MICRO FUEL CELLS

PRINCIPLES AND APPLICATIONS

Edited by

T.S. ZHAO
The Hong Kong University of Science & Technology
Hong Kong, China

AMSTERDAM • BOSTON • HEIDELBERG • LONDON
NEW YORK • OXFORD • PARIS • SAN DIEGO
SAN FRANCISCO • SINGAPORE • SYDNEY • TOKYO
Academic Press is an imprint of Elsevier

Academic Press is an imprint of Elsevier
30 Corporate Drive, Suite 400, Burlington, MA 01803, USA
525 B Street, Suite 1900, San Diego, California 92101-4495, USA
84 Theobald's Road, London WC1X 8RR, UK

Library of Congress Cataloging-in-Publication Data
Application submitted

British Library Cataloguing-in-Publication Data
A catalogue record for this book is available from the British Library.

ISBN: 978-0-12-374713-6

For information on all Academic Press publications
visit our Web site at www.elsevierdirect.com

Printed in the United States of America
09 10 11 9 8 7 6 5 4 3 2 1

Contents

Preface

Following the success of our previous book, *Advances in Fuel Cells*, this closely related volume continues to provide in-depth coverage of the newest, most important developments in the general field of fuel cells, this time specifically focusing on the pivotal area of Micro Fuel Cells. Each chapter in this book deals with a significant, emerging topic by beginning with fundamental physiochemical considerations and then proceeding in a logical fashion to the forefront of recent developments and future challenges. The contributing authors are a combination of young, upcoming experts and well-established leaders in the field.

A growing number of increasingly ubiquitous portable consumer electronics, such as personal digital assistants (PDAs), laptop computers, and cellular phones, demand small, lightweight power sources with high power density and energy capacity. Over the past few years, a number of different types of micro fuel cells have been developing at a rapid pace in order to meet this demand. This book pays particular attention to these recent developments, including electrolytes for long-life, ultra low-power direct methanol fuel cells, MEMS-based micro fuel cells, microfluidic fuel cells, micro tubular solid oxide fuel cells, enzymatic fuel cells, and glucose biosensors that mainly focus on diabetes management.

The editorial board expresses their appreciation to the contributing authors of this volume, who have maintained the high standards established in *Advances in Fuel Cells*. The editor is grateful to Dr. Zhenxing Liang and Dr. Erdong Wang for their assistance in preparing this book. Last, but not least, the editor acknowledges the efforts of the professional staff at Elsevier for providing invaluable editorial assistance.

T.S. Zhao
The Hong Kong University of Science & Technology
Hong Kong, China

About the Editor

Dr. T.S. Zhao is a Professor of Mechanical Engineering and Director of the Center for Sustainable Energy Technology at the Hong Kong University of Science & Technology (HKUST). As an internationally renowned expert in energy technology, he presently focuses his research on fuel cells, multiscale multiphase heat/mass transport with electrochemical reactions, and computational modeling. He has published many important papers in prestigious journals in the fields of mechanical engineering, physics, and fuel cells and has presented numerous plenary/keynote lectures at international conferences. In the international community, Dr. Zhao serves as Editor-in-Chief of *Advances in Fuel Cells,* Asian Regional Editor of *Applied Thermal Engineering*, and as a member of the Editorial Board for more than 12 international journals. He has received a number of recognitions for his research and teaching, including the Bechtel Foundation Engineering Teaching Excellence Award at HKUST in 2004, the Overseas Distinguished Young Scholars Award by the Natural Science Foundation of China in 2006, Fellow of the American Society of Mechanical Engineers (ASME) since 2007, and the Croucher Senior Fellowship award from the Croucher Foundation in 2008.

About the Contributors

Robert L. Arechederra was born in the United States in 1980. He received his M.S. in Chemistry from Saint Louis University in 2007 and continued to pursue his doctoral degree under the supervision of Dr. Shelley D. Minteer. His current research efforts are focused on creating artificial metabolic pathways for substrates on electrode surfaces for generating electricity.

Michael Beilke earned his B.S. in Chemistry at Missouri Southern State University in 2005. He also obtained a M.S. in Chemistry from Saint Louis University in 2007. He will be attending the Ohio State University in the fall of 2009 to work on a Ph.D. in Analytical Chemistry.

Frank Davis is a researcher within Cranfield Health, located at Cranfield University, UK. He initially graduated and gained a Ph.D. from Lancaster University, which was then followed by research positions at Manchester University, Sheffield University, and Gillette UKRDL. Dr. Davis has worked on biosensors based on enzymes and antibodies at Cranfield since 2002 and is co-author of over 90 publications.

Ned Djilali obtained B.S. and M.S. degrees in Aeronautics from Hertfordshire University and Imperial College. After completing a Ph.D. in Mechanical Engineering at the University of British Columbia, he first worked with the Advanced Aerodynamics Department of Bombardier (Canadair Aerospace Division) and then joined the University of Victoria. Dr. Djilali is a Professor in the Department of Mechanical Engineering, an active member of the Institute for Integrated Energy Systems (IESVic), and a Canada Research Chair in Energy Systems Design and Computational Modelling.

Marguerite Germain is a Ph.D. candidate in the Department of Chemistry at Saint Louis University. She started her collegiate career at Rockhurst University where she graduated with a B.A. in French. She is currently studying poly(methylene green) and developing a self-powered explosive biosensor that uses mitochondria as catalysts and pyruvate as fuel.

Séamus P. J. Higson holds a chair in Bio- and Electro-analysis and is currently Dean of the Faculty of Medicine and Biosciences at Cranfield University in the United Kingdom. Professor Higson also is author of a major Oxford University Press textbook, *Analytical Biotechnology*, and acts as Research Director of Microarray Ltd., a University spin-off company based on research from his laboratory.

Erik Kjeang, originally from Sweden, completed his Ph.D. in Mechanical Engineering at the University of Victoria, British Columbia, Canada, in 2007. His research demonstrated several unique microfluidic fuel cell architectures resulting in practical miniature devices with high performance and high efficiency, and was awarded with the Governor General's Gold Medal for Outstanding Dissertation. Since then, Dr. Kjeang has worked as a research engineer at Ballard Power Systems, Inc., and is presently an Assistant Professor at Simon Fraser University in Mechatronic Systems Engineering.

Tamara Klotzbach received her B.S. in Chemistry from Saint Louis University in spring of 2007. She is currently attending the Ohio State University where she is working toward a Ph.D. in Inorganic Chemistry under the direction of Professor Malcolm Chisholm. Her work focuses on using calcium catalysts for the ring opening polymerization of cyclic esters for the development of biodegradable polymers.

Paul A. Kohl is Regents' Professor and Hercules Inc./Thomas L. Gossage Chair in Chemical and Biomolecular Engineering at Georgia Institute of Technology. He received a Ph.D. in Chemistry at the University of Texas at Austin and was previously employed at AT&T Bell Laboratories. His research interests include electrochemistry and electronic materials.

Shelley D. Minteer received her Ph.D. in Analytical Chemistry from the University of Iowa in 2000 after receiving her B.S. in Chemistry at Western Illinois University in 1995. After getting her Ph.D., she took a position as an Assistant Professor of Chemistry at Saint Louis University in the fall of 2000. She was promoted to Associate Professor in 2005 and to Full Professor in 2008. Her research interests are in the area of enzymatic biofuel cells.

Michael J. Moehlenbrock was born in the United States in 1982. He received his M.S. in Chemistry from Saint Louis University in 2007. In 2007, he also began his pursuit of his Ph.D. in Chemistry under the supervision of Dr. Shelley D. Minteer. His current research efforts are focused on the miniaturization of immobilized enzymatic biofuel cells and biosensors and the examination of substrate channeling in metabolic complexes for use in biofuel cells.

William E. Mustain is currently an Assistant Professor of Chemical Engineering at the University of Connecticut in the Department of Chemical, Materials and Biomolecular Engineering. Dr. Mustain received his Ph.D. from the Illinois Institute of Technology in 2006. He then spent two years as a Postdoctoral Fellow at Georgia Institute of Technology. His areas of expertise are electrochemical engineering and electrocatalysis.

Tristan Pichonat was born in Migennes, France, in 1975. He received the M.S. in Image Processing and Medical Scanning in 1998 from the University of Dijon, Burgundy, and the M.S. in Optics, Laser, and Signal Processing from the University of Rouen in 1999. After working on micro accelerometers at the LPMO-CNRS in Besançon, he received a Ph.D. in Engineering Sciences in 2004 working on micro fuel cells. He then moved on to a post-doctorate position, working on micro resonators at the IEMN-CNRS, in Villeneuve d'Ascq. He currently works as a research engineer IEMN, in Villeneuve d'Ascq, on micro power sources for autonomous networks nodes supply.

Shruti Prakash was born in Patna, India. She did her B.S. in Chemical Engineering from University of Wisconsin, Madison. She received her Ph.D. in Chemical Engineering from Georgia Institute of Technology, Atlanta. She is currently employed by the Lawrence Berkeley National Laboratory in Berkeley, California.

David Sinton obtained his Ph.D. from the University of Toronto, focusing on microscale flow visualization in 2003. He is currently an Associate Professor in Mechanical Engineering at the University of Victoria, and member of the Institute for Integrated Energy Systems, University of Victoria (IESVic). Dr. Sinton leads an award- winning research program focused on the study and application of microfluidics and nanofluidics for biomedical and energy applications.

Daria Sokic-Lazic was born in Tuzla, Bosnia and Herzegovina, in 1980. She received her M.S. in Chemistry from Saint Louis University in 2008. In 2008, she also began her pursuit of Ph.D. in Chemistry under the supervision of Dr. Shelley D. Minteer. Her current research efforts are focused on enzymatic biomimics of metabolic pathways for use in biofuel cells.

Toshio Suzuki is a research scientist in the National Institute of Advanced Industrial Science and Technology (AIST), Nagoya, Japan, working on the development of micro tubular SOFCs. Before joining AIST, he worked as an Assistant Research Professor at the University of Missouri-Rolla, on single chamber SOFCs and low temperature (under 1000°C) ceramic processing for SOFC components (electrolyte and electrodes). He earned his M.S. in Applied Physics at Tohoku University, Sendai, Japan, and his Ph.D. in Ceramic Engineering at the University of Missouri-Rolla.

Becky L. Treu is a native Missourian who earned her Ph.D. in integrated and Applied Sciences with an emphasis in Chemistry from Saint Louis University in 2008. Her field of expertise is bioelectrochemistry, namely the isolation and purification of PQQ-dependent dehydrogenates for biofuel cell applications. From this research she has generated numerous publications and patents. She currently holds a position as a postdoctoral fellow at Missouri University of Science and Technology in the Graduate Center for Materials Research, researching environmentally friendly rare earth-based coating systems for aerospace applications.

Janice Wildrick finished her B.A. in Chemistry at University of Missouri-St. Louis after working in the legal and computer industries for 15 years. She joined Dr. Shelley Minteer's research group at St. Louis University in 2007 to contribute to the most compelling issue of our time-renewable energy. Ms. Wildrick's research focuses on the development and optimization of an air-breathing cathode.

CHAPTER

1

Electrolytes for Long-Life, Ultra Low-Power Direct Methanol Fuel Cells

Shruti Prakash, William E. Mustain, and Paul A. Kohl

[1]Department of Chemical and Biomolecular Engineering

[2]Georgia Institute of Technology

Portable direct methanol fuel cells are potentially excellent power sources for small electronic devices because of the high energy density of pure methanol. For example, wireless sensors are valuable in monitoring and control situations. The ability of wireless sensors to form self-assembled networks may provide rapid growth for the technology. In each of the small electronic applications, the cost, lifetime, size, and weight of the power source is a critical part of the value of the overall system. These devices may require tens of milliwatts for milliseconds to acquire or transmit data, and tens of microwatts for long periods in sleep mode. This style of operation (low intermittent power over a long time period) is far different from the power sources for transportation, high-power electronic devices, or electric power. The fuel cells must have very low energy losses, including low methanol permeability, and must allow the use of highly concentrated fuels. In this chapter, the existing fuel cell technologies are examined in light of these new requirements.

1.1 INTRODUCTION

The fuel cell market can be divided into different segments of our energy infrastructure based on power level and end use. These areas (with example power levels) are (i) stationary plug power (hundreds of megawatts), (ii) back-up power (tens to hundreds of kilowatts), (iii) traction power (portable supplies at 10 to 100 kW), (iv) small portable power (1 to 100 W) and (v) mini or micro power (10 μW to 1 W). Fuel cells have the potential to provide clean, efficient, sustainable power in all market segments. However, in each market segment, there are multiple energy conversion devices available. Heat engines (Rankine & Brayton cycles of stationary power and Otto & Diesel cycles for traction) are highly competitive in high power-density applications where device size and fuel infrastructure are not primary concerns. However, at smaller sizes, electrochemical devices become more competitive because they are simpler than heat engines, requiring no moving parts, while providing higher energy densities since they scale as a function of their surface area, not volume.

1.1.1 Potential Applications for Micro Fuel Cells

Energy sources for portable devices, such as electronic devices, are under continual pressure to decrease both consumed power and weight. The scaling of electronic devices has produced products with higher functionality (requiring more power) and smaller form-factors. The increase in popularity of portable electronics has continued to put pressure on increasing the energy density and lowering the cost of portable power sources. The average power can range from microwatts to watts, depending on function and duty cycle.

One growing market segment is the use of electronic sensors. Some devices communicate wirelessly and are deployed in locations where plug-in power is difficult or impossible to implement and a portable power source is essential to its implementation. Among these, industrial sensors are the largest market segment. Sensors such as temperature, pressure, gas composition, smoke, motion, humidity, and light are needed in mining, construction, utilities, manufacturing, transportation, and warehouse locations. A second area where wireless sensors will have an impact is in commercial buildings. The sensing and control of heating, ventilation, and air-conditioning systems will improve the energy efficiency and make buildings more environmentally benign. Also, wireless security systems can be easily added to commercial and residential properties.

A third technological area where small wireless sensors are valuable is in the implementation of automated meter readers, such as water and gas meters. The ability of wireless sensors to form self-assembled networks allows utilities to accurately monitor and bill communities at low cost. A fourth area of interest is in home automation. Wireless sensors can be used to monitor remote door opening as well as smoke, gas, flame, and water situations. Again, energy efficiency can be improved by monitoring and selectively controlling heating, air conditioning, and lighting.

Finally, wireless sensors, especially those forming networks, are valuable in environmental and homeland security situations. Simple sensors can be implemented for the detection of biological, chemical, and radiation hazards. In addition, environmental sensors can be used for weather forecasting, agricultural monitoring (fertilizer and water monitoring), and motion tracking at borders and other secure locations.

In each of these applications, the cost, lifetime, and size and weight of the power source are a critical part of the value of the overall system. Many of the monitor-style sensors, as described above, have low duty cycles (need to acquire data only occasionally) and require low power to operate due to their simple function. Devices may require tens of milliwatts for milliseconds to acquire or transmit data, and tens of microwatts for long periods in sleep mode. This style of operation—low, intermittent power over a long time period—is far different from the power sources for transportation, high-power electronic devices, or electric power. The general approach for low-power applications is for the fuel cell to provide the constant power for the sleep mode (e.g., tens of microwatts), and a secondary storage device, which is maintained by the fuel cell, provides the burst-power for events including sensing and transmission [1]. Thus, the design parameters for the fuel cell shift from the traditional high-power mode, to low-power and high energy efficiency. The parameters which affect energy efficiency will be explained in the following section, and the ability of existing PEM systems to provide high efficiency during ultra low-power operation will be the subject of the remaining sections of this chapter.

1.1.2 Direct Methanol Fuel Cells

In the direct methanol fuel cell (DMFC), liquid methanol is fed directly to the anode compartment of the PEM fuel cell. This provides several advantages over its hydrogen counterpart, namely that 1) liquid methanol has a higher volumetric energy density than hydrogen and 2) the generation, storage, transportation of methanol is facile. However, unlike the hydrogen oxidation, the direct oxidation of methanol requires a considerable amount of water in the anolyte since the water is oxidized along with methanol to form carbon dioxide (Eq. 1.1).

$$CH_3OH + H_2O \rightarrow CO_2 + 6H^+ + 6e^- \tag{1.1}$$

Equivalent to the PEM fuel cell, the proton in Eq. 1.1 is transported through the electrolyte membrane as a hydrated ion and the electrons travel through an external circuit (the device). The protons then meet molecular oxygen at the cathode, forming water via the acidic oxygen reduction reaction (Eq. 1.2).

$$O_2 + 4H^+ + 4e^- \rightarrow 2H_2O \tag{1.2}$$

This process is illustrated in Figure 1.1.

Theoretically, a 17 M methanol anode feed is possible. However, in conventional systems, the highest currents and powers are achieved with dilute methanol solutions in the 0.5 M to 2 M methanol range [2]. The concentration optimization is a balance of three effects. First, if the methanol concentration is high, permeation of fuel through the electrolyte is prohibitive. Second, if the methanol concentration is too low, the reaction kinetics reduces and mass transport of the methanol reactant to the anode limits the current

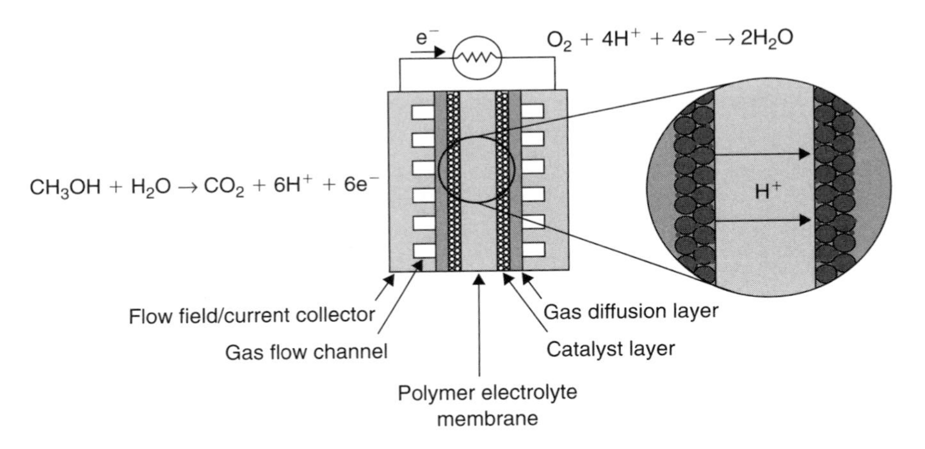

FIGURE 1.1 DMFC schematic.

density. Finally, up to 15 water molecules can be transported by electro-osmotic drag from the anode to the cathode for each methanol molecule oxidized [3]. This dilution of the fuel is highly undesirable because it decreases the energy density of the cell and can cause water management problems.

1.1.3 Energy Efficiency and Device Life

Fuel cells have very high theoretical energy densities because concentrated liquid fuels with high equivalence (e.g., six electrons from methanol) can be used, and the oxidizing agent (species to be reduced at the cathode) does not have to be carried within the cell if oxygen from air is reduced. The theoretical energy density of pure methanol is 6100 Whr/kg whereas lead-acid and nickel-cadmium batteries offer 30–85 Whr/kg, and lithium ion generally offers between 110–160 Whr/kg [4]. However, 6100 Whr/kg is not achievable because one cannot discharge a methanol fuel cell at the theoretical thermodynamic voltage and pure methanol is not acceptable in the PEM fuel cell since water must also be provided as a reactant

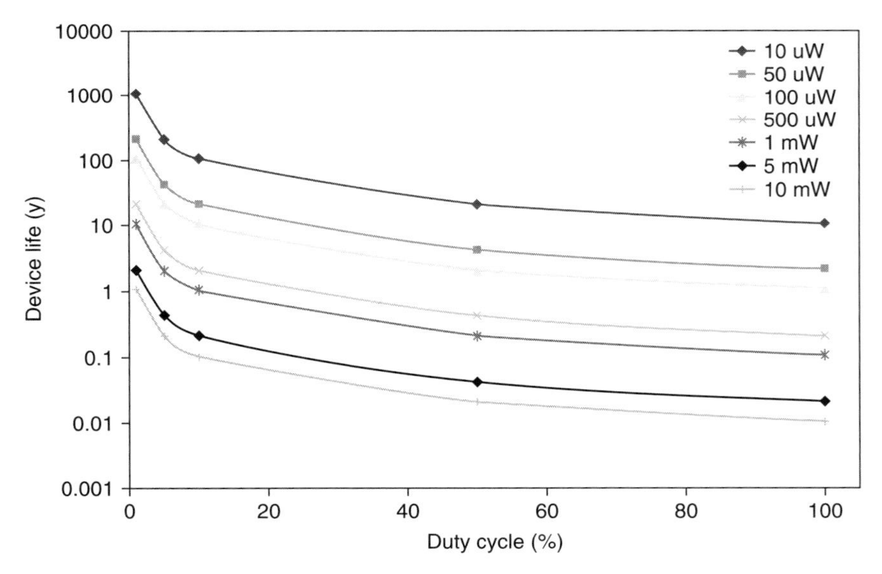

FIGURE 1.2 Maximum device life as a function of duty cycle for a 12 M methanol fuel cell carrying 1 cc^3 of fuel and discharged at 0.5 V.

at the anode, thus diluting the fuel. A more realistic goal may be a discharge voltage of 0.5V and use of 12M methanol. Under these conditions, one would have an energy density of over 1200Whr/kg. Figure 1.2 shows the lifetime of such a cell as a function of duty cycle and average power level. One milliliter of fuel theoretically could last 10 years at an average power of 100μW and 10% duty cycle.

However, there are two primary energy loss mechanisms which must be considered and mitigated in order to achieve long operating life. The first is the permeation of fuel, methanol in the direct methanol fuel cell, through the electrolyte. In this case, the mass transport of methanol from the anode to cathode results in not only a simple fuel loss, but the electrochemical performance of the cathode is decreased as well. Second, the ionic conductivity of the electrolyte is important as ohmic-type heating losses must be minimized.

An expression for the energy conversion efficiency of a fuel cell can be derived by considering the energy available relative to the energy delivered. The useful energy delivered from a fuel cell, E_U, is given in Eq. 1.3.

$$E_U = iV_{op} \tag{1.3}$$

where i is the fuel cell current and V_{op} is the operating voltage. Resistive loss caused by ionic transport through the proton exchange membrane, E_R, is expressed by Eq. 1.4:

$$E_R = \left(\frac{i^2 \rho \delta_1}{A_1}\right) \tag{1.4}$$

where ρ is the ionic resistivity of the electrolyte, δ_1 is ionic path length, A_1 is the electrochemically active area.

Fuel can be lost by permeation through the electrolyte, often referred to as methanol crossover. This loss, E_x, is given by Eq. 1.5.

$$E_x = \left(\frac{P_1 \Delta p A_2}{\delta_2}\right) nFV_{ocv} \tag{1.5}$$

where P_1 is the permeation coefficient of the membrane, Δp is the pressure drop across the membrane, A_2 is the exposed membrane area available for fuel transport through the membrane, δ_2 is the electrolyte thickness, n is the number of electrons transferred in the reaction, and F is Faraday's constant. It should be noted that generally $\delta_1 = \delta_2$; however, A_1 and A_2 need not be the same. Appropriately engineering the electrode structure may block the membrane from crossover loss while maintaining a large membrane area for low resistive losses (i.e., $A_2 < A_1$).

Combining Eqs. 1.3–1.5, the energy efficiency, ε, is given by Eq. 1.6.

$$\varepsilon = \frac{iV_{op}}{\left(\frac{i^2\rho\delta}{A_1}\right) + \left(\frac{P_1\Delta p A_2}{\delta}\right) nFV_{ocv} + iV_{op}} \tag{1.6}$$

Fuel loss from permeation through the PEM membrane is especially important in ultra low-power fuel cells compared to intermediate and high-power cells because 1) the rate of fuel consumption through electrochemical oxidation is orders of magnitude lower; 2) the ohmic loss is significantly reduced since it is a function of the square of the operating current; and 3) the electrolyte aspect ratio ($\alpha = A/\delta$) is high. That is, the current in the numerator of Eq. 1.6 is smaller (can be much smaller) than in high-power systems, making it more important to have tight control of losses (terms in the denominator of Eq. 1.6), especially the permeation losses.

It should be noted here that the device life is a significant function of the chemical stability of the components used. Though the chemical stability of the electrode-electrolyte interface will not be discussed in detail in this review, a few examples are listed below. The most common limitation observed in the DMFC is the chemical stability of the solid electrolyte membrane. In many cases, the polymer backbone is degraded by either heat treatment or contact with peroxide radicals generated at the cathode. Second, it has been found that the PtRu anode catalyst degrades during DMFC operation. Evidence suggests that the ruthenium is electrochemically oxidized, transported in its solvated ionic form through the electrolyte and electrodeposits on the cathode [5]. This leads to a loss of activity for both the anode and cathode. DMFC lifetime can also be affected by the corrosion of the carbon catalyst support at each of the electrodes. When the carbon support is oxidized, especially at the positive electrode, the platinum or other alloy particles peel away from the support and electrical contact is lost, resulting in an increase in electrode resistance [6]. This process is exacerbated at high currents, high temperature, and low relative humidity conditions.

1.2 PERFLUORINATED POLYMER PROTON EXCHANGE MEMBRANES

Several fluorinated membranes have been developed for use as proton exchange membranes in both the PEM and DMFC systems. The fluorinated polymers are typically synthesized by copolymerization of tetrafluoroethylene and a perfluorinated vinyl ether with sulfonyl acid fluoride. The sulfonyl fluoride groups, $-SO_2F$, are converted to sulfonic acid ($-SO_3^-H^+$) by consecutive soaking in hot sodium hydroxide (to yield $-SO_3^-Na^+$), hydrogen peroxide, and sulfuric acid. The resulting polymer contains a fluorocarbon backbone, which is hydrophobic, and perfluoroether side chains containing hydrophilic sulfonic acid ionic groups. The sulfonic acid groups form internal clusters so that small channels (ca. 4 nm) form. When hydrated, these channels provide ionic pathways for protons [7]. The general structure for the fluorinated membranes is presented as Figure 1.3.

The coefficients in the generalized structure, x, y and z, can be adjusted to tailor the various physical properties of the membrane, including its molecular weight, hydrophobicity, and acidity, which is commonly expressed as equivalent weight (grams of polymer/mole

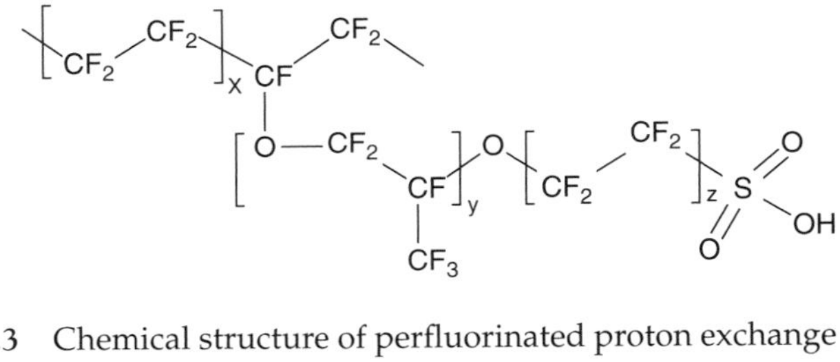

FIGURE 1.3 Chemical structure of perfluorinated proton exchange membranes.

H^+). For Nafion®, by far the most common proton exchange membrane at every power level, $y = 1$, $z = 1$, and x is adjusted between 6 and 10. The adjustment in x yields the various equivalent weight (EW) membranes that are available from DuPont, including Nafion 120 (EW = 1200), Nafion 117, 115, and 112 (EW = 1100), and Nafion 105 (EW = 1000).

Dow has also developed a fluorinated proton exchange membrane that has been extensively studied [8, 9]. In the Dow Hyflon membrane, x is varied between 3 and 10, y is equal to zero, and z is equal to one. In its most popular form, the resulting equivalent weight is around 800. Asahi Glass Company has marketed its Flemion membrane [10]. The Flemion membrane has a structure quite similar to the Dow membrane where x is varied between 3 and 10, y is around 0.1, and z is varied between 0 and 3; however, the equivalent weight of the Flemion membrane is much greater than the Dow membrane, around 1000. Asahi Chemical has reported the Aciplex-S proton exchange membrane with the same general structure to the other films [9]. In the Aciplex-S, the structure and equivalent weight is similar to Nafion, where x is varied between 2 and 14, y is 0.3, and z is either 1 or 2. Finally, two additional fluorinated membranes are popular, Neosepta-F (Tokuyama) and Gore Select (W L Gore and Associates Inc.), though their exact structures have not yet been reported. This is summarized in Table 1.1.

All of the perfluorinated electrolytes tend to have similar performance characteristics, namely enhanced ionic conductivity at room temperature (>10 mS/cm) and good chemical and electrochemical stability. However, they all also tend to suffer from high methanol crossover and poor ionic performance at low relative humidity conditions [11–14]. It appears that both of these are linked to the need

TABLE 1.1 Structure of various commercial fluorinated proton exchange membranes

Membrane	Manufacturer	x	y	z
Nafion	DuPont	6–10	1	1
Hyflon	Dow Chemical	3–10	0	1
Flemion	Asahi Glass	3–10	0.1	0–3
Aciplex-S	Asahi Chemical	2–14	0.3	1 or 2

for a high degree of solvation of the sulfonic acid groups to impart ionic conductivity. A detailed discussion on the ion transport and methanol permeability of perfluorinated Nafion and Nafion composite electrolytes is given in the following two sections.

1.2.1 Nafion®

Nafion was created by E.I. DuPont de Nemours and Company in the 1960s in response to the need for a more robust cation exchange membrane for the electrochemical production of chlorine and sodium hydroxide (chlor-alkali industry). In the 1990s, Nafion became the membrane of choice for both PEMFCs and DMFCs. This is mostly due to the perfluorination of the polymer backbone, which greatly improved the chemical stability of Nafion compared to its predecessors, though degradation through reaction with peroxide radicals generated during fuel cell persists [15]. Tang et al. showed that Nafion soaked in hydrogen peroxide solution in the presence of trace amounts of iron, chromium, and nickel ions degraded from the ends of the main chain and resulted in the loss of polymer repeat units and formation of voids in the membrane [15]. This process occurs more readily when hydrogen is used as the fuel because the humidified hydrogen can react directly with oxygen in the membrane.

It is well known that the ionic conductivity of Nafion significantly improves with membrane water content. Unfortunately, as the water content of the membrane increases, its permeability to methanol also increases. This is caused by an increase in both mass transport mechanisms, diffusion, and electro-osmotic drag. The crossover is influenced by the membrane aspect ratio, operating temperature, current density, cathode feed humidity, and methanol concentration in the fuel. Nakagawa et al. measured the crossover flux of methanol for Nafion 112, 115, and 117 [16]. The thinnest membrane, Nafion 112, had a crossover flux of $0.02\,g/m^2s$ using 4 M methanol at 297 K. If one were to convert this flux into an equivalent current (the current resulting from that oxidation of this flux of methanol), it would correspond to a crossover current of $38\,mA/cm^2$. This is especially important for low-power, long-life fuel cells because this crossover alone is orders of magnitude higher than the device operating current. Blum et al. measured the crossover current as a function of discharge voltage and methanol concentration. The maximum crossover current calculated was $25\,mA/cm^2$. Liu et al. note that the crossover current is approximately linear with methanol feed concentration at the anode [3]. The crossover decreased from $32\,mA/cm^2$ at 2 M to $11\,mA/cm^2$ at 4 M. Ling et al. measured the concentration of methanol reaching the cathode as a function of Nafion thickness and concentration of methanol at the anode [17]. Typically, the methanol reached 0.01 M at the cathode for Nafion 112 when 0.5 M fuel was used, and reached 0.14 M when 3 M fuel was used.

These experimental values can be used to estimate the methanol permeability coefficient through Nafion, which is estimated as 2.67×10^{-6} mol-cm/cm^2-day-Pa. Using this and the accepted value for the ionic conductivity of fully hydrated Nafion, 0.083 S/cm, we can estimate the efficiency of a $1\,cm^2$ active area micro DMFC with $100\,\mu A$ operating current discharged at 0.5 V and a membrane aspect ratio of 10^{-2} (Nafion 112) using Eq. 1.6. The fuel loss due to methanol permeability alone in this system is over 99.9%. This indicates that a micro DMFC operating with a conventional Nafion electrolyte would have an efficiency less than 0.1%. This agrees well with both Nakagawa and Liu, whose experimental values for the methanol crossover current estimate the fuel loss in a $100\,\mu A$ device as 99.7% and 99.1%, respectively [3, 16]. This is clearly unacceptable and alternative electrolytes with lower methanol permeability must be developed if micro DMFCs are to be commercially realized.

1.2.2 Nafion Composite Membranes for Direct Methanol Fuel Cells

Due to its high ionic conductivity under fully hydrated conditions, several groups have explored modified Nafion membranes for DMFCs. In this alternative approach, many research studies have proposed the modification of the perflurosulfonic acid matrix of Nafion by i) impregnation with acids that have low solubility, ii) solution casting with solubalized oxygenated acids, and iii) composites of nonconducting polymers with Nafion. Each of these proposed methods can potentially combat Nafion's shortcomings while providing proton conductivities similar to the host material.

Nafion/Silica Composites

As mentioned previously, the most serious challenge to using Nafion in ultra low-power DMFCs is its high methanol permeation rate. Methanol crossover through the proton exchange membrane reduces the kinetic efficiency of oxygen reduction at the cathode and leads to a loss in the lifetime of the device due to the diminution of methanol fuel. Therefore, it is essential to incorporate materials that can potentially reduce the crossover rate of methanol without compromising the conductivity of the proton exchange membrane. One approach is to impregnate Nafion membranes with hydrous tetravalent oxides such as SiO_2 (silica). Over years, tetravalent oxides have been known for their ability to conduct protons. The −OH terminated silicon end groups, caused by incomplete condensation of the silicate precursors, leads to ion transport in these oxides. According to Prakash et al., the proton conductivity of silicate glasses can be explained on the basis of proton hopping from one −OH site to another, much like the Grotthuss mechanism [18]. Results provided by the authors showed that plasma enhanced chemical vapor deposition of phosphorus-doped silicate glasses has conductivity in the range of 10^{-4} S/cm. Although this value is lower than the reported Nafion conductivity, the authors have proposed the inclusion of silica particles to act as a barrier for methanol transport. The resultant composite Nafion/SiO_2 membrane consists of a thin layer of phosphorus-doped silica glass (3–6 μm) to act as a methanol barrier resulting in an overall improvement of DMFC.

A more popular method of incorporating silica particles in the Nafion matrix is thorough sol-gel chemistry. Compared to plasma deposition, the wet chemical process provides a lower cost alternative for making glass. In sol-gel processes, precursors of hydrous tetravalent oxides (siloxane) are reacted in the presence of acid to form a continuous matrix of silica and are thereafter impregnated in Nafion matrix. Shao et al. have used a similar method of preparing Nafion-silica composite membranes where commercial Nafion membranes were immersed in silica sol-gel to allow impregnation of membranes with inorganic composites [19]. Adjemian et al. also fabricated sol-gel glass using TEOS (tetraethoxy ortho siloxane) [20]. The authors reported silica loading of 10% with TEOS while maintaining the mechanical rigidity of Nafion matrix. It has been shown that silica-impregnated Nafion membranes maintain their high conductivity with increasing temperature unlike Nafion. The authors explained this observation based on the water retention of the composite membrane. As a result, the composite membranes showed better conductivities than Nafion. At temperatures higher than 100°C, the conductivity of hybrid membranes ranged between 0.2–2 ohm-cm^2.

In a similar work done by Muaritz et al., Nafion membranes were impregnated with silica using a sol-gel synthesis [21] (see Figure 1.4). Unlike the previous study, where an acid catalyzed sol-gel reaction took place, the study conducted by the authors reports a base-catalyzed sol-gel reaction. In this case, the pH of the Nafion solution was adjusted between 8 and 13. The authors have suggested that in base-catalyzed sol-gel reactions, it is easier to control the amount of silica acquired by the Nafion matrix. As a result, the proton exchange membranes can be customized to have hydrophilic (silica) and hydrophobic (PFSA) regions within their matrix.

To study the behavior of these hybridized and customized Nafion-silica matrices, Adjemian et al. conducted a comparative study between commercial Nafion/silica and hybrid Nafion/silica membranes [22]. In this study, commercially available Nafion membranes were impregnated with silica by soaking them in a sol-gel solution of TEOS. The performance of this membrane was compared with recast PFSA composite with TEOS using solution

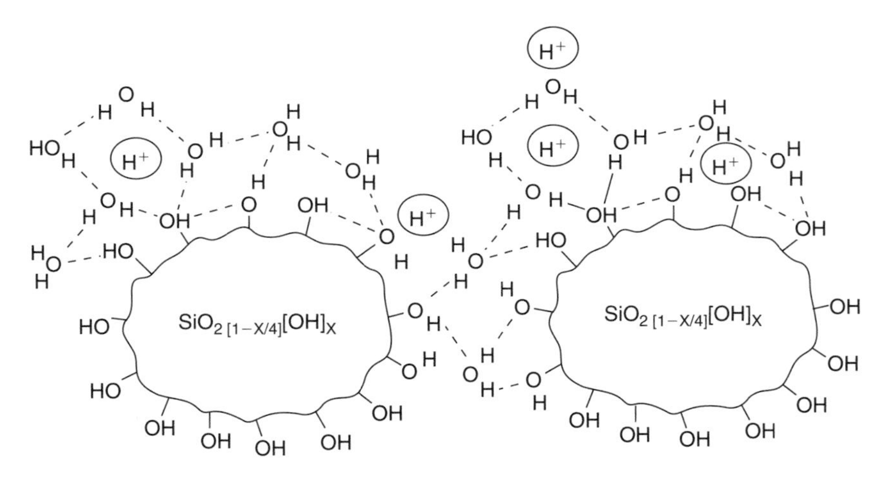

FIGURE 1.4 Proton hopping between the hydrophilic pathways through the silicate nanoparticles that are encapsulated within hydrophobic Nafion clusters [21].

casting. The authors have reported 50% lower resistivity for the recast membranes. It is believed that the low resistivity observed in the case of recast PFSA-silica hybrid membrane is a direct result of the continuous pathway available for faster proton transfer within their matrix.

With advances in nanotechnology and nano-scale understanding of materials, there has been interest in using this technology for fuel cells. Recently, incorporation of nano-sized silica particles for fabricating Nafion-silica hybrid membranes has drawn much attention. The silica nanoparticles are found to be well dispersed in the Nafion matrix and are both thermally and mechanically stable, even at high temperature. As studied by Shao et al., the nano silica particles are mixed ultrasonically with Nafion solution before being cast into Nafion/silica membranes [23]. In this work, the authors have reported that the conductivity of Nafion-silica composite membranes increases with increase in temperature due to their higher water uptake, much unlike the behavior of Nafion. The incorporation of silica particles not only improved the thermal, mechanical, and proton conductivity of the hybrid membrane, but it also lowered the methanol permeability. Not only were the Nafion-silica hybrid membranes more robust regarding their electrical, mechanical and thermal properties, they also showed decreased methanol permeability. Also, Nafion membranes that were recast from a solution mixture of 5 wt% Nafion inomer with dispersed silica particles showed a lower methanol crossover rate [24].

Further, it has been shown that optimization between conductivity and methanol permeability can be achieved by integrating hydrophilic silica particles in such a way so as to alter the transport properties of perflurosulfonic acid membranes [25]. According to the authors, the higher water uptake observed in the case of silica-Nafion composite membranes is mostly because of the interaction of sulfonic groups with hydrophilic sites of silica particles. The transport properties of sulfonic acid backbone are highly influenced by the formation of ionic clusters in the polymer matrix due to the increased water uptake. As a result, hydrophilic and hydrophobic clusters are formed in the polymer matrix that can potentially lower methanol crossover rate without reducing conductivity. The authors reported a resistance of 0.02 ohm-cm^2 which is lower than Nafion's 0.75–0.15 ohm-cm^2 [25].

Though the silica particles increase the composite film conductivity compared with Nafion alone, they alone do not have significant conductivity. However, they can be functionalized in order to improve their conductivity [26] (see Figure 1.5). In the reported work, silica particles were functionalized to form sulfonated silica-nanoparticles (S-SNP). The composite membranes were thereafter solution cast from a dispersion of S-SNP and Nafion. It was shown that membranes formed with functionalized silica particles cross-linked in Nafion matrix showed higher conductivity than pristine Nafion at all temperatures. At room temperature, the measured conductivity of pristine Nafion was reported to be around 4 mS/cm while that for the composite membranes was in the range of 5–6 mS/cm and was also reported to exhibit lower methanol crossover rates.

While most popular methods of incorporating silica structures in Nafion membrane involve incorporating nanoparticles of silica in the Nafion matrix by either sol-gel or by impregnation of Nafion membranes with silica particles, a completely different and novel approach has been suggested by Chen et al. [27]. In this work, a blending procedure is proposed where polysiloxane structures form covalent bonds with Nafion the matrix. According to the authors, one of the biggest advantages of this methodology is that the resulting hybrid membrane retains the desired properties from each of the components in the blend. It is believed that this cross-linking framework allows better mixing of perfluorinated hydrophobic moieties with the extrinsic species without compromising on proton conductivity and mechanical stability. In this work, the authors have explored the usage of triply crosslinked hybrid

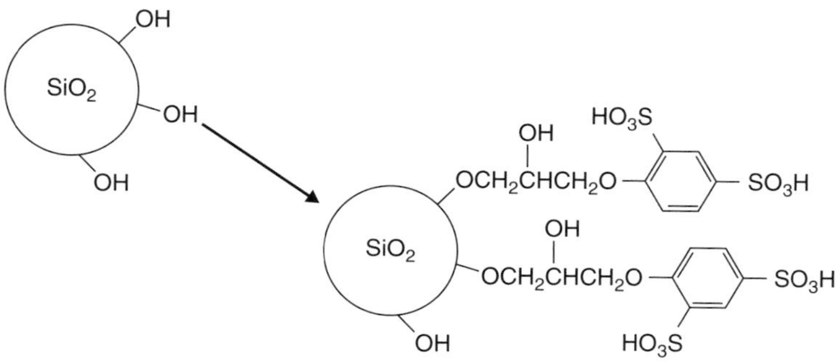

FIGURE 1.5 Proposed structure of sulfonated-silica nanoparticles (S-SNP) [26].

membrane that includes Nafion with 4,4′-methylenedianiline (MDA) and 3-glycidoxypropyltrimethoxy silane (GPTMS) [27]. In the resulting hybrid membrane, the polymer framework facilitates the interaction between Nafion and polysiloxane, allowing them to form a continuous network. The best performing hybrid polymer network showed better water bonding within the matrix that allowed continuous pathway proton mobility, leading to a conductivity of 0.034 S/cm. This conductivity is 25% lower than their reported conductivity of Nafion 117 (0.045 S/cm); however, the methanol permeability was three orders of magnitude lower than Nafion with 15 wt.% polysiloxane [27]. Despite this quite significant decrease in the methanol permeability, only very slight improvements in the device efficiency are realized using Eq. 1.6. A 100 μA, 0.5 V DMFC with a Nafion-polysiloxane composite electrolyte (aspect ratio = 1.0) still shows a fuel loss of 97%. This is a significant improvement over Nafion alone; however, it is unlikely that such an electrolyte can be utilized in an ultra low-power DMFC.

Nafion-Hetropoly Acid Composites

Hetropoly oxygenated acids, like phosphotungstic acid (PWA), have been known to have high proton conductivity (0.02–0.1 S/cm). However, since they easily dissolve in water, their use as an electrolyte membrane for fuel cell applications is limited. To overcome this, hetropoly acids have been immobilized on silica particles before being incorporated in Nafion the matrix through an in situ micro emulsion process [28]. The resulting membrane was homogenous with well-distributed inorganic regions. The authors have reported proton conductivities comparable to pristine Nafion and have explained the proton conductivity based on the Grotthuss mechanism, where interconnected hydrophilic channels are formed due to the well dispersed PWA, providing a continuous pathway for proton transfer. It was also observed that the methanol permeability for composite membranes was around half that of Nafion, which is not significant enough for consideration in an ultra low-power device.

In a similar work done by Shao et al., PWA/SiO_2 particles were ultrasonically mixed with 5% Nafion solution to form electrolyte membranes by solution casting [23]. They reported lower conductivity than Nafion at low temperatures and similar conductivity at higher temperature for composite membranes, and a lower activation energy has been observed for the composite membranes. However, at low relative humidity, the composite membranes showed higher conductivities than Nafion membranes. At 110°C and 70% RH, the conductivity of recast Nafion/SiO_2/PWA membranes was observed to be 2.67×10^{-2}S/cm in comparison to 8.13×10^{-3}S/cm for Nafion.

In another approach, Nafion/SiO_2/PWA composite membranes were prepared by immersing Nafion 117 membranes in TEOS solution followed by treatment with PWA solution [29]. In this study, the authors have reported lower methanol permeability for Nafion/SiO_2/PWA composite membrane. The reported conductivity for Nafion/SiO_2/PWA composite membrane was 24 mS/cm.

Hydrated oxides, such as $Ta_2O_5.nH_2O$, have also been reported to exhibit proton conductivity. Yang et al. in their work have suggested incorporation of nanoparticles of these pentavalent oxides into the fuel cell electrolyte as well as on the active electrode layers [30]. It was observed that the inclusion of nanosized particles of $Ta_2O_5.nH_2O$ facilitated water retention at intermediate temperatures and hence demonstrated steady cell performance. The observed conductivity was found to be higher with the composite membranes at all temperatures than Nafion by itself. The authors have explained that, at low temperatures, vehicle mechanism dominates the proton conduction in these membranes, which is facilitated by the presence of large numbers of physically absorbed water molecules on the surface of the membrane. Nafion, on the other hand, does not have such absorbed water and hence fails to demonstrate similar conductivities. The presence of $Ta_2O_5.nH_2O$ particles provides strong hydrogen bonding in composite membranes, which leads to faster proton conduction compared to Nafion.

Nafion-Zirconium Composites

Zirconium phosphates are Brønsted acids due to their ability to donate protons. Unlike hetropoly acids, zirconium phosphates can immobilize themselves directly in the polymer matrix, ensuring adequate anchoring. This lessens the probability of leaching out as is often observed in the case of polymer/hetropoly acid composite membranes. Along with their moderate proton conductivities, $\sim 10^{-3}$S/cm, these inorganic oxides are known for their high thermal stability and are hygroscopic. Because of this increased structural stability, these inorganic additives have become a popular additive in the design of Nafion composite membranes. These materials incorporate a molecular sieve-type material within the polymer matrix such that methanol transport through the polymer can be hindered and provide higher stiffness to the Nafion backbone.

In studies conducted by Bauer et al., Nafion-zirconium phosphate composite membranes were formed by immersion of Nafion in $ZrOCl_2$ and H_3PO_4 solution [31]. The resulting membranes showed higher mechanical stiffness than Nafion with low water content; however, the stiffness was similar to Nafion at high water content. In a similar study, Yang et al. incorporated zirconium phosphate in Nafion matrix by carrying out ion exchange such that protons are replaced by zirconium ions which are subsequently reacted in place by phosphoric acid [32]. According to the authors, water uptake was higher for the composite membranes and did not depend significantly on the temperature. It was further observed that at low water activity, Nafion 115 showed higher conductivity, while at high water activity, the proton conductivity of the composite membrane was higher. It is suggested that, with increasing activity of water in the membrane, the sulfonic acid groups dissociate themselves from the matrix, increasing the proton concentration, which leads to an increase in conductivity. Furthermore, at elevated temperatures, the performance of fuel cells with composite membrane was an order of magnitude higher than that of Nafion 115 membrane. It was also observed that the effect of MEA resistance for composite membrane was lower than Nafion both at low and high water activity.

In another study conducted at Princeton University, two different Nafion/zirconium composite membranes were fabricated [33]. In the first case, like those previously shown, Nafion membrane was immersed in $ZrOCl_2$ solution followed by immersion in phosphoric acid to anchor ion-exchanged zirconium ions in the matrix. The second approach taken by the authors involved solution casting of Nafion solution mixed with sulfonated zirconia particles. The authors observed no significant difference between Nafion and Nafion/sulfonate zirconia particles mostly due to nonhomogeneous distribution of sulfonated zirconia in the polymer matrix.

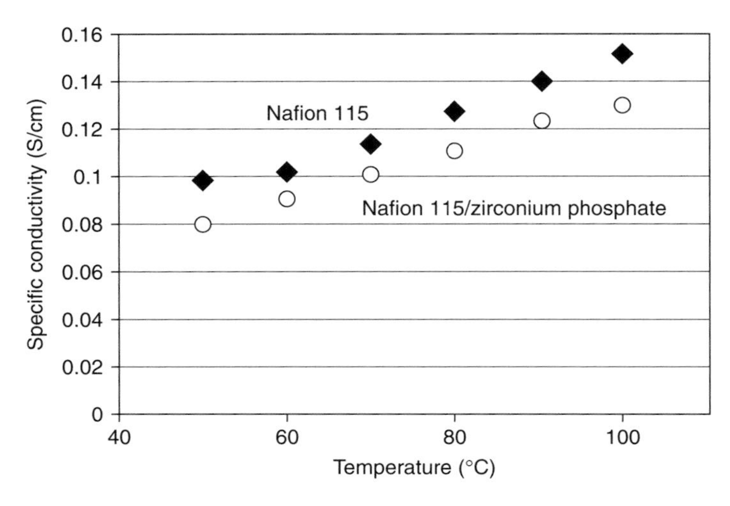

FIGURE 1.6 A comparative study of specific conductivity (S/cm) as a function of temperature between Nafion 115 and Nafion 115/zirconium phosphate membranes [33].

While Nafion/zirconium phosphate formed a continuous pathway bridging different ionic clusters between the Nafion backbone, particles of zirconia in Nafion/sulfonated zirconium were discontinuous and did not provide any added advantage over Nafion. This is shown in Figure 1.6, where the conductivity of the composite membrane was lower than that of Nafion.

It was also desired that the incorporation of these solid inorganic particles in the polymer matrix could potentially lead to the formation of a barrier layer to the transport of methanol through the polymeric membrane. In order to investigate the performance of the DMFC with a composite membrane fabricated with Nafion/zirconium phosphate, studies were conducted by Bauer et al. and it was observed that zirconium phosphate particles imbedded in Nafion matrix act as a diffusion barrier for methanol transport and also provide stronger mechanical structure to the membrane [34]. In comparison to Nafion, which showed a diffusion coefficient of $1.1 \times 10^{-5}\,cm^2/s$ for methanol, Nafion/zirconium phosphate composite membranes showed diffusion coefficients of methanol in the range of $7 \times 10^{-6}\,cm^2/s$ with 35 wt% zirconium phosphates. Like others, Bauer et al. have also suggested the strong interaction between ionic clusters of Nafion backbone and the embedded zirconium particles, therefore forming a continuous pathway for proton hopping.

Nafion/PTFE Composites

One of the initial efforts made in the modification of Nafion was motivated by its high cost. One suggested cost reduction method is to simply reduce the amount of Nafion in the film, while maintaining desired mechanical rigidity and proton conductivity. To this effect, Nafion can be incorporated into a microporous structure of compatible frameworks formed by fluorine-free or partially fluorinated membranes. By doing so, the framework could provide high rigidity without loading an excessive amount of Nafion and reduce the overall thickness of the membrane electrode assembly. This could result in significant cost reduction and higher area conductance. Few of such networks include Celgard (polypropylene and polyethylene resins), Gore-Select membranes, microporous network of poly(vinylidene fluoride) (PVDF), poly(tetrafluoro ethylene) (PTFE), and microporous silicon membrane. The adhesion of Nafion membrane with the microporous support can be increased by the addition of various surfactants.

In a study conducted by Nouel et al., Nafion-impregnated microporous support membranes were investigated [35]. In this investigation, Nafion/Celgard membranes showed conductivity in the range of 21–27 mS/cm, while that for Nafion/PTFE ranged around 104 mS/cm, which was lower than Nafion 112 (144 mS/cm) reported in [35]. Tang et al. report a more reasonable conductivity for Nafion/PTFE composite membranes ranging between 47–61 mS/cm in comparison to Nafion 112 at 91 mS/cm [36]. It was also observed that the overall performance of the impregnated membranes improved with chemical alteration of PTFE (making it more hydrophilic). In a very similar approach taken by Tang et al., the proton conductivity of Nafion/PTFE hybrid membranes was found to be around 81 mS/cm, which was significantly higher than Nafion (30 mS/cm) [37]. Along with higher conductivity, the authors also report that the new membranes exhibited very high chemical durability, mostly because of strong grafting of Nafion in the microporous support structure.

Although many research studies have reported the effect of using microporous-supported Nafion on the conductivity, not many have been reported regarding its effect on methanol permeability. Huang et al. fabricated similar Nafion/PTFE hybrid membranes, but they treated this composite membrane with an additional hybridization process where they coated the Nafion/PTFE membrane with a layer of silicate via a sol-gel process of TEOS [38]. The OH terminated ends in the silica matrix allow proton hopping from one site to another, and have often been associated with proton conductivity through the Grotthuss mechanism [18, 39, 40]. Thus, this additional layer could potentially act as a barrier to methanol transport and still be able to conduct protons. In the work done by Huang et al., conductivities in the range of 3×10^{-3} S/cm were obtained as compared to the conductivity of 1.01 mS/cm for Nafion 117 [38]. It was also observed that the addition of a silica layer did not adversely affect the conductivity of the composite Nafion/PTFE matrix. Furthermore, in methanol crossover measurements, it was observed that the methanol crossover rate for silica coated Nafion/PTFE hybrids was lower than Nafion 112 membranes, though the reduction was modest.

Nafion/Imidazole Composites

In most of the Nafion composite membranes discussed previously, the presence of bound water allows proton conduction via the conductive pathways though ionic clusters of polymer backbone. Water molecules grafted in the polymer backbone can easily hydrogen bond with neighboring polar sites, thus increasing the mobility of protons within the polymer. Because of the high volatility of water, which leads to dehydration of membranes at higher temperature, few studies have been carried out to replace water with a nonaqueous solvent in the polymer matrix that will be able to provide proton carrier characteristics and can also act as a barrier for methanol transport. In the case of Nafion/imidazole membranes, proton conduction is carried out by nonvolatile imidazole solvent which provides protonated and nonprotonated nitrogen functions that can act as proton donors and acceptors.

One of the biggest challenges in the design of the Nafion/imidazole membrane is to overcome the absorption of imidazole on platinum [33]. Therefore, the performance of Nafion/imidazole hybrid membranes showed a strong dependence on membrane preparation technique. It was elucidated that Nafion impregnated with molten imidazole showed high conductivities (10 mS/cm) at temperatures above 90°C. However, they lost their mechanical rigidity at low temperatures. In another fabrication method, the authors used a recast Nafion and imidazole membrane and observed conductivities in the order of 100 mS/cm at elevated temperatures. Since, at high temperatures and anhydrous conditions, water is lost from Nafion, its conductivity declines. At elevated temperatures, proton transfer is facilitated by the presence of donor and acceptor sites along with the realignment of the imidazole chain [41, 42].

In order to address the issue of catalyst poisoning by imidazole and mechanical rigidity at room temperatures, Fu et al. prepared Nafion/imidazole composite membranes that were doped by phosphoric acid, which showed slightly better conductivities than Nafion at all temperatures [43]. Suppression of poisoning of Pt catalyst was shown to be somewhat possible with the new membrane; however, the overall performance of the fuel cell was observed to be weaker than Nafion cells. Alternatively, replacing Pt with Pd-Co-Mo catalyst showed better tolerance to the hybrid membrane [43].

Nafion Composites with Other Additives

Modifications of the Nafion backbone to address methanol crossover and high temperature operation have led to the introduction of very novel additives in the polymer matrix, although their overall performance in fuel cell conditions is yet to be clearly defined. Jia et al. impregnated Nafion membranes with poly(1-methylpyrrole) by *in situ* polymerization [44]. This lowered the methanol crossover by 50% compared to untreated Nafion, though the conductivity was slightly compromised.

Lin and coworkers observed that Nafion blends with Teflon-fluorinated Ethylene-Propylene (FEP) prepared by melt-processing and hot pressing resulted in membranes that showed significant reduction in the methanol crossover rate along with high thermal, mechanical, and chemical rigidity [45]. Because of the inclusion of FEP in the polymer matrix, methanol experiences a more tortuous pathway for its transport. Apart from this, Teflon-FEP blend also provides a structural support and minimizes membrane degradations due to swelling. At 60% loading of FEP blend, the diffusion coefficient of methanol through hybrid membrane was observed to be $4.36 \times 10^{-9} cm^2/s$ in comparison to $3.6 \times 10^{-6} cm^2/s$ for Nafion membrane. In DMFC operating with 10M methanol solution, it was observed that a 100 μm thick Nafion-FEP membrane outperformed Nafion 117 mostly due to the reduced crossover rate. This is confirmed by Eq. 1.6, which shows efficiencies as high as 11% for the micro DMFC, a significant improvement over Nafion, though still too low for serious consideration.

Lee et al. have also addressed the issue of methanol crossover by introducing functionalized montmorillonite (MMT) in Nafion matrix [46]. At 5% loading of Nafion matrix with functionalized MMT, it was observed that methanol permeability decreased by approximately 30% compared to Nafion 115, without a compromise in the conductivity.

In another class of proton conducting materials, recent attention has been given to fullerene acids. Sony first developed polyhydroxy hydrogen sulfated fullerenes (PHSF), $C_{60}(OH)_n(OSO_3H)_{12-n}$ ($n \sim 6$), which showed conductivities on the order of mS/cm range. Since these materials are incapable of withstanding high temperature due to chemical instability, Wang et al. have suggested incorporating them in Nafion matrix [47]. Addition of super acids in proton conducting sulfonated membranes like Nafion can help improve the proton conductivity without having to compromise on the membrane structure. Novel fullerene/Nafion composite membranes ($HC_{60}(CN)_3$-$C_{60}(TEO)_5$-Nafion) were fabricated where $C_{60}(TEO)_5$ was used as a dispersant to improve the uptake of hydrophobic fullerene in Nafion. At low humidity conditions, the composite fullerene membranes showed conductivity more than three times that of Nafion alone. Unfortunately, studies pertaining to methanol permeability are yet to be conducted to fully verify their performance in DMFC applications.

1.3 NON-NAFION POLYMER PROTON EXCHANGE MEMBRANES

Though several advances have been made in Nafion-based proton exchange membranes, no candidate materials have surfaced which provide acceptable efficiencies for micro direct methanol fuel cells and it seems highly unlikely that the Nafion family of fluorinated polymers will ever be able to provide proton exchange membranes with sufficiently low methanol permeability. With this in mind, many groups have developed nonfluorinated membranes with the goal of decreasing the methanol permeability. Several approaches have been tried for the synthesis of low methanol crossover polymer membranes for the direct methanol fuel cell including the blending of ionic and non-ionic polymers, sulfonating poly ketones, sulfones, imides and their derivates, acid-doping of polybenzimidazole, and forming organic/inorganic composites. In this section, we will summarize the work that has been done with non-fluorinated polymeric materials for use as proton exchange membranes in direct methanol fuel cells.

1.3.1 Polyvinyl Alcohol Blends

Poly(vinyl alcohol) (PVA), Figure 1.7, has been used as the base material for several DMFC electrolyte investigations. PVA has received a significant amount of attention due to its high selectivity in the separation of water and ethanol [48, 49]. The low permeability of ethanol through the PVA film in the presence of copious amounts of water suggests that this polymer may act as an effective barrier for other low molecular weight polar compounds, including methanol. Though the decreased methanol permeability of PVA is promising, PVA is not an efficient ion conductor, limiting its usefulness as a proton exchange membrane. Therefore, PVA has been blended with various other materials in order to increase the ionic conductivity while maintaining reduced methanol crossover through the film.

The most straightforward PVA blend for fuel cell applications is PVA/Nafion. DeLuca and Elabd have recently investigated PVA/Nafion blends for the DMFC [50]. Though fuel cell data was not presented, the ionic conductivity and methanol permeability was investigated with a 50:50 blend composition over a wide range of annealing temperatures. It was found that the conductivity of the prepared films was nearly independent of annealing temperature

n
OH

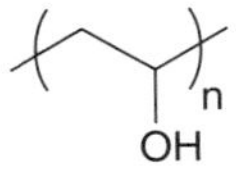

FIGURE 1.7 Structure of poly(vinyl Alcohol)

and quite high at 20 mS/cm, nearly equivalent to their measured value for Nafion 117 (26 mS/cm). On the other hand, the methanol permeability of the prepared films was a strong function of the annealing temperature, varying by more than an order of magnitude. The lowest measured permeability was found with an annealing temperature of 230°C and was only three times lower than Nafion.

PVA blended with polystyrene sulfonic acid (PSSA) has also been investigated as possible proton exchange membranes with reduced methanol permeability. The ionic conductivity and methanol permeability has been studied over a wide range of PSSA content and annealing temperatures [51]. The best performing membrane was achieved at 17 wt% PSSA and an annealing temperature of 177°C. In this case, the methanol permeability was reduced by half compared with Nafion, though the ionic conductivity was compromised at only 4 mS/cm. It was believed that membrane swelling contributed to both the lower conductivity of the polymer film and the modest reduction in the methanol crossover. In order to address this, a ternary component, maleic acid, has been added to the PVA/PSSA blend. It was found that the membrane swelling was significantly reduced and the methanol permeability was decreased by an order of magnitude and an ionic conductivity of 95 mS/cm was demonstrated [52].

Poly(2-acrylamido-2-methyl-1-propanesulfonic acid) (PAMPS), stabilized with poly(vinylpyrrolidone), has also been added to PVA films [53, 54]. PAMPS is particularly promising due to its ionic performance in other devices including humidity sensors and Li-ion cells [55, 56]. Also, its water content per sulfonic acid group is significantly lower than the fluorinated membranes [57, 58]. Qiao et al. reported an ion exchange capacity of 1.61 meq/g for a 50:50 polymer film. This yielded an excellent proton conductivity of 88 mS/cm. Also, methanol permeation was reduced by three times compared with Nafion. The chemical stability was also found to be quite good in hydrogen peroxide solutions of various compositions [53]. Walker demonstrated ternary membranes of PVA/PAMPS and poly(2-dydroxyethyl methacrylate) (PHEMA) [54]. The ternary system demonstrated lower conductivity than the PVA/PAMPS films, with experimental values ranging from 5 to 15 mS/cm between 20 and 80°C. This may be due to the lower

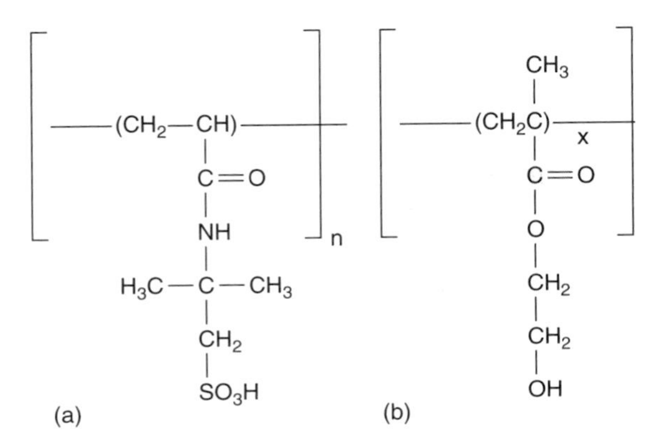

FIGURE 1.8 Structure for PHEMA and PAMPS additives for lower methanol crossover proton exchange membranes.

exchange capacity of the dangling hydroxyl group on PHEMA when compared to the sulfonic acid group in PAMPS; the structures for both PHEMA and PAMPS are shown in Figure 1.8. On the other hand, the methanol-water selectivity of the film was increased nearly tenfold when compared with Nafion films.

1.3.2 Sulfonated Poly (Ether Ketone)s

One of the most popular substitute materials for PEMFCs and DMFCs is sulfonated poly(ether ether ketone) (SPEEK). SPEEK membranes have shown acceptable thermal, mechanical, and chemical stability. It is also easy to control their degree of sulfonation, which allows researchers to limit water uptake, a key parameter in diffusive methanol crossover. Kruer et al. found that the electro-osmotic drag, and hence, methanol crossover is greatly reduced in SPEEK films [59–61]. It was postulated that this is due to the reduced separation between hydrophobic and hydrophilic regions in the polymer backbone as well as the increased distance between the sulfonic acid groups in the structure.

Yang and Manthiram obtained SPEEK membranes by dissolving poly(ether ether ketone) in concentrated sulfuric acid at room temperature with vigorous agitation [62]. The resulting sulfonated monomer was solvent exchanged with N,N-dimethylacetamide and cast on an inert substrate. The ion exchange capacity (IEC) was controlled by the sulfonation time. The water uptake was a strong

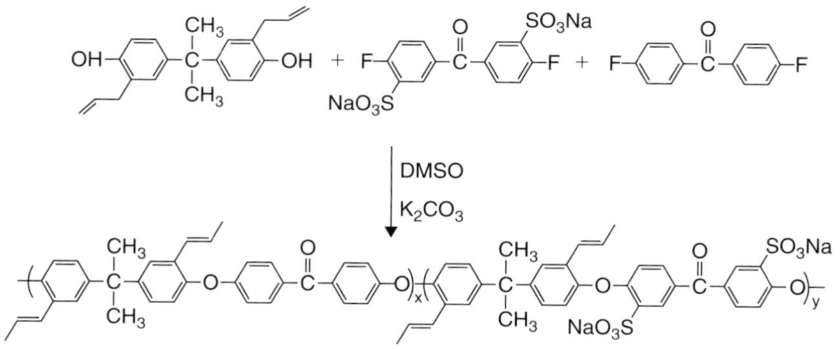

FIGURE 1.9 Synthesis of SPEEK polymers [64].

function of the IEC. The highest proton conductivity, measured at 80°C and 100% relative humidity, was 7 mS/cm. Despite reduced electrolyte conductivity, performance curves in a DMFC (2 M methanol) using the SPEEK film were superior to Nafion. Reduced methanol crossover was cited as the reason for the performance improvement, though no permeability data is available. However, Li et al. have reported conductivity values for SPEEK as high as 40 mS/cm and methanol permeabilities of 5.7×10^{-7} cm^2/s [63], which agreed well with another study by Gil [64].

In order to further reduce the methanol permeability of SPEEK electrolytes, Zhong et al. crosslinked the sulfonated polymer with benzophenone [65]. The SPEEK polymer was prepared via nucleophilic substitution of diallyl bisphenol A, 4,4-difluorobenzophenone and 5,5-carbonyl-bis(2-fluorobenzenesulfonate) in DMSO (Figure 1.9) and then UV crosslinked (Figure 1.10).

It was found that crosslinking significantly reduced the water uptake and methanol permeability. The methanol diffusion coefficient was found as 8.5×10^{-8} cm^2/s, which is an order of magnitude lower than SPEEK alone and 50 times smaller than Nafion 117 without a compromise in the conductivity.

SPEEK blends with PVDF have also been reported [66]. The blend was achieved by mixing SPEEK and PVDF in N,N-dimethylformamide for 30 minutes and solution cast on glass. PVDF was chosen due to inherent hydrophobicity. The methanol permeability was reduced below 10^{-9} cm^2/s, a significant decrease compared to both Nafion and pure SPEEK films. The ionic conductivity

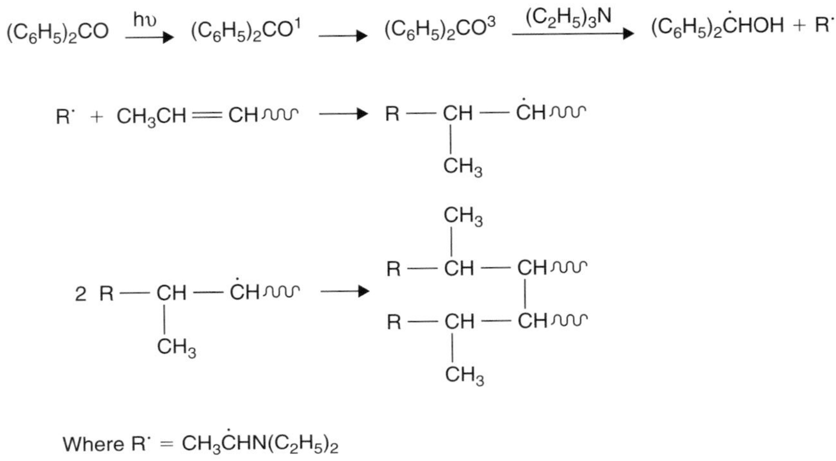

$$(C_6H_5)_2CO \xrightarrow{h\upsilon} (C_6H_5)_2CO^1 \longrightarrow (C_6H_5)_2CO^3 \xrightarrow{(C_2H_5)_3N} (C_6H_5)_2\dot{C}HOH + R^{\cdot}$$

$$R^{\cdot} + CH_3CH{=}CH\sim \longrightarrow R{-}CH(CH_3){-}\dot{C}H\sim$$

$$2\ R{-}CH(CH_3){-}\dot{C}H\sim \longrightarrow \begin{array}{l} R{-}CH(CH_3){-}CH\sim \\ \qquad\qquad\quad | \\ R{-}CH(CH_3){-}CH\sim \end{array}$$

Where $R^{\cdot} = CH_3\dot{C}HN(C_2H_5)_2$

FIGURE 1.10 UV crosslinking mechanism for SPEEK polymer films [65].

was also maintained comparable to neat SPEEK electrolytes. This is a significant reduction in the methanol loss due to fuel permeation and micro DMFC efficiencies around 27% can be achieved at high aspect ratio films, $\delta/A = 1\,\text{cm}^{-1}$. However, at reasonable aspect ratios, < 0.2, the fuel loss due to methanol permeation remains above 93% as shown in Figure 1.11.

Several ketone variations have been proposed in the literature including sulfonated poly(ether ketone ketone) [67], poly(aryl ether ether ketone) [68], poly(arylene ether ketone) [69], poly(phathalazinone ether ketone) [70] and poly(benzoxazole ether ketone) [71]. Poly(ether ether sulfone) [72], poly(arylene ether sulfone) [73], poly(phathalazinone ether sulfone ketone) [74], poly(arylene ether benzonitrile) [75] and poly(arylene ether 1,3,4-oxadiazole) [76] polymer networks have also been proposed. In all cases, however, the overall efficiency was higher than Nafion, though lower than SPEEK.

1.3.3 Sulfonated Poly (Phenylene Oxide)

Guan and coworkers have recently reported a novel sulfonated poly(phenylene oxide) (SPPO) electrolyte for the direct methanol fuel cell [77–80]. In the investigations, PPO was purchased, and dissolved in chloroform and sulfonated introducing a 5 wt%

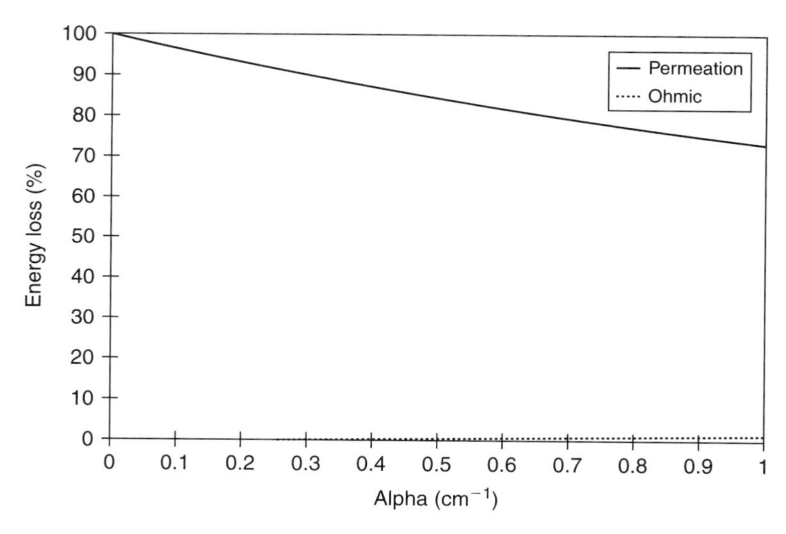

FIGURE 1.11 Energy loss for a SPEEK-PVDF electrolyte micro DMFC as a function of the membrane aspect ratio operating at 100 μA, 0.5 V.

chlorosulfonic acid solution dropwise. The ion exchange capacity for the SPPO was reported to be as high as 2.83 meq/g, which is three times higher than the accepted value for Nafion 112, though the conductivity is only 0.4 mS/cm. The methanol permeability for neat SPPO was not reported.

SPPO has been combined with a phospho-silicate sol and then solution cast to yield SPPO-glass composite electrolytes. The SPPO-glass composite electrolytes exhibited outstanding performance when compared to Nafion for both the ionic conductivity and methanol permeation experiments. The ionic performance was a modest function of the phospho-silicate glass composition, ranging from 17 to 216 mS/cm, though all of the films prepared with 2.83 meq/g ion exchange capacity showed a conductivity of at least 76 mS/cm in the fully hydrated state. However, the films showed a significant decrease in the ionic performance at decreased relative humidity. Finally, the SPPO-glass films showed a methanol permeability coefficient only four times lower than Nafion 112.

1.3.4 Polybenzimidazoles

Polybenzimidazole (PBI) membranes have received a considerable amount of attention in recent years as a possible proton

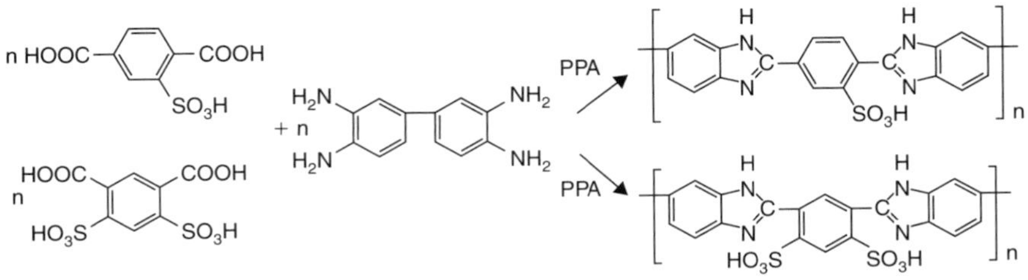

FIGURE 1.12 Preparation of sulfonated PBI films [86].

exchange membrane. PBI films generally have good mechanical strength, outstanding chemical and thermal stability (>300°C), and high proton conductivity under low relative humidity conditions (nearly two orders of magnitude greater than Nafion).

Sulfonated PBI films have been synthesized in polyphosphoric acid by several methods [81–83] but the most common is polycondensation of sulfoterephthalic acid and disulfoisophthalic acid with 3,3-diaminobenzidine [81, 84, 85]. The mechanism for the polycondensation is showed in Figure 1.12.

The ionic conductivity of the films has been shown to increase with increased sulfonation and was comparable to Nafion at 80°C [84]. However, the methanol permeability of the films was sufficiently high as to yield no advantage over Nafion for use in the DMFC.

The proton conductivity of phosphoric acid (PA) doped PBI films has also been characterized [87–91]. Unlike the sulfonated PBI films, the phosphoric acid groups are not directly bonded to the polymer backbone. Instead, it has been suggested that the low charge density anion is immobilized and linked to the structure by a strong hydrogen-bond network. This is clearly seen in Figure 1.13.

It has also been hypothesized that this network provides the pathway for proton transport. The authors have observed that the association between the immobilized anion and PBI reduces the effective dissociation constants for phosphoric acid by nearly two orders of magnitude [87]. This gives a pKa for the first proton dissociation of 1.44, which is significantly less than the sulfonic acid group. This is apparent in the ionic conductivity of the films, which was 10^{-4}S/cm at a relative humidity of 35% measured by

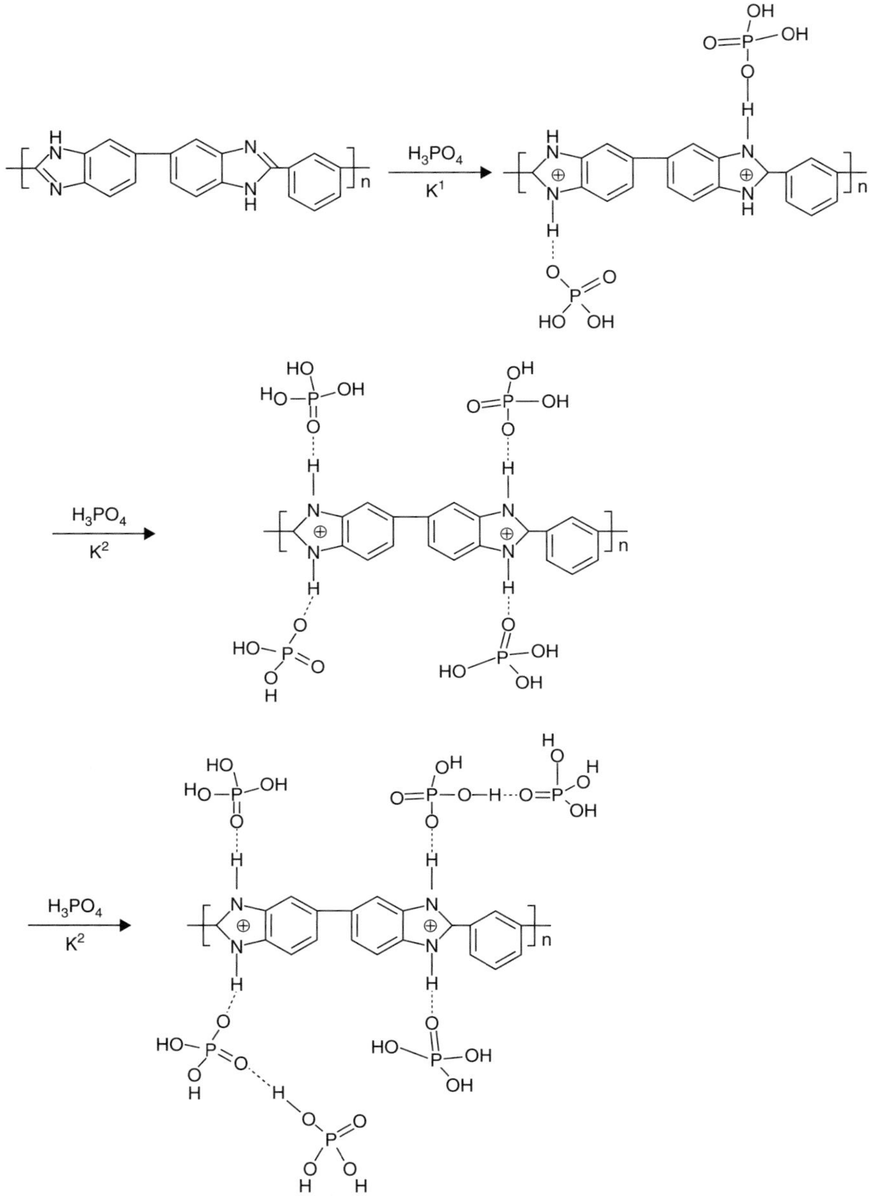

H_3PO_4

K^1

H_3PO_4

K^2

H_3PO_4

K^2

FIGURE 1.13 Proposed association of phosphoric acid in PBI films [87].

Liu et al. At more fully hydrated conditions, proton conductivities as high as 40 mS/cm have been reported [89]. Also, the methanol permeability of the PBI/PA films has been estimated as nearly two orders of magnitude lower than the fluorinated polymers [89]. This should make the PBI/PA films a promising alternative to Nafion in the high and intermediate power regime; however, its application in ultra low-power DMFCs is unlikely as the energy loss due to fuel permeation would be 97% at reasonable form factors. Also, there is still much uncertainty about the long-term mobility of the anion during operation.

1.3.5 Polyimides

Among the aromatic electrolytes, sulfonic acid functionalized polyimides are considered among the most promising candidates. The polyimide backbone provides the membranes with good thermal, chemical, and mechanical stability. It is also responsible for low moiety permeation that has been observed for several gases including hydrogen and oxygen. Like other polymers, the sulfonic acid functionality provides the ionic conductivity and water uptake. Several different polyimides have been prepared [92–103] and their general structure is shown in Figure 1.14.

The polyimide which -Ar- contained in its structure has shown very promising performance for both their conductivity and methanol permeability.

Li and coworkers have recently reported the electrochemical performance of 2,2-bis(3-sulfobenzoyl) benzidine (BSBB) [96]. The electrolyte monomer was synthesized by reaction of 2,2-dibenzoyl benzidine with fuming sulfuric acid. The resulting monomer was copolymerized with 1,4,5,8-naphathlenetetracarboxylic dianhydride to give the proton-conducting membrane. The resulting

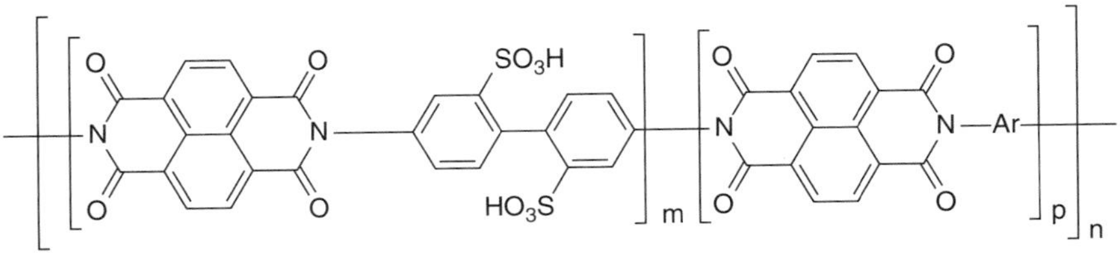

FIGURE 1.14 General structure for the polyimide electrolytes [86].

conductivity for the BSBB film was 125 mS/cm at room temperature, measurably higher than Nafion under identical conditions. Also, Hu et al. synthesized a sulfonated diamine bearing sulfophenyl pendant groups, 2,2-(4-sulfophenyl) benzidine (BSPhB) [97]. The BSPhB film had an ion exchange capacity of 1.5–2.8 meq/g. A BSPhB film with an ion exchange capacity of 1.77 meq/g had proton conductivities of 120 and 260 mS/cm at 60 and 120°C, respectively. Unfortunately, the methanol permeability was not measured in either case.

Both the ionic conductivity and methanol permeability of 2,2-bis (p-aminophenoxy)-1,1-binaphthyl-6,6-disulfonic acid (BNDADS) copolymerized with 1,4,5,8-naphthalenetetracarboxylic dianhydride (NTDA) and 4,4-diaminodiphenyl ether (DADE) [98–103]. The BNDADS content was adjusted between 30 and 80 wt%. The lowest methanol permeability was observed with the 30 wt% BNDADS polymer and was around two orders of magnitude lower than Nafion with comparable conductivity at room temperature. However, the membrane with the highest selectivity of conductivity to methanol permeability had a BNDADS content of 40 wt%. The proton conductivity of the electrolyte was 50 mS/cm at 20°C and 93 mS/cm at 80°C. The methanol permeability was an order of magnitude lower than Nafion.

1.3.6 Non-Nafion Polymer Organic/Inorganic Composites

Composite membranes featuring a blend of organic polymer electrolyte and a solid inorganic material has been the subject of much research in recent years. Several materials have been explored including silica [104–107], titanium dioxide [34, 107], zirconium dioxide [107–109], zirconium phosphate [110], boron orthophosphate [111], and phosphotungstic acid [29, 105]. It is believed that the introduction of these materials will decrease the methanol crossover of the electrolyte by increasing the diffusion path length of the methanol clusters. Also, the water retention of the films may be improved since the water is more tightly bound to the hydrophilic metal oxide than the organic material that contains both hydrophilic and hydrophobic regions.

Silica has been one of the most extensively studied filler materials for composite films due to its ease in preparation and incorporation into the polymer film via the sol-gel process. In general, the silica-composite proton exchange membranes show a significant decrease in the methanol permeability [104–106, 112]. Reductions in the methanol permeability of 50–90% with SiO_2/SPEEK membranes have been reported, depending on the silica preparation technique [107]. This is consistent with that observed by both Zhang and Thangamuthu [104, 112]. Also, in each case the ionic conductivity of the film was reduced by nearly half, leading to only modest increases in the membrane efficiency. The decrease in both the methanol and ionic transport in the film is most likely due to the strong bonding between the silica particles and water in the film, which leads to a reduction in the water mobility through the film.

Another popular metal oxide additive is zirconium dioxide [107–109]. Unlike SiO_2, zirconium dioxide is not a simple filler in the sense that it only provides an impermeable hydrophilic material within the organic network. It is believed that the ZrO_2 forms a large heteropoly anion with the proton-conducting sulfonic acid group. In this case, the stationary anionic group is surrounded by a self-assembled structure of metal oxide polyhedrons. A scanning electron micrograph showing finely dispersed ZrO_2 in SPEEK is shown in Figure 1.15.

This self-assembled proton transport complex should lead to an improvement in the proton conductivity when compared to the silica films since the effective charge radius is increased. This was confirmed by Nunes and coworkers where the proton conductivity of a ZrO_2/SPEEK membrane was reduced by only 30% while the water flux was reduced by 28 times [107]. Also, methanol permeabilities 60 times lower than the uncomplexed SPEEK membranes were reported. Silva and coworkers observed similar behavior for ZrO_2/SPEEK films with a sulfonation degree of 87% [108, 109]. Promising DMFC performance was also reported. Polarization curves for the ZrO_2/SPEEK showed lower overall performance and peak power densities than Nafion, though the open circuit voltages were higher and the CO_2 emission from the cathode compartment was lower.

Phosphotungstic acid (PWA) is another common additive in organic–inorganic hybrid composites. PWA is added in order to increase the ionic conductivity of the electrolyte, which is typically

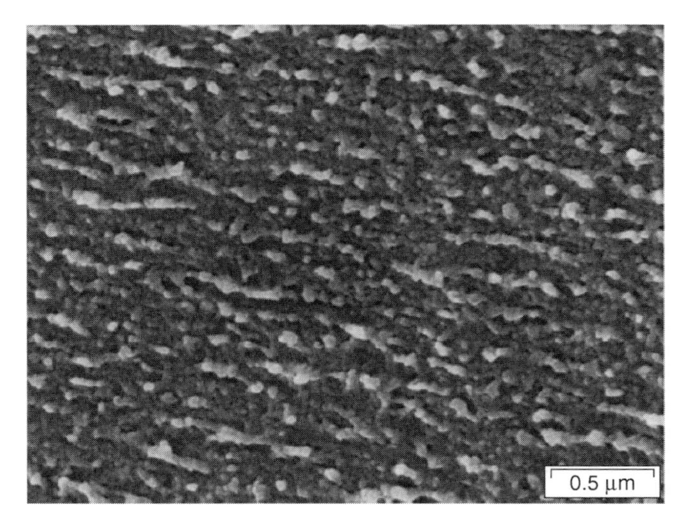

FIGURE 1.15 Scanning electron micrograph of a 5% ZrO_2/SPEEK composite proton exchange membrane [108].

reduced compared to the neat polymer film. PWA has been added to several systems, including Nafion-SiO_2 and PVA-SiO_2 composites [23, 104, 105, 113]. One concern with PWA and other unbound additives, including phosphoric acid, is their mobility in the film. It is likely that over time they will deplete by simple solvation and leaching when exposed to water and methanol. For the Nafion-SiO_2-PWA case, the conductivity of the electrolyte was restored to the value for Nafion at 100 RH, though the performance was significantly better than Nafion at low relative humidity. Unfortunately, the methanol crossover is not reported. However, Shao et al. showed that the water uptake for the composite was 50% higher than Nafion and that the hydrogen crossover through the film was 20% higher [23, 113]. From this, it is a reasonable conclusion that the methanol crossover will be higher than Nafion as well, decreasing the film selectivity. The PVA-SiO_2-PWA composite is an interesting system as neither PVA nor SiO_2 exhibits any significant ionic mobility on their own. Despite this, Xu has reported ionic conductivity values of 17 mS/cm with a PVA:SiO_2:PWA ratio of 2:1:2 [105]. This is in disagreement with Lin, who reported limited conductivity, 2×10^{-4} S/cm, with a 1:0:1 composite [114]. The methanol permeability was shown to be more than two orders of magnitude lower than Nafion 115.

1.4 INORGANIC MEMBRANES

Some attention has recently been given to the design of novel inorganic proton exchange membranes. Successful design of such membranes could potentially overcome the drawbacks of the polymer-based electrolytes by exhibiting comparable proton conductivities and ultra-low methanol crossover at reasonable cost.

One approach has been the use of transition metal phosphates, silicates, or super acids such as phosphotungstic acid as the proton-conducting medium. They all have the ability to transfer protons by either the Grotthuss or vehicle mechanism. According to the Grotthuss mechanism, protons hop from one site to another based on the strength of hydrogen bonding. In the vehicle mechanism, protons diffuse through the bonded water in the matrix, which is prevalent in polymer-based materials. The Grotthuss mechanism, being a surface transport mechanism, dominates most of the observed protons, transport through these materials. Both mechanisms are shown in Figure 1.16.

These inorganic materials with adequate proton conductivity are ideal candidates for proton exchange membranes by virtue of their low cost, high reliability, stable performance, and easy synthesis route over a wide range of thicknesses. In addition, they can also act as a methanol barrier layer to increase device efficiency in DMFCs. However, one of the challenges associated with these membranes is their inability to form free standing membranes. As a result, their application in fuel cell devices is limited. Recently, the synthesis of these inorganic proton exchange membranes by means of wet sol-gel chemistry has been successful in forming high quality films on various substrates. Consequently, sol-gel processing of inorganic materials has attained increased interest in the area of proton-conducting membranes. One of the added benefits of sol-gel processing is that it allows customized fabrication of silicate materials to control the physical and chemical properties. This is achieved by optimizing the processing conditions or by addition of a functionalized group in their matrix.

1.4.1 Silicate Glasses

Since the early 1980s, silicate glasses have drawn attention as a possible proton-conducting medium. Incomplete oxidation of

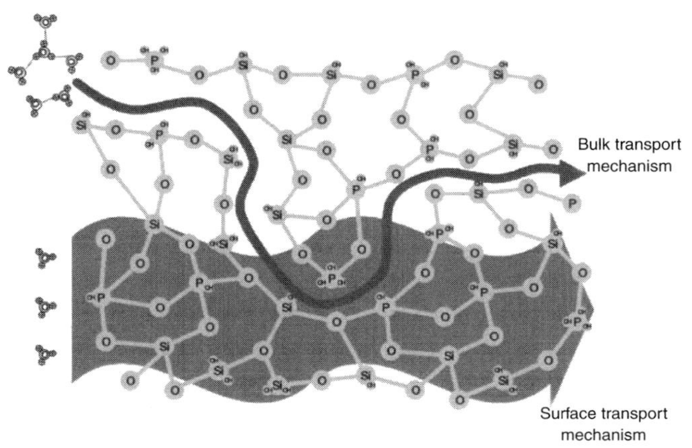

(a) The contribution of bulk transport mechanism is smaller than that of surface transport for proton conductivity in the glass membrane

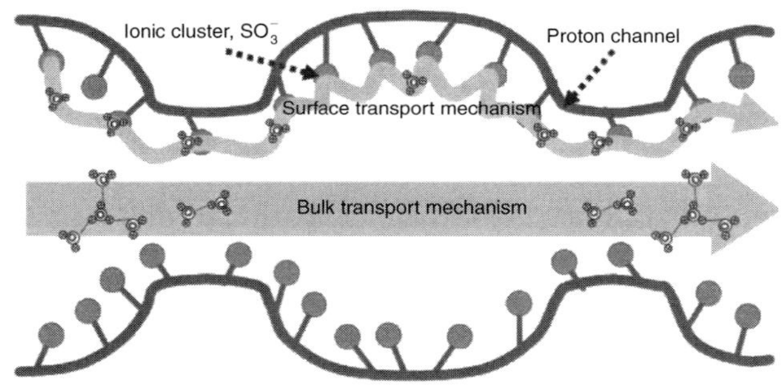

(b) The contribution of bulk transport mechanism is larger than that of surface transport for proton conductivity in Nafion membrane

FIGURE 1.16 Proton transport mechanisms for (a) glass (b) and Nafion electrolytes [115].

silicate glasses can lead to the formation of defect sites in the form of −OH terminated silanol sites within the otherwise closely packed tetrahedral SiO_2 structure. As a result, high mobility is imparted to the protons along the −OH terminated end groups, which travel under the influence of hydrogen bonding from one defect site to another. Therefore, silicate glasses can exhibit moderate to high proton conductivity [18].

Abe et al. have several publications which divulge the proton-conducting mechanism in these silicate glasses and their hybrids [116–119]. According to them, low temperature fabrication of these glasses allows the formation of −OH terminated defect sites in

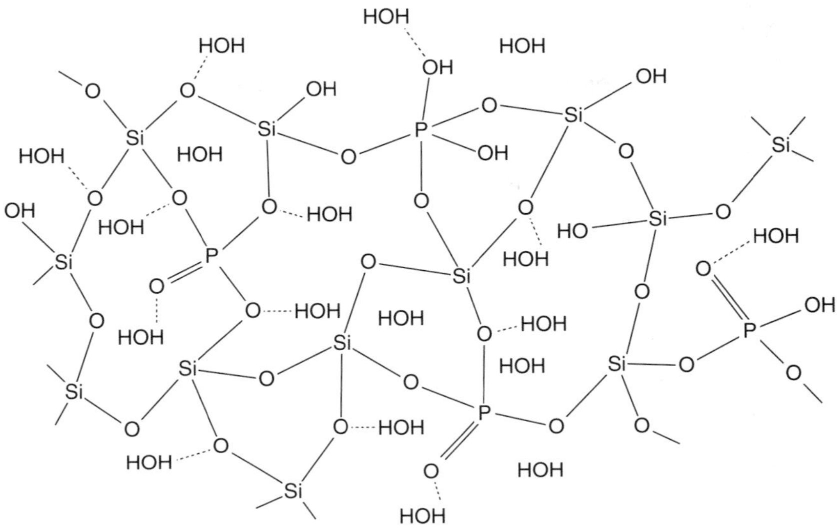

FIGURE 1.17 Proposed structure of SiO_2-P_2O_5 network [121].

comparison to the high temperature processing conditions. Consequently, the sol-gel process of fabricating glasses is more favorable than the traditional melt quenching technique. In an attempt to impart higher conductivities to silicate glasses, addition of acidic groups like phosphates, zirconium, and titanium oxides are of particular interest.

1.4.2 Phospho-Silicate Glasses

In a study conducted by Abe et al., the authors have indicated that the presence of group V elements such as phosphorus in Si-O-Si improves the conductivity of glass networks by allowing the formation of Si-O-P-OH groups [120]. The resulting matrix, shown in Figure 1.17, exhibits high ion exchange capacity by the virtue of high free volume and pore wall surface area. Since P-OH bonds are more acidic in nature, their inclusion in glass matrix can potentially provide higher mobility to protons, as has been explained in the work done by Prakash et al. [18]. In their study, phosphor-silicate glasses were synthesized by plasma-enhanced chemical vapor deposition (PECVD) with silane, phosphine, and nitrous as the reaction precursor. The resulting films showed conductivities around 10^{-4} S/cm and the methanol permeability was not reported. Also, since the plasma deposited glass membranes were amorphous in nature, fabricating free-standing films of significant area was not difficult.

Alternatively, Nogami and coworkers have explored phospho-silicate glasses prepared by a sol-gel synthesis route [116, 122–126]. In one of their earlier works, silicate glasses with varying Si:P ratios were formed [123]. The resulting P_2O_5-SiO_2 glass membranes were porous in nature and showed higher content of OH groups in comparison to pure SiO_2 glasses. Moreover, conductivities as high as 10^{-2} S/cm were reported [126], which is a significant improvement over dense SiO_2 films (10^{-9} S/cm). Also, Wang et al. investigated silicate glass performance in the presence of non-ionic and cationic surfactants and concluded that three-dimensionally oriented pores showed easier pathways for proton conduction. The highest observed conductivity for phospho-silicate glass membranes is 2×10^{-2} S/cm with 5 wt% P_2O_5 in the hybrid membrane [126].

Even though the phospho-silicate proton exchange membranes have shown adequate conductivity at room temperature, it has been difficult for researchers to obtain films that were crack-free, which has limited their ability to investigate the methanol permeability through the films. High mechanical glass films are difficult to obtain due to the high rate of water loss during high-temperature processing. To combat this, Tung et al. devised a water vapor management system for the sol-gel method for the fabrication of phospho-silicate glasses, whereby the environment was saturated by water vapor during the gelation period [115, 121]. In this method, the authors hydrolyzed the precursors for the sol-gel reaction in ethanol and water followed by hydrolization in formamide to control pore formation. Not only was this process successful in fabricating crack-free membranes with conductivities of 9.35×10^{-3} S/cm which were less sensitive to humidity, the resulting membrane also showed a considerable reduction in methanol permeability, 2.1×10^{-9} cm^2/s, compared to the measured methanol permeability of 1.57×10^{-6} cm^2/s for Nafion 117 membrane. Furthermore, they observed that membranes with lower content of P_2O_5 showed more cracks and had fewer numbers of bonding water/OH. The high conductivity and low methanol crossover of the phosphor-silicate electrolytes are significant improvements over all the polymer films previously discussed.

1.4.3 P_2O_5-ZrO_2-SiO_2 and P_2O_5-TiO_2-SiO_2 Glasses

In an attempt to modify the structure of the inorganic, solid state electrolytes, some studies have focused on the performance of phospho-silicate glass with additional oxides embedded in their matrix. Titanium and zirconium oxides have been identified as possible additives because of their ability to form a defined molecular structure with chemisorbed water. The presence of water in the matrix extends the formation of ionic clusters, thus increasing the ion exchange capacity. Additionally, the presence of these metallics in the matrix can further restrict methanol transportation through the electrolyte membrane. The titanium and zirconium oxides are mixed with proton-conducting phospho-silicate glass via the sol-gel process. Uma et al. have studied the performance of phospho-silicate glasses with zirconium and titanium oxide additives (P_2O_5-TiO_2-SiO_2 and P_2O_5-ZrO_2-SiO_2) [127, 128]. The authors observed that the resulting glass matrix was noncrystalline in nature and that the proton conductivity showed a significant dependence on the porosity and the activated pore size. For $9P_2O_5$-$6TiO_2$-$85SiO_2$, it was observed that the conductivity was 3.6×10^{-2}S/cm at 90°C and 30% RH. On the other hand, $5P_2O_5$-$4ZrO_2$-$91SiO_2$ showed conductivities in the order of 10^{-2}S/cm and exhibited a similar dependence on relative humidity and temperature. In both of these glass membranes, proton conduction was identified to take place by the Grotthuss mechanism and protons transported through the electrolyte material by hopping from one negatively charged site to the other. As has been previously mentioned, P_2O_5 is more acidic than silica, and so are zirconium and titanium oxides. As a result, inclusion of more of these makes the overall proton conductivities of silicate glasses increase due to their higher water adsorption properties. Unfortunately, the methanol permeability of these materials is yet to be reported.

1.4.4 Inorganic/Organic Nano Composite Membranes

In order to provide increased mechanical rigidity to the surface structure of silica glasses, 3-glycidoxypropyl trimethoxy silane (GPTMS, $C_9H_{20}O_5Si$) has often been mixed with sol-gel derived SiO_2 glasses. GPTMS was chosen because it has been speculated that the epoxy linkage would be effective for mechanical rigidity, while the silane groups could potentially provide a pathway for proton conduction by forming −OH terminated end groups in the

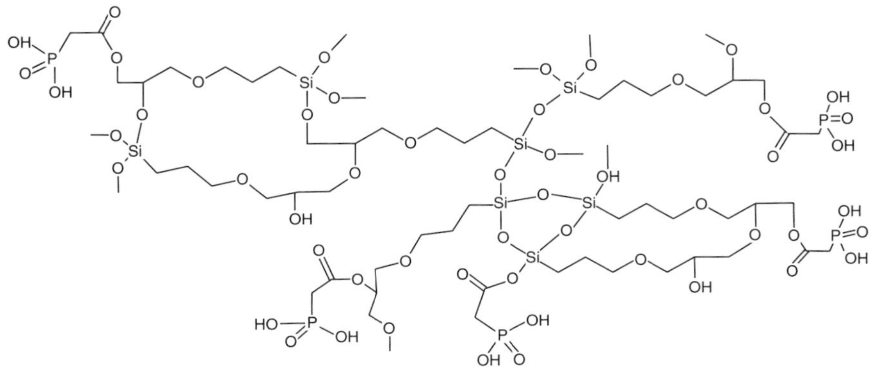

FIGURE 1.18 Proposed structure of GPTMS and MPTS hybrid membrane [132].

Si-O-Si matrix [129]. Furthermore, in an attempt to create a more conductive structure, sulfonic acid groups have been introduced into the glass matrix with 3-mercaptopropyl trimethoxy silane (MPTS $C_6H_{16}O_3SSi$), whose thiol group can be oxidized to sulfonic acid. The performance of MPTS-GPTMS composites (Figure 1.18) has been reported by Park et al. [130]. It was observed that the resulting hybrid glass exhibited excellent mechanical stability and showed no cracks even after treatment in either an ultrasonic bath or boiling water. Also, the conductivity of the films was found to be directly proportional to the MPTS content. At room temperature and 50 mol% MPTS, the conductivity was 76 mS/cm, which was increased to 92 mS/cm at 100% RH. Also, preliminary experiments on MPTS-GPTMS composite films in our group indicate that the methanol permeability may be lower than 10^{-10} mol-cm/cm^2-day-Pa [131]. This yields the most significant improvement in device performance of any electrolyte reported for micro DMFCs. This is clearly observed in Figure 1.19, where efficiencies as high as 80% are possible at high aspect ratios.

In a slightly different approach, Kato et al. mixed GPTMS with phosphonoacetic acid (PA) in sol-gel reaction to create a linkage between carboxyl groups of PA and the epoxy groups of GPTMS with an added advantage of a phosphorus group in the cross-linked matrix [132].The resulting matrix can be readily formed into crackless free-standing films. It was observed that with increasing phosphorus content, the conductivity of the films increased along with

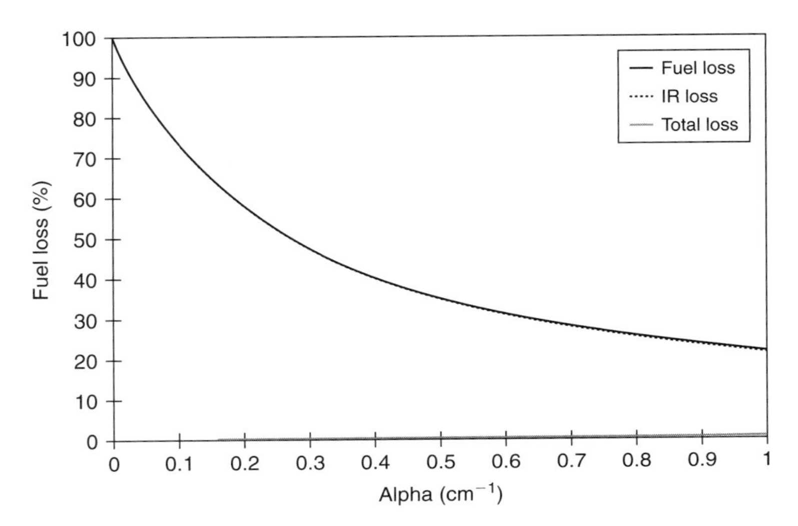

FIGURE 1.19 Calculated efficiency for a 100 μA, 0.5 V micro DMFC with a 50:50 GPTMS-MPTS composite proton exchange membrane.

film brittleness. At about equal GPTMS and PA content, the highest conductivity observed without compromising the mechanical rigidity of the film was approximately 10^{-3} S/cm. Unfortunately, the methanol permeability was not reported.

1.5 CONCLUSIONS

There is no doubt that significant advances have been made in recent years in reducing the permeability of methanol in polymer-based proton exchange membranes. Several new membrane types have been developed which have both increased ionic conductivity and reduced methanol permeability compared with Nafion. A summary of the ionic conductivity and methanol permeability of state-of-the-art electrolytes and their comparison with Nafion 117 is shown in Figure 1.20 [133].

It is clear from Figure 1.20 that many polymer electrolytes have been prepared with methanol permeabilities nearly two orders of magnitude lower than Nafion with minimal compromise in the conductivity. Several of these have the potential for exceptional behavior at intermediate and high power scales. This is demonstrated in Figure 1.21, where the DMFC efficiency as a function of electrolyte

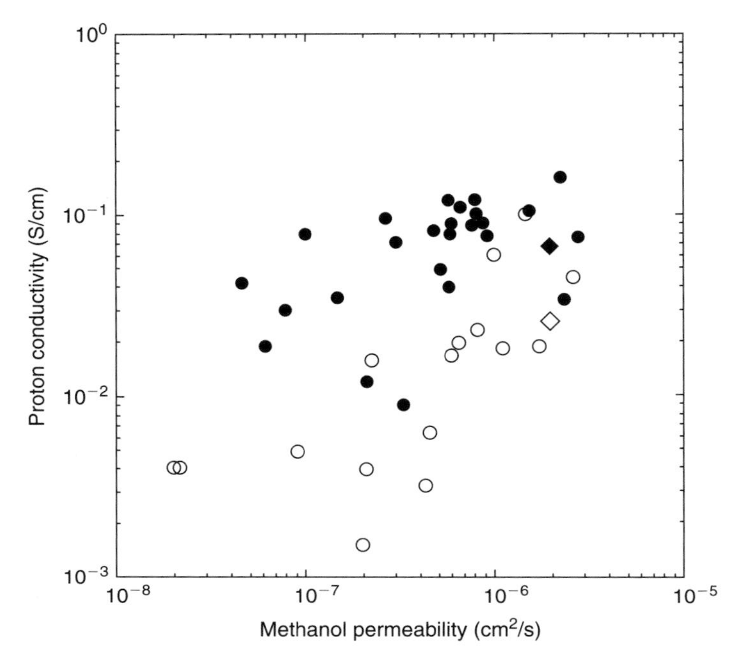

FIGURE 1.20 Proton conductivity and methanol permeability for polymer-based proton exchange membranes [133]. ◇ and ◆ denote the reference values for Nafion. ○ and ● represent the films where the conductivity was measured in-plane and through-plane, respectively.

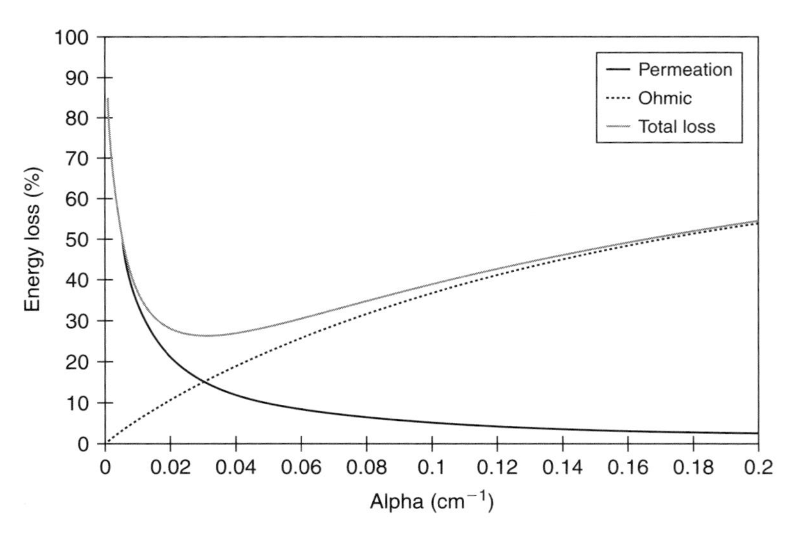

FIGURE 1.21 Energy loss for a SPEEK-PVDF electrolyte DMFC as a function of the membrane aspect ratio operating at 50 mA/cm^2, 0.5 V.

aspect ratio is shown for a fuel cell operating at $50\,mA/cm^2$. The highest efficiency realized is 72% at an aspect ratio of 0.024, which is reasonable. However, as was shown for the same material in Figure 1.11, even the best performing alternative polymer membrane falls far short of what is required for ultra low-power applications. Unfortunately, this is most likely intrinsic to polymer films and cannot be overcome. In polymer films, due to the solvated proton transport mechanism, both the conductivity and permeability are strong functions of the water content. This means that a change in one will always be mirrored to some extent in the other. This is clear from Figure 1.20, where the relationship between conductivity and methanol permeability is essentially linear over a wide range of electrolyte chemistries. Therefore, it seems highly unlikely that polymer films will ever be able to meet the requirements of an ultra low-power device and future work should be focused on finding electrolyte materials with higher film density and lower water content.

To this end, novel inorganic electrolytes have been prepared with ultralow methanol permeability, often more than an order of magnitude lower than their polymer counterparts. Also, sulfonic acid functionalized glasses have been prepared with proton

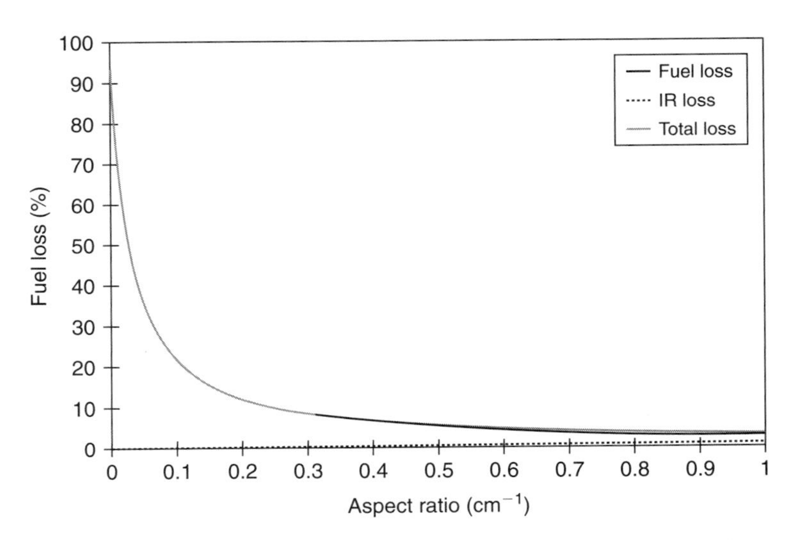

FIGURE 1.22 Energy loss for a micro DMFC with a methanol permeability coefficient of 10^{-11} mol-cm/cm^2-day-Pa and conductivity of 7 mS/cm operating at 100 μA and 0.5 V.

conductivities comparable to Nafion with methanol permeabilities more than four orders of magnitude lower. These electrolytes are also poised to provide low power with quite high efficiency, which was shown in Figure 1.19 for a GPTMS-MPTS composite membrane. Therefore, glass-based proton exchange membranes appear most promising among all of the candidates in the ultra-low power regime. However, much work remains and future work should focus on further reducing the methanol permeability, even if this somewhat compromises the ionic conductivity. This is illustrated in Figure 1.22, which shows what the energy efficiency of a $100\,\mu A$, $0.5\,V$ micro DMFC would be if the methanol permeability were reduced by an order of magnitude compared to the GPTMS-MPTS composite and the ionic conductivity was compromised by an order of magnitude as well. Not only would the efficiency be increased over the entire range of form factors, the energy loss at reasonable aspect ratios ($<0.1\,cm^{-1}$) would be less than 20%.

References

[1] G. Rincon-Mora, Harvesting microelectronic circuits, in: S. Priya, D.J. Inman (Eds.) Energy Harvesting Technologies, Springer, 2009.
[2] D.R. Chu, R.Z. Jiang, Electrochim. Acta 51 (2006) 5829–5835.
[3] F.Q. Liu, G.Q. Lu, C.Y. Wang, J. Electrochem. Soc. 153 (2006) A543–A553.
[4] H.W. Cooper, Chem. Eng. Prog. 103 (2007) 34–43.
[5] P. Piela, C. Eickes, E. Brosha, F. Garzon, P. Zelenay, J. Electrochem. Soc. 151 (2004) A2053–A2059.
[6] A. Blum, T. Duvdevani, M. Philosoph, N. Rudoy, E. Peled, J. Power Sources 117 (2003) 22–25.
[7] K.A. Mauritz, R.B. Moore, Chem. Rev. 104 (2004) 4535–4585.
[8] A. Ghielmi, P. Vaccarono, C. Troglia, V. Arcella, J. Power Sources 145 (2005) 108–115.
[9] M. Wakizoe, O.A. Velev, S. Srinivasan, Electrochim. Acta 40 (1995) 335–344.
[10] X.Z. Du, J.R. Yu, B.L. Yi, M. Han, K.W. Bi, Phys. Chem. Chem. Phys. 3 (2001) 3175–3179.
[11] S. Hikita, K. Yamane, Y. Nakajima, Jsae Rev. 22 (2001) 151–156.
[12] V.M. Barragan, C. Ruiz-Bauza, J.P.G. Villaluenga, B. Seoane, J. Power Sources 130 (2004) 22–29.
[13] A.V. Anantaraman, C.L. Gardner, J. Electroanal. Chem. 414 (1996) 115–120.
[14] D. Chu, R.Z. Jiang, J. Power Sources 80 (1999) 226–234.
[15] H. Tang, S. Peikang, S.P. Jiang, F. Wang, M. Pan, J. Power Sources 170 (2007) 85–92.
[16] N. Nakagawa, M.A. Abdelkareem, K. Sekimoto, J. Power Sources 160 (2006) 105–115.
[17] J. Ling, O. Savadogo, J. Electrochem. Soc. 151 (2004) A1604–A1610.

[18] S. Prakash, W.E. Mustain, S. Park, P.A. Kohl, J. Power Sources 175 (2008) 91–97.
[19] P.L. Shao, K.A. Mauritz, R.B. Moore, J. Polym. Sci. Part B-Polym. Phys. 34 (1996) 873–882.
[20] K.T. Adjemian, S.J. Lee, S. Srinivasan, J. Benziger, A.B. Bocarsly, J. Electrochem. Soc. 149 (2002) A256–A261.
[21] K.A. Mauritz, J.T. Payne, J. Membr. Sci. 168 (2000) 39–51.
[22] K.T. Adjemian, S. Srinivasan, J. Benziger, A.B. Bocarsly, J. Power Sources 109 (2002) 356–364.
[23] Z.G. Shao, P. Joghee, I.M. Hsing, J. Membr. Sci. 229 (2004) 43–51.
[24] P.L. Antonucci, A.S. Arico, P. Creti, E. Ramunni, V. Antonucci, Solid State Ionics 125 (1999) 431–437.
[25] P. Dimitrova, K.A. Friedrich, B. Vogt, U. Stimming, J. Electroanal. Chem. 532 (2002) 75–83.
[26] Y.H. Su, T.Y. Wei, C.H. Hsu, Y.L. Liu, Y.M. Sun, J.Y. Lai, Desalination 200 (2006) 656–657.
[27] W.F. Chen, P.L. Kuo, Macromolecules. 40 (2007) 1987–1994.
[28] H. Kim, H. Chang, J. Membr. Sci. 288 (2007) 188–194.
[29] W.L. Xu, T.H. Lu, C.P. Liu, W. Xing, Electrochim. Acta 50 (2005) 3280–3285.
[30] B. Yang, A. Manthiram, J. Electrochem. Soc. 151 (2004) A2120–A2125.
[31] F. Bauer, M. Willert-Porada, Solid State Ionics 177 (2006) 2391–2396.
[32] C. Yang, S. Srinivasan, A.B. Bocarsly, S. Tulyani, J.B. Benziger, J. Membr. Sci. 237 (2004) 145–161.
[33] C. Yang, P. Costamagna, S. Srinivasan, J. Benziger, A.B. Bocarsly, J. Power Sources 103 (2001) 1–9.
[34] F. Bauer, M. Willert-Porada, J. Membr. Sci. 233 (2004) 141–149.
[35] K.M. Nouel, P.S. Fedkiw, Electrochim. Acta 43 (1998) 2381–2387.
[36] H.L. Tang, M. Pan, S.P. Jiang, X. Wang, Y.Z. Ruan, Electrochim. Acta 52 (2007) 5304–5311.
[37] H.L. Tang, M. Pan, F. Wang, P.K. Shen, S.P. Jiang, J. Phys. Chem. B 111 (2007) 8684–8690.
[38] L.N. Huang, L.C. Chen, T.L. Yu, H.L. Lin, J. Power Sources 161 (2006) 1096–1105.
[39] C.W. Moore, J. Li, P.A. Kohl, J. Electrochem. Soc. 152 (2005) A1606–A1612.
[40] J. Li, C.W. Moore, D. Bhusari, S. Prakash, P.A. Kohl, J. Electrochem. Soc. 153 (2006) A343–A347.
[41] K.D. Kreuer, A. Fuchs, M. Ise, M. Spaeth, J. Maier, Electrochim. Acta 43 (1998) 1281–1288.
[42] W. Munch, K.D. Kreuer, W. Silvestri, J. Maier, G. Seifert, Solid State Ionics 145 (2001) 437–443.
[43] Y.Z. Fu, A. Manthiram, J. Electrochem. Soc. 154 (2007) B8–B12.
[44] J. Nengyou, C.L. Mark, H. John, Q. Zhigang, G.P. Peter, Electrochem. Solid State Lett. 3 (2000) 529–531.
[45] J. Lin, J.K. Lee, M. Kellner, R. Wycisk, P.N. Pintauro, J. Electrochem. Soc. 153 (2006) A1325–A1331.
[46] W. Lee, H. Kim, T.K. Kim, H. Chang, J. Membr. Sci. 292 (2007) 29–34.
[47] H.B. Wang, R. DeSousa, J. Gasa, K. Tasaki, G. Stucky, B. Jousselme, F. Wudl, J. Membr. Sci. 289 (2007) 277–283.

[48] G. Li, W. Zhang, J. Yang, X. Wang, J. Colloid Interface Sci. 306 (2007) 337–344.
[49] M.C. Burshe, S.A. Netke, S.B. Sawant, J.B. Joshi, V.G. Pangarkar, Sep. Sci. Technol. 32 (1997) 1335–1349.
[50] N.W. DeLuca, Y.A. Elabd, J. Membr. Sci. 282 (2006) 217–224.
[51] H. Wu, Y.X. Wang, S.C. Wang, J. New Mater. Electrochem. Syst. 5 (2002) 251–254.
[52] M.S. Kang, J.H. Kim, J. Won, S.H. Moon, Y.S. Kang, J. Membr. Sci. 247 (2005) 127–135.
[53] J.L. Qiao, T. Hamaya, T. Okada, Polymer 46 (2005) 10809–10816.
[54] C.W. Walker, J. Electrochem. Soc. 151 (2004) A1797–A1803.
[55] Y. Sakai, M. Matsuguchi, N. Yonesato, Electrochim. Acta 46 (2001) 1509–1514.
[56] J. Travas-Sejdic, R. Steiner, J. Desilvestro, P. Pickering, Electrochim. Acta 46 (2001) 1461–1466.
[57] J.P. Randin, J. Electrochem. Soc. 129 (1982) 1215–1220.
[58] S.Y. Huang, R.D. Giglia (1984), US Patent 4,478,991.
[59] K.D. Kreuer, J. Membr. Sci. 185 (2001) 29–39.
[60] K.D. Kreuer, M. Ise, A. Fuchs, J. Maier, J. De Phys. Iv 10 (2000) 279–281.
[61] K.D. Kreuer, Solid State Ionics 97 (1997) 1–15.
[62] B. Yang, A. Manthiram, Electrochem. Solid State Lett 6 (2003) A229–A231.
[63] X.F. Li, C.J. Zhao, H. Lu, Z. Wang, H. Na, Polymer 46 (2005) 5820–5827.
[64] M. Gil, X.L. Ji, X.F. Li, H. Na, J.E. Hampsey, Y.F. Lu, J. Membr. Sci. 234 (2004) 75–81.
[65] S.L. Zhong, X.J. Cui, H.L. Cai, T.Z. Fu, C. Zhao, H. Na, J. Power Sources 164 (2007) 65–72.
[66] J. Wootthikanokkhan, N. Seeponkai, J. Appl. Polym. Sci. 102 (2006) 5941–5947.
[67] S. Vetter, B. Ruffmann, I. Buder, S.R. Nunes, J. Membr. Sci. 260 (2005) 181–186.
[68] C.J. Zhao, Z. Wang, D.W. Bi, H.D. Lin, K. Shao, T.Z. Fu, S.L. Zhong, H. Na, Polymer 48 (2007) 3090–3097.
[69] X.Y. Shang, S.M. Fang, Y.Z. Meng, J. Membr. Sci. 297 (2007) 90–97.
[70] F.C. Ding, S.J. Wang, M. Xiao, Y.Z. Meng, J. Power Sources 164 (2007) 488–495.
[71] J.H. Li, H.Y. Yu, J. Polym. Sci. Part a-Polym. Chem. 45 (2007) 2273–2286.
[72] Z. Wang, X.F. Li, C.J. Zhao, H.Z. Ni, H. Na, J. Appl. Polym. Sci. 104 (2007) 1443–1450.
[73] K.B. Wiles, C.M. de Diego, J. de Abajo, J.E. McGrath, J. Membr. Sci. 294 (2007) 22–29.
[74] X.M. Wu, G.H. He, S. Gu, Z.W. Hu, P.J. Yao, J. Membr. Sci. 295 (2007) 80–87.
[75] M. Sankir, Y.S. Kim, B.S. Pivovar, J.E. McGrath, J. Membr. Sci. 299 (2007) 8–18.
[76] X.Y. Shang, D. Shu, S.J. Wang, M. Xiao, Y.Z. Meng, J. Membr. Sci. 291 (2007) 140–147.
[77] W. Lu, D.P. Lu, J.H. Liu, C.H. Li, R. Guan, Polym. Adv. Technol. 18 (2007) 200–206.
[78] D.P. Lu, W. Lu, C.H. Li, J.H. Liu, R. Guan, Polym. Bull. 58 (2007) 673–682.

[79] S.F. Yang, C.L. Gong, R. Guan, H. Zou, H. Dai, Polym. Adv. Technol. 17 (2006) 360–365.
[80] I. Cabasso, Y. Yuan, C. Mittelsteadt, Blend Membranes Based on Sulfonated Poly (phenylene oxide) for Enhanced Polymer Electrochemical Cells, (1999), US Patent 5,989,742.
[81] K. Uno, K. Niume, Y. Iwata, F. Toda, Y. Iwakura, J. Polym. Sci. Part a-Polym. Chem. 15 (1977) 1309–1318.
[82] T.D. Dang, N. Venkatasubramanian, D.R. Dean, G.E. Price, F.E. Arnold, Abstr. Pap. Am. Chem. Soc. 214 (1997) 110-POLY.
[83] J.A. Asensio, S. Borros, P. Gomez-Romero, J. Polym. Sci. Part a-Polym. Chem. 40 (2002) 3703–3710.
[84] C. Hasiotis, V. Deimede, C. Kontoyannis, Electrochim. Acta 46 (2001) 2401–2406.
[85] W.F. D'Alelio, W.E. Gibbs, R.L. Van Deusen (1970), US Patent 3,536,674.
[86] A.L. Rusanov, D. Likhatchev, P.V. Kostoglodov, K. Mullen, M. Klapper, Inorg. Polym. Nanocompos. Membr. 179 (2005) 83–134.
[87] R.H. He, Q.F. Li, J.O. Jensen, N.J. Bjerrum, J. Polym. Sci. Part a-Polym. Chem. 45 (2007) 2989–2997.
[88] Y.F. Liu, Q.C. Yu, Y.H. Wu, Electrochim. Acta 52 (2007) 8133–8137.
[89] J.S. Wainright, J.T. Wang, D. Weng, R.F. Savinell, M. Litt, J. Electrochem. Soc. 142 (1995) L121–L123.
[90] R. He, J.O. Jensen, N.J. Bjierrum, Q. Li, Fuel Cells 4 (2004) 147–159.
[91] S.L.-C.H. Shih-Wei Chuang, J. Polym. Sci. Part A: Polym. Chem. 44 (2006) 4508–4513.
[92] Z.M. Qiu, S.Q. Wu, Z. Li, S.B. Zhang, W. Xing, C.P. Liu, Macromolecules 39 (2006) 6425–6432.
[93] X.X. Guo, J.H. Fang, T. Watari, K. Tanaka, H. Kita, K.I. Okamoto, Macromolecules 35 (2002) 6707–6713.
[94] J.S. Wainright, M.H. Liu, Y. Zhang, C.C. Liu, R.F. Savinell, Electrochem. Soc. Proc. 489 (2000) 14.
[95] R. Jin, Z. Wang, Z. Cui, W. Xing, L. Gao, Yuhan Li, J. Polym. Sci. Part A: Polym. Chem. 45 (2007) 222–231.
[96] N.W. Li, Z.M. Cui, S.B. Zhang, W. Xing, J. Membr. Sci. 295 (2007) 148–158.
[97] Z.X. Hu, Y. Yin, H. Kita, K.I. Okamoto, Y. Suto, H.G. Wang, H. Kawasato, Polymer. 48 (2007) 1962–1971.
[98] C. Genies, R. Mercier, B. Sillion, N. Cornet, G. Gebel, M. Pineri, Polymer 42 (2001) 359–373.
[99] Y. Zhang, M. Litt, R.F. Savinell, J.S. Wainright, J. Vendramini, Polym. Prepr. 41(2), (2000), 1561.
[100] K. Miyatake, H. Zhou, H. Uchida, M. Watanabe, Chem. Commun. (2003), 3 368–369.
[101] Y. Yin, J.H. Fang, T. Watari, K. Tanaka, H. Kita, K. Okamoto, J. Mater. Chem. 14 (2004) 1062–1070.
[102] B.R. Einsla, Y.S. Kim, M.A. Hickner, Y.T. Hong, M.L. Hill, B.S. Pivovar, J.E. McGrath, J. Membr. Sci. 255 (2005) 141–148.
[103] T. Yasuda, K. Miyatake, M. Hirai, M. Nanasawa, M. Watanabe, J. Polym. Sci. Part a-Polym. Chem. 43 (2005) 4439–4445.

[104] R. Thangamuthu, C.W. Lin, Solid State Ionics 176 (2005) 531–538.
[105] W.L. Xu, C.P. Liu, X.Z. Xue, Y. Su, Y. Lv, W. Xing, T.H. Lu, Solid State Ionics 171 (2004) 121–127.
[106] V.K. Shahi, Solid State Ionics 177 (2007) 3395–3404.
[107] S.P. Nunes, B. Ruffmann, E. Rikowski, S. Vetter, K. Richau, J. Membr. Sci. 203 (2002) 215–225.
[108] V.S. Silva, B. Ruffmann, H. Silva, Y.A. Gallego, A. Mendes, L.M. Madeira, S.P. Nunes, J. Power Sources 140 (2005) 34–40.
[109] V.S. Silva, J. Schirmer, R. Reissner, B. Ruffmann, H. Silva, A. Mendes, L.M. Madeira, S.P. Nunes, J. Power Sources 140 (2005) 41–49.
[110] V. Baglio, A. Di Blasi, A.S. Arico, V. Antonucci, P.L. Antonucci, F. Serraino Fiory, S. Licoccia, E. Traversa, J. New Mater. Electrochem. Syst. 7 (2004) 275–280.
[111] M.H.D. Othman, A.F. Ismail, A. Mustafa, J. Membr. Sci. 299 (2007) 156–165.
[112] X.W. Zhang, J. Electrochem. Soc. 154 (2007) B322–B326.
[113] Z.G. Shao, H.F. Xu, M.Q. Li, I.M. Hsing, Solid State Ionics 177 (2006) 779–785.
[114] C.W. Lin, R. Thangamuthu, C.J. Yang, J. Membr. Sci. 253 (2005) 2–31.
[115] S.P. Tung, B.J. Hwang, J. Mater. Chem. 15 (2005) 3532–3538.
[116] Y. Abe, G.M. Li, M. Nogami, T. Kasuga, L.L. Hench, J. Electrochem. Soc. 143 (1996) 144–147.
[117] Y. Abe, H. Shimakawa, L.L. Hench, J. Non-Cryst. Solids 51 (1982) 357–365.
[118] M. Nogami, Y. Abe, Phys. Rev. B 55 (1997) 12108–12112.
[119] Y. Abe, H. Hosono, Y. Ohta, L.L. Hench, Phys. Rev. B 38 (1988) 10166–10169.
[120] Y. Abe, H. Hosono, O. Akita, L.L. Hench, J. Electrochem. Soc. 141 (1994) L64–L65.
[121] S.P. Tung, B.J. Hwang, J. Membr. Sci. 241 (2004) 315–323.
[122] M. Nogami, R. Nagao, W. Cong, Y. Abe, J. Sol-Gel Sci. Technol. 13 (1998) 933–936.
[123] K. Makita, M. Nogami, Y. Abe, J. Ceram. Soc. Jpn. 106 (1998) 396–401.
[124] M. Nogami, K. Miyamura, Y. Abe, J. Electrochem. Soc. 144 (1997) 2175–2178.
[125] C. Wang, Y. Abe, T. Kasuga, M. Nogami, J. Ceram. Soc. Jpn. 107 (1999) 1037–1040.
[126] C. Wang, M. Nogami, Y. Abe, J. Sol-Gel Sci. Technol. 14 (1999) 273–279.
[127] T. Uma, S. Izuhara, M. Nogami, J. Eur. Ceram. Soc. 26 (2006) 2365–2372.
[128] T. Uma, A. Nakao, M. Nogami, Mater. Res. Bull. 41 (2006) 817–824.
[129] Y. Park, M. Nagai, Solid State Ionics 145 (2001) 149–160.
[130] P. Yong-il, J. Moon, K. Hye Kyung, Electrochem. Solid State Lett. 8 (2005) 191–194.
[131] H. Kim, S. Prakash, W.E. Mustain, P.A. Kohl, Inorganic glass proton exchange membranes, 212th Meeting of the Electrochemical Society, Washington, D.C, 2007.
[132] M. Kato, S. Katayama, W. Sakamoto, T. Yogo, Electrochim. Acta 52 (2007) 5924–5931.
[133] N.W. Deluca, Y.A. Elabd, J. Polym. Sci. Part B-Polym. Phys. 44 (2006) 2201–2225.

CHAPTER

2

MEMS-Based Micro Fuel Cells as Promising Power Sources for Portable Electronics

Tristan Pichonat
IRCICA, CNRS FR 3024
IEMN, CNRS UMR 8520,
Villeneuve d'Asq, France

This chapter presents a review of miniature fuel cells fabricated thanks to microtechnologies. First it introduces the reader to micro-electro-mechanical systems (MEMS) technologies and what they can bring to miniature fuel cells. Then the reasons for miniaturizing fuel cells and the specifications required by this miniaturization are discussed. It shows what kinds of fuel cells may fit to these specifications and which fuels can be employed to supply them. Despite a common use of the same kinds of technologies and global structure, the review highlights the diversity of choices and possibilities that research has to face and the large range of performances proceeding from them. The actual commercial developments in the domain are also briefly described. It finally details, as an example, the complete fabrication process of a particular microfabricated fuel cell based on a silane-grafted porous silicon membrane as the proton-exchange membrane instead of a common ionomer such as Nafion®.

2.1 INTRODUCTION

Over the last few years, a large number of electronic devices, such as computers, cellular phones, and camcorders, have become portable with the miniaturization of electronic components. Such equipment requires appropriate power supplies, which have to be more and more effective, especially in terms of power density and lifetime, in order to be able to provide for the increasing number of functionalities of these portable applications (such as Internet on cellular phones, for example).

In this strong growth market, fuel cells represent promising power sources for these applications. Indeed, even if they are generally associated with much larger applications such as vehicles and power plants, they can be of any size, given their modularity. This technology enables us to conceive of very small size fuel cells, involving only a few thin cells with a small effective area.

Thirty years ago, only the nickel cadmium (NiCd) technology was available for the supply of portable electronics. The last twenty years have seen the emergence of three new technologies on the market: nickel metal hydride (NiMH) since the late 1980s, lithium-ion (Li-Ion) since 1991, and lithium-polymer (LiPo) since 1996. The main battery manufacturers have been searching since to increase the lifetime of batteries, but despite their efforts batteries remain heavy, bulky, of limited charge, and, above all, polluting. Actually, the major problem for batteries is that their energy is stored in heavy and voluminous sources such as metal oxides and graphitic materials [1–3].

Compared with batteries, fuel cells can produce electrical energy from much lighter materials such as gases, alcohols, and hydrides by catalytic electrochemical processes. They theoretically provide energy densities 3 to 5 times higher than their competing batteries (Figure 2.1). Autonomies of 30 days in stand-by mode and more than 20 hours in communication are projected for a cellular phone within the same volume. Moreover, contrary to present batteries, which suffer from low autonomy combined with still frequent and long-time recharging, the autonomy of fuel cells only depends on the size of the fuel tank and refilling is immediate by changing the fuel tank. This technology is also theoretically nonpolluting. For these reasons, research in this area is growing rapidly, and all the more so given the fact that portable electronics is not the only market miniature fuel cells can claim. Indeed, military departments are also eager for long-time operation and lightweight batteries. The health services may also be interested in micro biofuel cells as implantable power sources for devices such as pacemakers or glucose sensors. This latter kind of micro fuel cell and its development has been recently reviewed by Davis et al. [5].

However, miniaturizing fuel cells is still a challenge, with many issues to consider such as fabrication techniques, cost, and customer safety [6]. Actually, fuel cells are two-part systems—the fuel

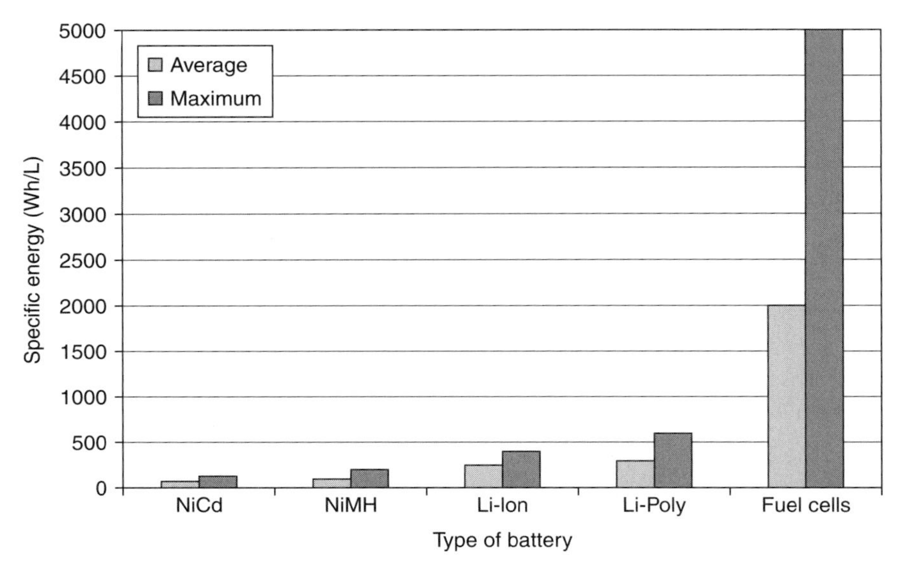

FIGURE 2.1 Comparison between specific energies of fuel cells and competing batteries, *after* [2, 4].

as the energy source and the fuel cell as the energy converter—both of which have to work as one. If the ability to store and efficiently manage the fuel will mostly influence the total volume and the autonomy of the fuel cell, the capability of efficiently transforming the specific energy of the fuels into electrical energy will depend on the effectiveness of catalytic reactions on the electrodes of the fuel cell. The cost is then bound, among other factors, to the catalyst itself, mostly platinum, even if in very small amounts. Moreover using fuel cells in portable applications implies methods to provide a high level of safety. This means, given that fuel cells are not generally sealed since they require oxygen from air, that the fuel has to be isolated to avoid any risk of accidental direct oxidation with air or user contact if toxic [2]. Other issues such as management of the heat and water produced are also challenges that technology developers have to deal with.

Focusing on the miniaturization, two main research axes are possible for the global conception of small fuel cells: either a reduction in the size of existing fuel cells, or a search for new materials and structures usable for their fabrication. Although the first solution seems to be the easiest, it involves the miniaturization of all the

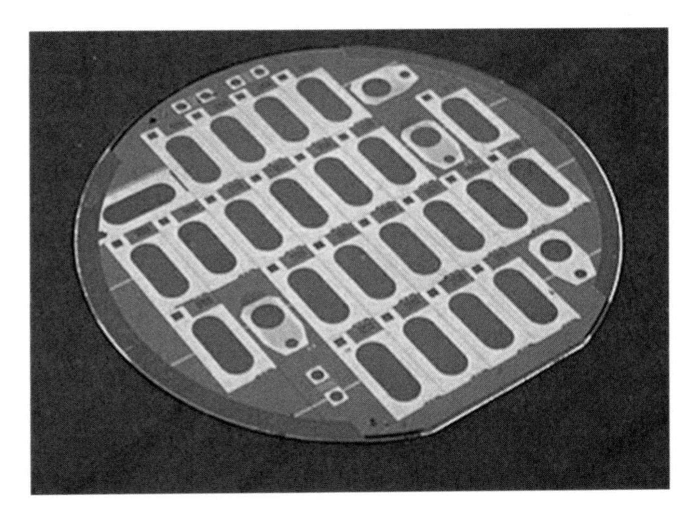

FIGURE 2.2 Fuel cells from the French CEA microfabricated on a silicon wafer.

auxiliary components and the difficulty in adapting "macrofabrication" techniques to miniature sizes.

Therefore, research rather tends to apply itself to the development of size-adapted methods with suitable materials.

One of the technological ways to miniaturize fuel cells is to make use of standard microfabrication techniques mainly used in microelectronics and, more especially, the fabrication of micro- and nano-electro-mechanical systems (MEMS/NEMS). More and more papers show a growing interest in developing MEMS-based fuel cells, either directly with silicon substrates (Figure 2.2), or adapting the methods to other substrates such as metals or polymers. These techniques enable notably mass fabrication at low cost (very large number of devices on a very small area), which could lead to a reduction in the global cost of miniature fuel cells.

The significant issues concerning the miniaturization of fuel cells are developed here, in particular showing which kinds of fuel cells may fit the requirements of portable electronics and which kinds of fuels may be appropriate for the supply of small fuel cells. Selected research contributions in the domain are reviewed, and a short survey of commercial developments of micro fuel cells is provided. Finally, the complete fabrication process of a micromachined fuel cell based on a silane-grafted porous silicon membrane is described as an example.

2.2 MINIATURIZATION

2.2.1 Microelectronics and MEMS

Historically, microfabrication techniques were first developed to meet the requirements of microelectronics, but they also have allowed the emergence and the development of a new research field in which mechanical elements can be manufactured and actuated with electrical signals at a micro- and even nanometer scale. To describe this emerging research field, R. T. Howe and others proposed the expression *micro electro mechanical systems*, or MEMS, in the late 1980s [7]. MEMS is not the only term used to describe this field; it is also called "micromachines," for instance in Japan, and more broadly referenced as "microsystem technology" (MST) in Europe.

These submillimeter devices are machined using specific techniques globally called *microfabrication technology*. This definition also includes microelectronic devices, but in addition to electronic parts, MEMS also features mechanical parts like holes, cavities, channels, cantilevers, or membranes. This particularity has a direct impact on their manufacturing processes, which must be adapted for thick layer deposition and deep etching and must introduce special steps to free the mechanical structures. Moreover, many MEMS are now not only based on silicon but are also manufactured with polymer, glass, quartz, or thin metal films.

These MEMS are now part of our daily life, with many applications such as inkjet heads, accelerometers for crash air-bag deployment systems in automobiles, micromirrors for digital projectors (the best example is the DLP® (for Digital Light Processing), presently the only all-digital display chip by Texas Instruments), and also pressure sensors for automotive and medical applications. Further applications to come could be lab-on-chip for medical analysis and radiofrequency MEMS for telecommunications (switches, filters, etc.). Two examples of MEMS fabricated at the IEMN (Villeneuve d'Ascq, France), a microgripper and a micromotor, are shown in Figures 2.3 and 2.4.

Besides the many applications previously cited, there are three main factors influencing the use of MEMS technology today:

- **Miniaturization of existing devices**, such as the production of silicon-based gyroscopes, reducing previously existing voluminous devices to a microchip;

FIGURE 2.3 SEM view of an electrostatic microgripper, *after* [8].

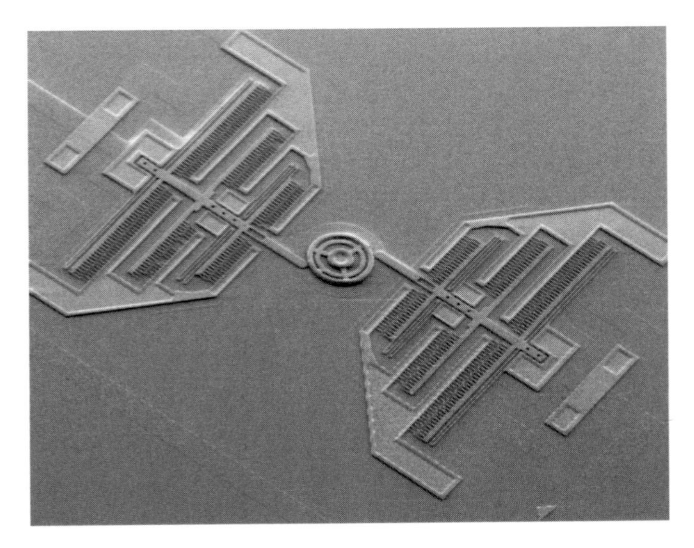

FIGURE 2.4 SEM view of an electrostatic micromotor, *after* [9].

- **Development of new devices** based on principles that do not work at larger scale, such as the electro-osmotic effect in microchannels;
- **Development of new tools** to interact with the micro-world.

The first two factors are the same as the motivations for the conception of miniature fuel cells. The key idea here is also to reduce the size to gain volume and thus enable low cost mass fabrication, without denigrating the performance of the system. Furthermore, the materials and techniques used for fabricating MEMS are compatible with the fabrication of fuel cells, and provide, for instance, easy ways to conceive microchannels to efficiently bring the fuel and the oxidant to the membrane-electrode assembly; but primarily these techniques can offer to the fuel cell an appropriate and dedicated micromachined support which may eventually be integrated with other microelectronic or MEMS devices.

Microfabrication and related MEMS techniques will not be described here. Short introductions already exist [7, 10] and far more information and accurate descriptions have been compiled by Sze [11], Senturia [12], or Madou [13].

2.2.2 Miniaturizable Fuel Cells

The first section in this chapter discussed the fact that miniaturization can bring several limitations to the conception of small fuel cells (materials, flows, techniques, etc.); additional restrictions come from the devices which they are intended for. Portable electronics also have their own requirements regarding the power to supply, the packaging, the working temperature, and the fuels to use. Thus, the manufacturing of miniature fuel cells will be determined by the final applications.

As it seems wiser to provide a fuel cell functioning at no higher temperatures than 80 or even 100°C for the user's safety, the choice of the type of fuel cell to use in portable devices may be limited to low temperature fuel cells such as PEMFC (for Proton Exchange Membrane Fuel Cell, or sometimes Polymer Electrolyte Membrane Fuel Cell) and DMFC (for Direct Methanol Fuel Cell). The later section titled "Reformed Hydrogen Fuel Cells" will nevertheless show that efforts are also currently focused on micro reformed methanol fuel cells and miniature solid-oxide fuel cells (SOFCs) functioning at high temperature (up to 350 and 500°C, respectively).

2.3 MICROFABRICATED FUEL CELLS

Various solutions have been reported in the literature in the domain of MEMS-based miniature fuel cells (FCs) with power densities of the miniature FCs ranging from a few tens of μW cm^{-2} up to several hundreds of mW cm^{-2}. The basic components are generally the same, namely the electrodes with flow fields and current collectors, diffusion layers, catalyst layers, and the proton exchange media. Concerning the architecture, two basic design approaches are currently employed: the classic bipolar design where all the components of the micro fuel cell are stacked together (Figure 2.5) and where fuel and oxidant are separated by the membrane electrodes assembly (MEA), and the planar design (Figure 2.6) where the fuel and oxidant channels are interdigitated and both electrodes are on the same single side [14]. Bipolar design ensures the separation of fuel and oxidant but requires all components to be fabricated separately and then assembled together. The planar design is more suitable for a monolithic integration but requires a larger surface area to deliver similar performance.

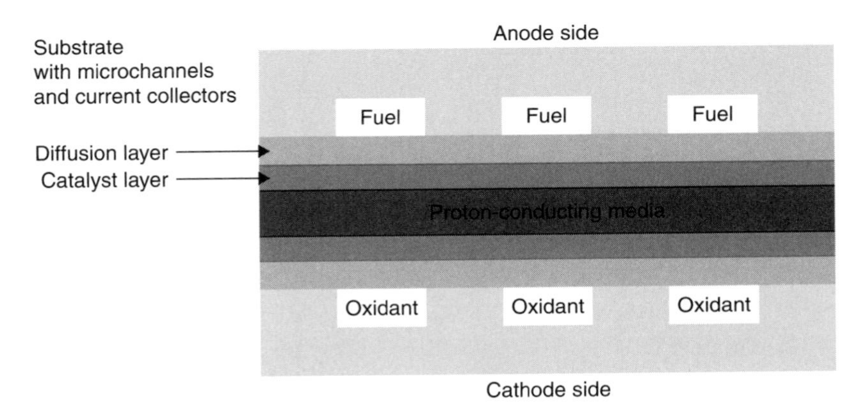

FIGURE 2.5 Bipolar design of micro fuel cell, *after* [14].

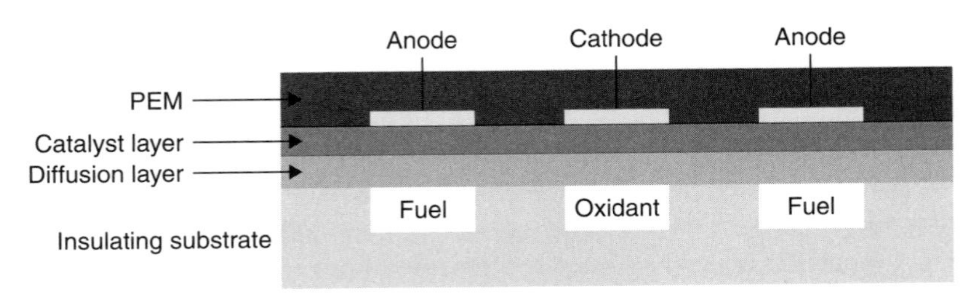

FIGURE 2.6 Planar design of micro fuel cell, *after* [14].

The micromachined parts of the micro fuel cell are generally the electrode plates and the fuel delivery system. The latter is often achieved through microchannels or flow fields. Nguyen et al. made a review of the different designs for the fuel delivery systems [14], illustrated in Figure 2.7.

Concerning the usable fuels for the supply of miniature fuel cells, the more investigated are presently hydrogen and methanol. Methanol is a low cost liquid fuel with a high energy density, which can be directly oxidized in the fuel cell. However, due to the poor kinetics of the charge transfer reaction at the heart of the fuel cell, direct methanol fuel cells need substantial amounts of noble metals as catalyst, increasing the cost. Moreover, concentrated methanol cannot be used as it leaks through the membranes of fuel cells, so diluted methanol has to be used, which drastically reduces energy density. On the other hand, hydrogen is the basic fuel, but is far too dangerous to be stored directly as compressed gas and too complicated to be stored as a liquid. Intermediate storages such as hydrides (chemical or metal) and methanol (for reformed methanol fuel cells or RMFC) are under consideration, though it adds complexity to the overall fuel cell [2]. Fuels such as

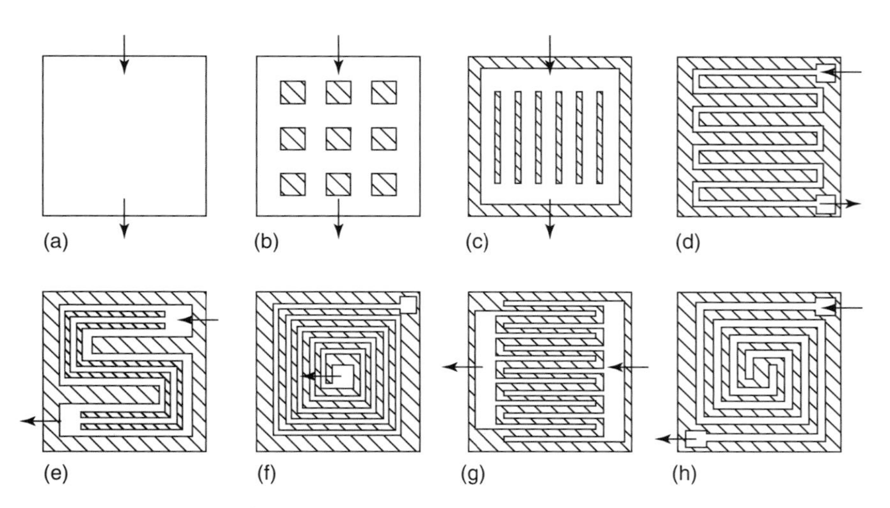

FIGURE 2.7 Typical fuel delivery designs: (a) direct supply, (b) with distribution pillars, (c) parallel microchannels, (d) serpentine microchannel, (e) parallel/serpentine microchannel, (f) spiral microchannel, (g) interdigitated microchannel, (h) spiral/interdigitated microchannel, *after* [14].

ethanol [15] or formic acid [16, 17] are also currently considered as interesting alternatives. In the following sections the different FCs will be indexed according to the different fuels used. FCs using hydrogen directly will be globally called hydrogen-fed FCs, the ones using a reformer will be called reformed hydrogen FCs, the ones using methanol directly (without reforming) direct methanol FCs or DMFCs, and the others will be presented as alternative fuel cells (ethanol, formic acid). This latter part will also be dedicated to alternative technologies, especially the ones using microscale effects such as laminar flow fuel cells.

2.3.1 Hydrogen-Fed Fuel Cells

Hydrogen-fed micro FCs are theoretically among the most simple and effective PEMFCs, except for the problem of fuel storage. If attention is given to finding appropriate materials for storing large amounts of hydrogen, there is currently no perfect solution for portable applications, and research tends to focus more on developing hydrogen FCs than on how these FCs could be supplied in a realistic way.

Concerning the FCs themselves, if the global structure is almost the same for all the FCs presented here—typically a membrane-electrodes assembly (MEA) surrounded on both sides by a gas diffusion layer (GDL)/flow fields and current collector, either on the same substrate, or not—different technological choices can be made during the fabrication.

For instance, concerning the basic substrate to use, a compromise has to be made, especially between the easiness of substrate patterning/etching, the easiness of assembly, the solidity of the whole system, and the easiness of cell interconnection.

As the basic material for MEMS technologies, silicon (Si) is also the most common material encountered in MEMS-based FCs, but is not the only one. Recently, solutions have been proposed using polymer or metal as the FC basic substrate instead of Si. Nevertheless, Si is still the leading material thanks to its properties and the microfabrication techniques associated with it, which are now well-known and mastered. Another advantage of Si-based FCs may also be to facilitate the possible integration of the FCs with other electronic devices on the same chip.

Silicon-Based

Since 2000, Meyers et al. (Lucent Technologies) have proposed two alternative designs using Si: a classical bipolar using separate Si wafers for the cathode and the anode and a less effective monolithic design that integrated the two electrodes onto the same Si surface [18, 19]. In the bipolar design, both electrodes were constructed from conductive Si wafers. The reactants were distributed through a series of tunnels created by first forming a porous silicon (PS) layer and then electropolishing away the Si beneath the porous film. The FC was completed by adding a catalyst film on top of the tunnels and finally by casting a Nafion® solution. Two of these membrane-electrode structures were fabricated and then sandwiched together. A power density of 60 mW cm^{-2} was announced for the bipolar design with H_2 supply.

Unlike Meyers and Maynard, Lee et al. (Stanford Univ.) proposed a "flip-flop" μ-FC design (Figure 2.8) where both electrodes were present on both sides [20]. If this design does provide ease of manufacturing by allowing in-plane electrical connectivity, it complicates the gas management. Instead of electrons being routed from front to back, gasses must be routed in crossing patterns, significantly complicating the fabrication process and sealing. Peak power achieved in a four-cell assembly was still 40 mW cm^{-2}.

A variant of this design was reported by Min et al. (Tohoky Univ., Japan) who proposed two structures of μ-PEFC using microfabrication techniques [21], the "alternatively inverted structure" and the "coplanar structure." These structures used Si substrates with porous SiO_2 layers with platinum (Pt)-based catalytic electrodes and gas feed holes, glass substrates with micro-gas channels, and a polymer membrane (Flemion® S). In spite of a reported

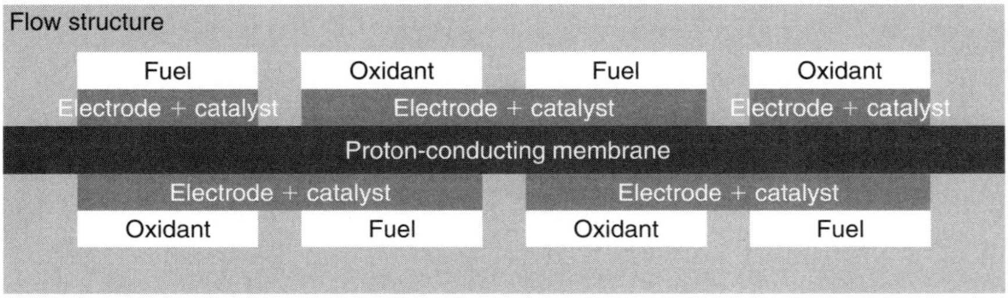

FIGURE 2.8 "Flip-flop" fuel cell design, *after* [20].

enhancement [22], the FC showed poor results (only 0.8 mW cm^{-2} for the alternatively inverted structure).

Starting from a classical miniature FC consisting of an MEA between two micromachined Si substrates but sandwiching Cu between layers of gold, Yu et al. (Hong Kong Univ.) were able to decrease the internal resistance of the thin-film current collectors. This involved an increase in the performance of the FC, achieving a peak power density of 193 mW cm^{-2} with H_2 and O_2 [23].

Xiao et al. proposed a silicon/glass micro fuel cell with catalyst layers formed by Pt sputtering on ICP etched silicon columns (Figure 2.9) and integrated gold-based micro current collectors patterned on both silicon and glass surfaces [24]. The complete process on silicon included notably the use of thermal oxide patterned by RIE, dry etching of feedholes and microcolumns, wet etching of fuel reservoirs and metallization for current collectors. The glass substrate (Pyrex Corning 7740) was wet etched by HF to form the chamber and inlet/outlet. The two substrates were then bonded by anodic bonding at 320°C and 400 V. The bonded pair was sputtered with TiW (adhesion layer) and Pt to produce the catalyst layer on the microcolumns. This process enabled an increase in the catalyst area with 3D columns. To complete the fuel cell, the fabricated electrodes were bonded with a Nafion® 117 membrane at 135°C.

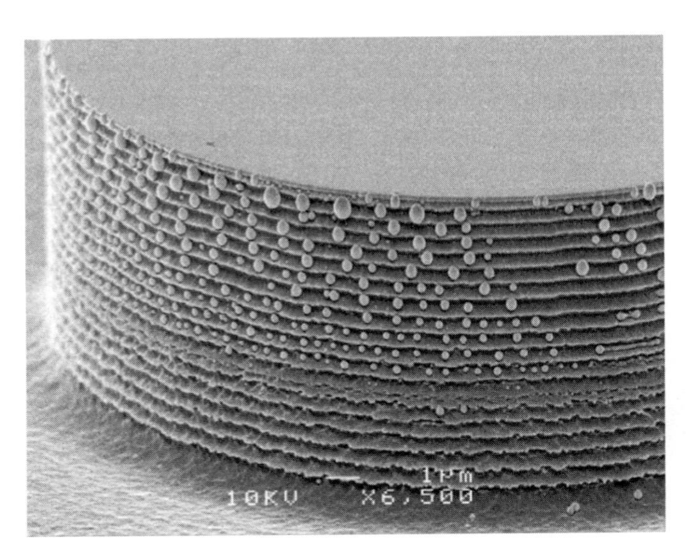

FIGURE 2.9 Platinum particles on ICP etched rings, *after* [25].

Two prototypes were achieved providing current densities of 7.1 and 0.1 mW cm^{-2} with H_2 and methanol respectively for the first one, and improved performances of 13.7 and 0.21 mW cm^{-2} for the second one, which benefited from a new etching process [25].

Chen et al. recently reported the fabrication and characterization of a high-power self-breathing PEMFC with optimally designed wet KOH etched flow fields and electrodes [26]. Ni/Cu/Au layers are used for current collecting, the 1.5 μm-thick in-between layer instead of a thick Au layer allowing a reduction in the fabrication cost. The MEA is composed of a classical Nafion® 1135 membrane with Pt-alloy sprayed carbon paper on both sides. The two silicon electrodes are sandwiched and pressed at RT with the MEA, and then sealed with epoxy. A base-chip formed by drilled Pyrex glass anodically bonded with KOH eched silicon acted both as H_2 inlet/outlet manifolds and as a support for a six-cell pack. An average power density of 144 mW cm^{-2} was measured with H_2 and air breathing, which was one of the highest performances currently reported.

Zhang et al. proposed a six-cell PEMFC planar stack combined with a small H_2 storage canister [27]. Each cell was made of a homemade MEA (Nafion® 112 membrane with Pt loading on each side) sandwiched between silicon-based flow field plates. These flow fields were fabricated using a classical wet etching process (KOH solution) and forming pin-type patterns with two square through holes for the hydrogen plates and through hole patterns for the oxygen plates (Figure 2.10). The plates were made electrically

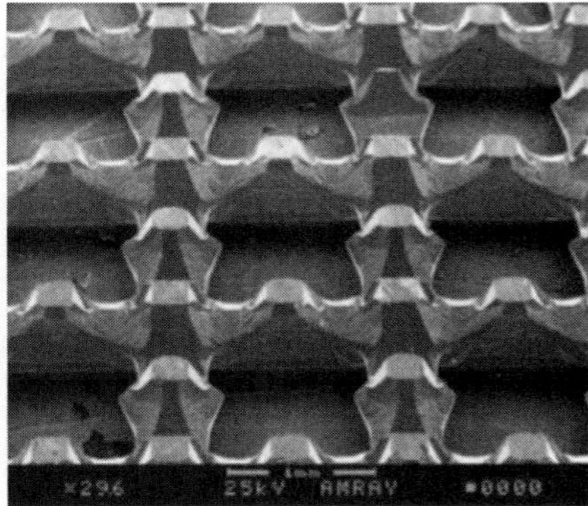

FIGURE 2.10 SEM pictures of flow fields, *after* [27].

conductive by sputtering Ti/Pt/Au metal layers on their surfaces. Six cells were then serially interconnected in the same plane with a fuel distributor fabricated on another piece of wet etched silicon anodically bonded to ultrasonically drilled Pyrex glass (Figure 2.11). Thanks to a homemade AB5-type H_2 storage canister and air-breathing conditions at RT, the stack reached a peak power of 0.9W at 250 mA cm^{-2} and an average power density of 104 mW cm^{-2} for a single cell. These performances tended to decrease with power increase and a power assistance may be needed to maintain their stability.

Metal-Based

R. Hahn and the Fraünhofer IZM have chosen foil materials for their prototypes of self-breathing planar PEMFCs (one example is seen in Figure 2.12) [28]. The dimensions were 1 cm^2 with 200 μm thickness. Stable long-term operation was achieved at 80 mW cm^{-2} at varying ambient conditions with dry H_2 fuel. They used a commercially available MEA further processed in their laboratory. Micro patterning technologies were employed on stainless steel foils for current collectors and flow fields and the assembled technology was adapted from microelectronics packaging. In their design, the interconnection between cells was performed outside the membrane area, which reduced sealing problems. The complete

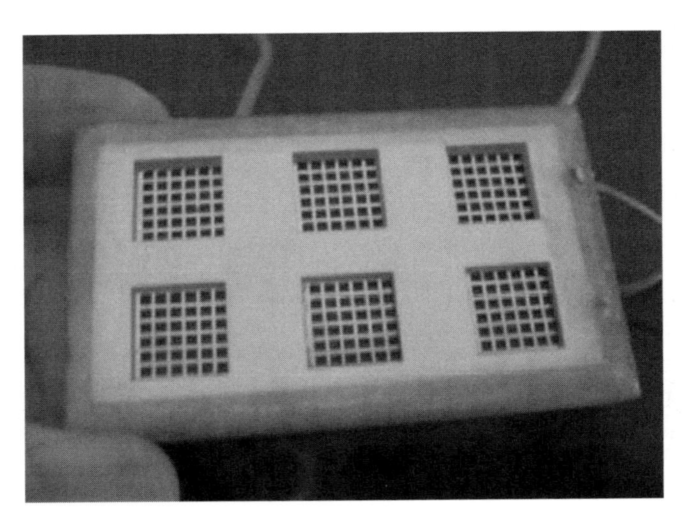

FIGURE 2.11 Assembled six-cell stack, *after* [27].

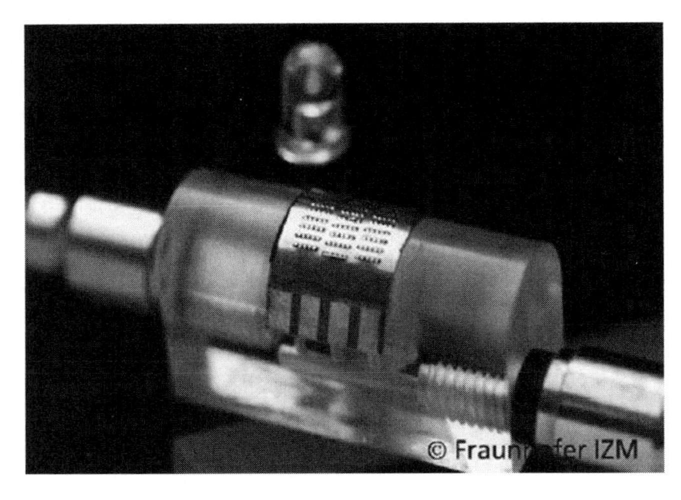

FIGURE 2.12 Fraünhofer IZM prototype of miniature fuel cell powering a LED, *after* [28].

FC consisted of only three layers: the current collectors with integrated flow fields on top and bottom and the patterned MEA in-between. The current collector foils with integrated flow field of the electrodes consisted of Au-electroplated and microstructured (micro patterning, wet etching, laser cutting, RIE) sandwiched metal-polymer foils. The commercial MEA was structured using RIE and laser ablation to avoid possible internal bypasses.

Müller et al. (IMTEK) also used micromachined metal foils to form the flow fields of their μ-FC [29]. Using Gore MEA, they were able to form very thin, high-power density stacks. They demonstrated both uncompressed and compressed FC designs. The uncompressed one had a peak power density of 20 mW cm^{-2} and the five-cell compressed stack achieved 250 mW cm^{-2}.

More recently a 280μm-thick titanium-based PEMFC was presented by N. Wan et al. [30]. Microchannels were fabricated on a Ti substrate using a photochemical wet etching process (etching solution composed of HF and HNO_3) (Figure 2.13). The etched Ti substrate was coated with a Ru layer (for electrical conduction), and then with a microporous carbon-PTFE composite layer under the form of an ink. The MEA was homemade by spraying an ink (Pt/C, isopropanol, and 5% Nafion® dispersion) onto a Nafion® 112 membrane sandwiched between two treated Ti substrates

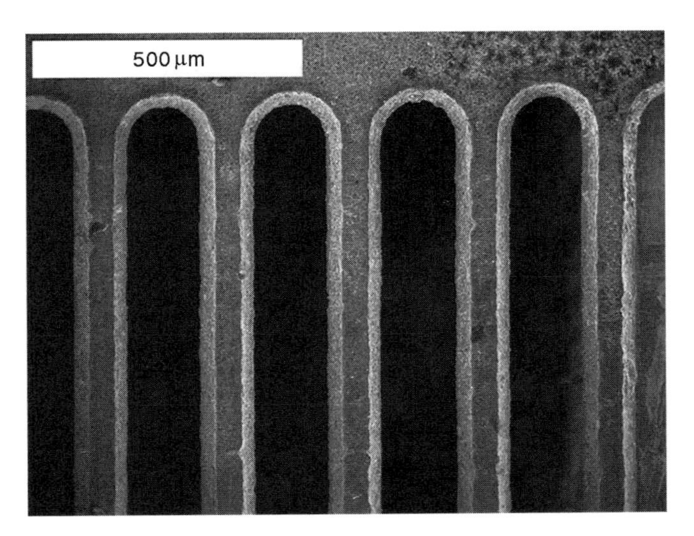

FIGURE 2.13 SEM image of microchannels in a Ti substrate, *after* [30].

(hot-pressing). This PEMFC showed high mechanical strength, good mass transport, and light weight. It also demonstrated good performances at RT, either under self-breathing conditions with 120 mW cm^{-2}, or with a forced flow of reactant with a peak power density of 220 mW cm^{-2} due to the faster mass transport, with no visible decrease of performances after 200 h operation.

Polymer-Based

Chan et al. (Nanyang Univ.) reported the fabrication of a polymeric μ-PEMFC (Figure 2.14) developed on the basis of micromachining of PMMA by laser [31]. The microchannels for fuel flow and oxidant were ablated with a CO_2-laser. The energy of the laser beam has a Gaussian distribution, so the cross-section of the channel also has a Gaussian shape. A 40-nm gold layer was then sputtered over the substrate surface to act both as the current collector and corrosion protection layer. The Gaussian shape allows gold to cover all sides of the channel. In this μ-FC, water generated by the reaction was utilized for gas humidifying. The flow channel has a spiral shape, which enables the dry gas in the outer spiral line to become hydrated by acquiring some of the moisture from the adjacent inner spiral line. The MEA consists of a Nafion® 1135 membrane with a hydrophobic carbon paper and a diffusion layer

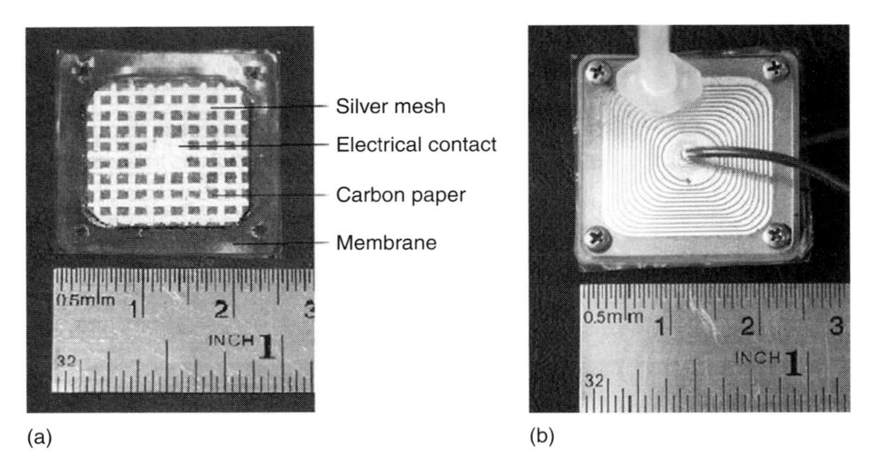

FIGURE 2.14 Assembled polymeric micro fuel cell, *after* [31]. (a) open view and (b) assembled view.

(carbon powder and PTFE) on one side and coated with a catalyst layer (ink with Nafion® 112 solution and Pt carbon) on the other side. Silver conductive paint was printed on the other side of the carbon paper to increase its conductivity and to contribute to current collection. In the final step, the two PMMA substrates and the MEA were bonded together using an adhesive gasket. H_2 was supplied by hydride storage, the air by an air pump with a constant flow rate of 50 mL min^{-1}, and O_2 flow with a constant flow rate of 20 mL min^{-1}, all tests were performed at room temperature. A high-power density of 315 mW cm^{-2} at 0.35 V has been achieved when O_2 is supplied at the cathode side, 82 mW cm^{-2} with air.

Another PMMA-based micro PEMFC has been developed by Hsieh et al. The technology included excimer laser lithography for patterning the flow field in the PMMA structure [32]. The channels were about 400 μm wide and 200 μm deep with a rib spacing of 50 μm for both anode and cathode. Electrically conductive regions acting as current collector for electrodes were defined by sputtering 200-nm-thick copper film on the flow-field plates. The MEA was based on Nafion® −117 membrane with thin sputtered Pt films on both sides. The assembly was then completed by 2-mm PMMA gaskets on both sides and sealed by surface mount technology or hot-pressing. A power density of 25 mW cm^{-2} at 0.65 V was obtained at room temperature, when supplied by H_2 and O_2. Further developments using high aspect ratio UV-curable epoxy SU-8 instead of PMMA and silver as current collector were reported with similar performances (30 mW cm^{-2} with H_2 and forced air) [33].

A complete PEMFC system consisting of a PDMS substrate with micro flow channels upon which the MEA was vertically stacked has been developped by Shah et al. [34, 35]. PDMS microreactors were fabricated by employing micromolding with a dry etched silicon master. The PDMS spincoated on micromachined Si was then cured and peeled off from the master. The MEA consisted of a Nafion® 112 membrane where Pt was sputtered through a Mylar® mask. Despite an interesting method, this FC gave poor results (0.8 mW cm^{-2}).

Other Solutions

W. Wan et al. presented another composite solution which used porous polytetrafluoroethylene (PTFE) as the support for the proton-conducting electrolyte [36]. An expanded PTFE substrate was sandwiched on both sides with very thin porous metal sheets made of titanium coated with a Ru layer and fabricated by photolithography and wet etching techniques. An electrolyte dispersion (DE 2021 from DuPont) was then impregnated into the pores of the PTFE membrane and the metal sheets, dried and protonized by immersing in boiling 5% sulfuric acid solution. To complete the MEA (Figure 2.15), a catalyst ink made of Pt-loaded carbon, alcohol, and 5% Nafion® solution was directly brush-coated onto both surfaces of electrolyte membrane and dried. Performances of 80 mW cm^{-2} were reached with H_2/air supply. The metal sheets served here both as flow fields and current collectors.

An original approach was recently proposed by Park and Madou. It used carbonized machined polyimide (CirlexR from Dupont, made from Kapton®) as their microfluidic plates [37]. The technique they currently called C-MEMS (for carbon-MEMS) consists of micromachining the polyimide and then pyrolyzing it at high temperature (900°C) to make micromachined carbon (Figure 2.16). The goal is to reproduce the advantages of graphite used in large fuel cells (such as high electrical conductivity, resistance to corrosion, high thermal conductivity, chemical compatibility, etc.) but is difficult to repeat on micro fuel cells due to the brittleness of the material and its high cost. The structure of the complete fuel cell

FIGURE 2.15 View of the integrated composite MEA, *after* [36].

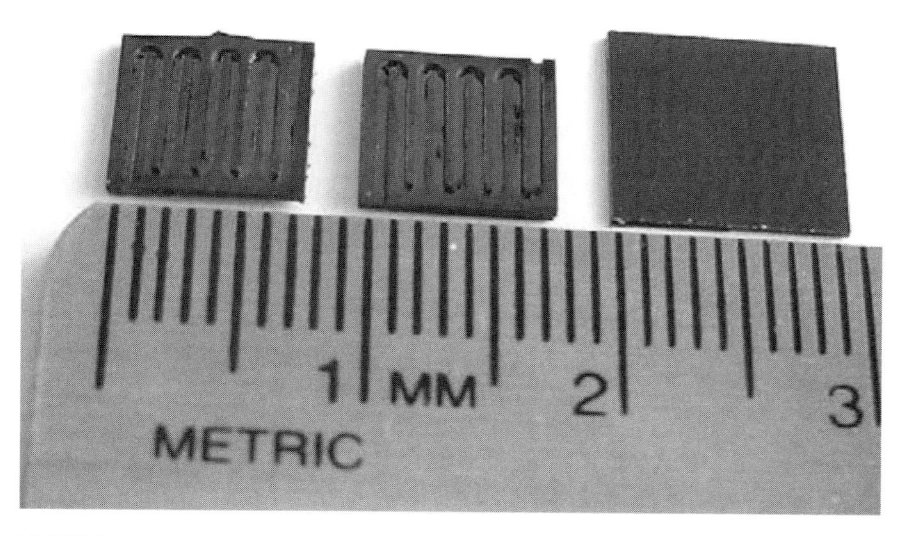

FIGURE 2.16 Carbonized fluidic plates shown to the left of $1\,cm^2$ Cirlex® square, *after* [37].

was as follows: carbon microfluidic flow channels on each side of a commercial MEA made of a Nafion® 115 membrane and 20% Pt-loaded carbon paper from Electrochem. The FC was sealed with epoxy and the membrane was larger than the microfluidic plates in order to hydrate the inner membrane from the external part.

A power density of 1.21 mW cm^{-2} was achieved. The major drawback of this technology is the cost of the polyimide, which remains expensive compared to other materials.

2.3.2 Direct Methanol Fuel Cells

Methanol would currently be the ideal candidate for supplying miniature fuel cells. Actually, this fuel has several advantages over H_2, especially its high volumetric energy density and its ease of storing and refilling. However, its use in a direct way involves new problems, particularly the amount of catalyst needed and the crossover of methanol through the proton exchange membranes. Still, extensive research on DMFCs is needed to attempt to overcome these problems.

Yen et al. [38] (Univ. of California/Pennsylvania State Univ.) presented a bipolar Si-based micro DMFC with a MEA (Nafion® 112 membrane) integrated and micro-channels 750 μm wide and 400 μm deep fabricated using Si micromachining (Deep RIE). This μ-DMFC with an active area of 1.625 cm^2 was characterized at near room temperature, showing a maximum power density of 47.2 mW cm^{-2} at 60°C, then 1M methanol was fed but only 14.3 mW cm^{-2} at room temperature. Since then, Lu et al. [39] have enhanced the performance of the μ-DMFC to a maximum power density of 16 mW cm^{-2} at room temperature and 50 mW cm^{-2} at 60°C with 2 M and 4 M methanol supply with a modified anode backing structure enabling a reduction in methanol crossover (Figure 2.17). Using the same structure as previously quoted, they have recently replaced the silicon, judged as too fragile for compressing and sealing with the MEA, with very thin stainless steel plates as bipolar plates with the flow field machined by photochemical etching technology [40]. A gold layer was deposited on the stainless steel plates to prevent corrosion. This enhanced fuel cell reached 34 mW cm^{-2} at room temperature and 100 mW cm^{-2} at 60°C.

Yao et al. (Carnegie Mellon Univ.) have been currently working on a room temperature DMFC to produce a net output of 10 mW for continuous power generation [41]. Their work focused on the design of the complete system, including water management at the cathode, micropumps and valves, CO_2 gas separation, and other fluidic devices. A passive gas bubble separator removed CO_2 from

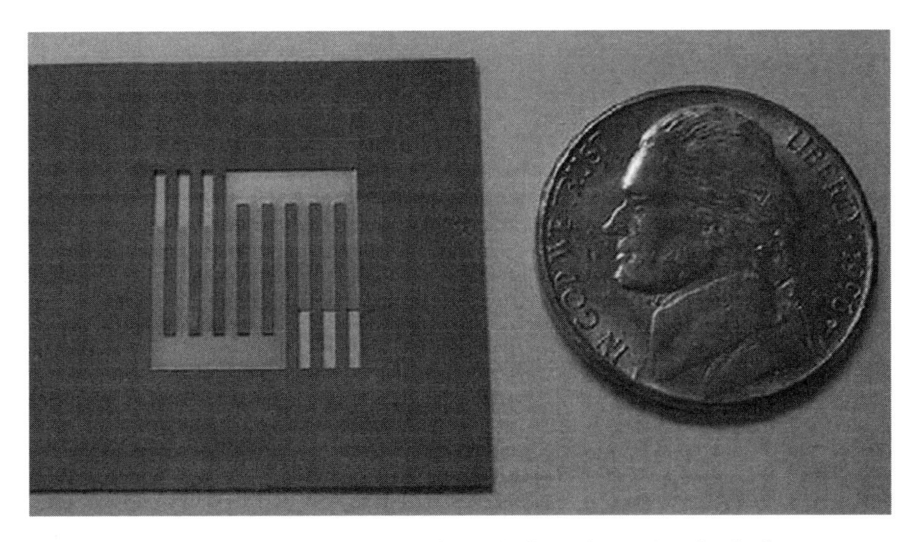

FIGURE 2.17 Picture of a silicon wafer with flow channels, *after* [39].

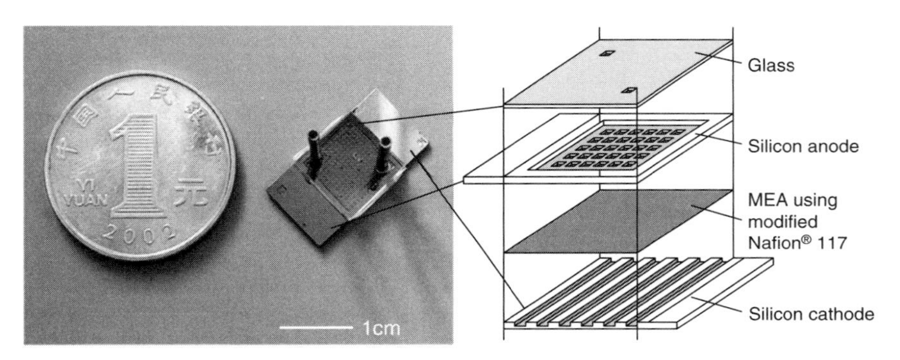

FIGURE 2.18 Photograph and schematic views of the final micro DMFC, *after* [43].

the methanol chamber at the anode side. The back planes of both electrodes were made of Si wafers with an array of etched micro-sized holes. Nano-tube catalysts were fabricated on the planes. A 3% methanol solution at the anode and the air at the cathode were driven by natural convection instead of being pumped. A micro-pump sent water back to the anode side. With 25 mW cm^{-2}, the total MEA area around 1 cm^2 could provide enough power to a 10 mW microsensor along with the extra power needed for internal use, such as water pumping, electronic controls, and process conditioning (see Figure 2.18).

Liu et al. also applied the MEMS technology to the fabrication of micro DMFCs [42, 43]. The process included the use of two silicon substrates (respectively, anode and cathode) sandwiching a MEA using a modified Nafion®. First, both wafers were patterned by photolithography and wet etched. Then, gold was deposited on both to form the electrodes using a PVD method. The fuel channels through the anode silicon substrate were made by laser beam, and silicon-glass anodic bonding was employed to complete the anode side. After the assembly of the silicon substrates and the MEA, epoxy resin was used as a sealant around the edges of the MEA and electrodes, which completed the μ-DMFC. The Nafion® membrane was modified by X-ray radiation and palladium deposition to reduce the methanol crossover. This modification allowed performances of 4.9 mW cm^{-2}, whereas with an unmodified membrane, power density was only 2.5 mW cm^{-2}.

Instead of Si or metal, SU-8 epoxy was the solution adopted by Cha et al. to create microchannels by simple UV photolithography and lift-off from a glass substrate [44]. Nafion® 115 was used as the PEM and sandwiched between two polymer chips with microchannels and two layers of photosensitive polymers. Microchannels were 300 μm wide and 100 μm deep. Current collectors were formed by sputtering Pt on polymer chips.

Carbon black, Pt black, and Pt-Ru catalyst were sprayed in the microchannels before assembling with the membrane. The maximum power density observed with this all-polymer μ-DMFC was 8 mW cm^{-2} with 2 M methanol as fuel and oxygen feeding.

2.3.3 Reformed Hydrogen Fuel Cells

Micro-reformed hydrogen FCs (μ-RHFCs) are actually PEMFCs supplied by hydrogen coming from the reforming of a hydrocarbon (generally methanol) or a chemical hydride (such as sodium borohydride, $NaBH_4$). According to the reformer feed, the reforming operation can be made either at a low operating temperature ($NaBH_4$) or at a relatively high temperature (between 200 and 300°C for methanol steam reforming). It adds complexity to the basic PEMFC system, especially with methanol high temperature reforming, where the proton exchange membrane has to tolerate high temperature, but it allows the use of hydrogen without storing it directly.

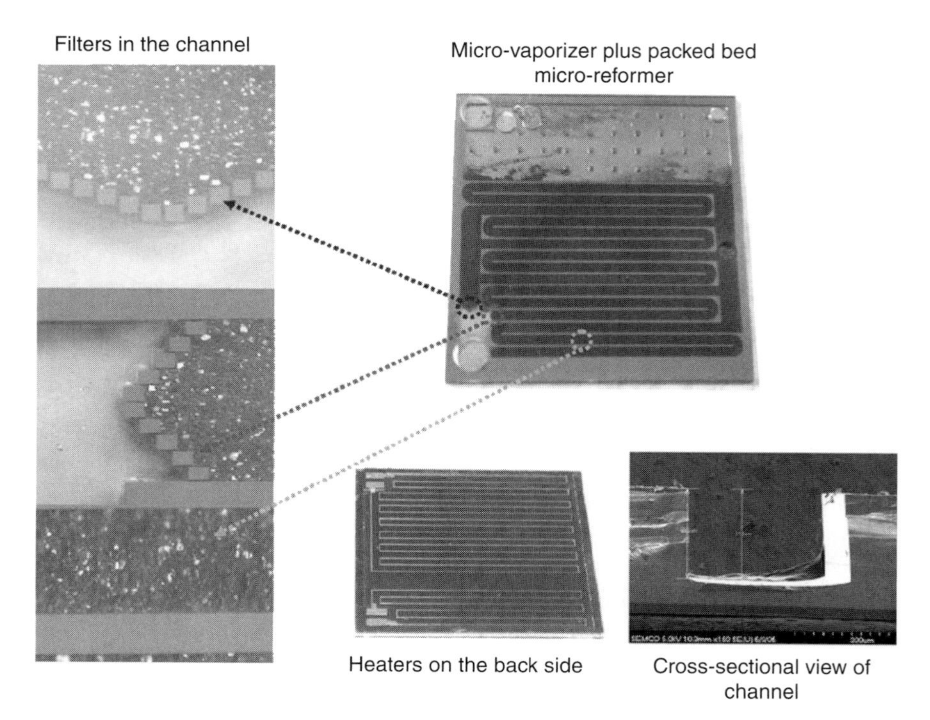

FIGURE 2.19 Micro-reformer developed in Samsung Electro-mechanics, *after* [46].

Concerning the methanol-based RHFCs, literature [45, 46] mentions two types of micro reformers: micro-packed reactors and catalyst coated micro-channel reactors. Pattekar et al. proposed a Si-based micro-packed reactor with microchannels fabricated by deep RIE in 1 mm-thick Si substrates [47]. They used a commercial $Cu/ZnO/Al_2O_3$ catalyst, a Pt resistance-temperature device for temperature sensing and Pt line for along the microchannels as the heater. They obtained 88% methanol conversion at a 1:1.5 molar ratio of methanol and water with a feed rate of $5\,cm^3\,h^{-1}$. Kundu et al. also contributed to this kind of micro-reformer with a Si-based serpentine micro-channeled reactor with high performance and durability test of the catalyst ($Cu/ZnO/\ Al_2O_3$) [48] (see Figure 2.19).

Catalyst-coated microchannel reactors are the second option for methanol steam micro-reforming. They are far more studied because they present several advantages relative to the previous technique—a lower pressure drop and less channeling of gas [46]. Extensive literature reports catalyst coated ($Cu/ZnO/Al_2O_3$) microchannels made of stainless steel [49], aluminum, or silicon. For example, Kawamura et al. reached a 100% conversion of methanol at 250°C using silicon substrate and a homemade catalyst ($Cu/ZnO/Al_2O_3$), with a volume of the micro-reformer less than 1 cm^3 [50]. The key points for high performance and durability are the good uniformity and the amount of the catalyst loading.

The use of $NaBH_4$ for the production of hydrogen seems more attractive since the reforming is made at low temperature (25–40°C). A stabilized aqueous solution of $NaBH_4$ is actually a safe, simple, and compact source of high-purity hydrogen [51]. The hydrolysis reaction is exothermic and proceeds according to the following equation:

$$NaBH_4 + 2H_2O \xrightarrow{\text{catalyst}} 4H_2 + NaBO_2 \tag{2.1}$$

Stabilized $NaBH_4$ solutions, which have a high pH, do not generate significant amounts of hydrogen under ambient conditions. Therefore, a catalyst such as Pt, Ru, or Pt/Ru is required and when this is removed from the $NaBH_4$ solution, hydrogen generation stops. This catalyst can be loaded on carbon support or other supports such as $LiCoCO_2$ or ion-exchange resins. As shown in Eq. 2.1, the reaction also produces boron oxide, which is hazardous and solid, thus risking the clogging of catalyst activity. However, it can be dissolved into the $NaBH_4$ solutions if low concentrations of $NaBH_4$ (10 to 15%) and/or additives such as ethylene glycol are used [46, 51–53]. Literature also reports the use of edible acids as catalyst. For instance, for a few years Seiko Instruments has been involved in the fabrication of micro fuel cells using malic acid solution as the catalyst for hydrogen production from $NaBH_4$ [54].

2.3.4 Alternative Fuels

Direct Formic Acid Fuel Cells

An interesting metallic miniature FC structure from the University of Illinois has been reported and proposed by Ha et al. [16]. They described the design and performance of a passive air breathing direct formic acid FC (DFAFC). The MEA was fabricated in house with catalyst inks (Pt black powder for cathode and Pt-Ru black for anode, Millipore water, and 5% Nafion® solution) directly painted on a Nafion® 117 membrane. The current collectors were fabricated from titanium (Ti) foils electrochemically coated with gold. The miniature cell at 8.8 M formic acid produced a maximum power density of 33 mW cm^{-2} with pieces of gold mesh inserted between the current collector and the MEA on both sides of the MEA. With Pd black used as the catalyst at the anode side [17], they recently obtained a maximum power density of 177 mW cm^{-2} at 0.53 V for their passive DFAFC with 10 M formic acid (see Figure 2.20).

Through the works of Gold and Chu et al., the University of Illinois also proposed another solution consisting of using nanoporous silicon as the basic material for the proton-conducting membrane [55–57]. A first study by Gold et al. had already shown the relevance of using porous silicon loaded with sulfuric acid as the membrane of the fuel cell, though the assembled fuel cell exhibited poor results (a few hundred μW cm^{-2}) [55]. Further evolutions proved to be more effective. Starting from a standard silicon wafer, 100 μm-thick silicon membranes were micromachined using deep

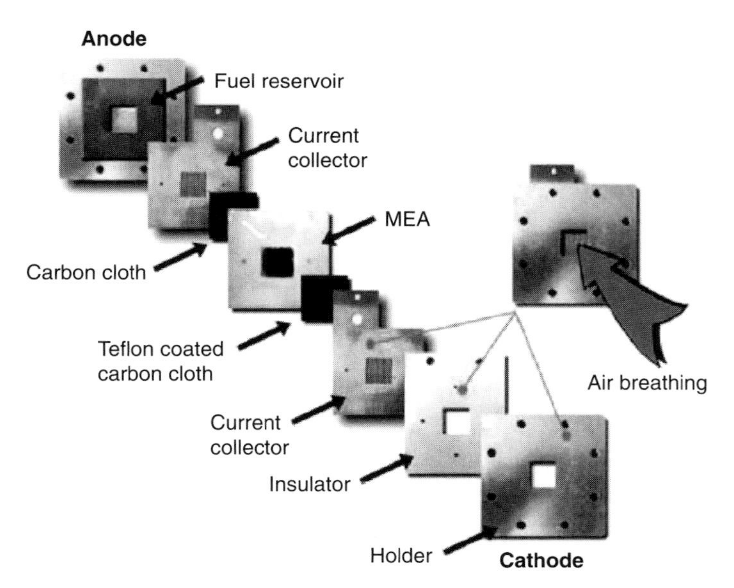

FIGURE 2.20 Structure of the passive air breathing direct formic acid fuel cell, *after* [17].

RIE (with silicon nitride masking) and then anodized in an ethanoic-HF bath to convert them to porous silicon membranes. Insulators (made of TiO_2) and catalyst layers (ink made of Pd black (anode), Pt black (cathode), 5% Nafion® solution and deionized water) were finally painted onto the porous structures to complete the membrane electrodes assembly. Supplied with a 5 M formic acid and 0.5 M sulfuric acid fuel solution, this fuel cell achieved a current density of 34 mW cm^{-2} at 120 mA cm^{-2} [56] with the sulfuric acid acting as the proton-conductor. An improved fabrication process (similar to the one described in the Example section later in this chapter) has been recently proposed with a membrane wet etched on both sides with a KOH solution, silicon nitride insulation layers on both sides grown by LPCVD and a backside RIE after anodization to open up the nanopores [57]. In order to partially oxidize the pore walls, the porous membrane was immersed in a 37% hydrochloric acid solution during one night. Actually, the porous membrane only acts here as a separator that limits fuel crossover and does not need to be proton-conducting. A 5-nm thick gold palladium alloy film was deposited on top of the catalyst layer of both anode and cathode sides using sputtering. Current collectors were then formed by painting a gold ink around and on top of the catalyst films. The performances were also improved up to 94 mW cm^{-2} at 314 mA cm^{-2} with the same fuel supplying conditions as previously.

Yeom et al. (Univ. of Illinois) reported the fabrication of a monolithic Si-based microscale MEA consisting of two Si electrodes, with catalyst deposited directly on them, supporting a Nafion® 112 membrane in-between [58]. The electrodes were identically gold-covered for current collecting, and were also covered with electrodeposited Pt black.

The electrodes and the Nafion® membrane were sandwiched and hot-pressed to form the MEA. The complete fuel cells were tested with various fuels: H_2, methanol, and formic acid and reached 35 mW cm^{-2}, 0.38 mW cm^{-2}, and 17 mW cm^{-2} respectively at room temperature with forced O_2. More recently, performances with formic acid as fuel were increased up to 28 mW cm^{-2} with electrodeposited palladium (Pd)-containing catalyst at the anode [59]. An all-passive system (passive 10 M formic acid and quiescent air) was also recently announced with interesting first results (12.3 mW cm^{-2}) [60].

Ethanol

Aravamudhan et al. (Univ. of South Florida) presented a FC powered by ethanol at room temperarure [15]. The electrodes were fabricated using macroporous Si technology. The pores developed acted both as micro-capillaries/wicking structures and as built-in fuel reservoirs, reducing the size of the FC. The pore sizes dictated the pumping/priming pressure in the FC. The PS electrode thus eliminated the need for an active external fuel pump. The structure of the MEA consisted of two PS electrodes sandwiching a Nafion® 115 membrane. Pt was deposited on both the electrodes, micro-columns to act both as an electrocatalyst and as a current collector. The FC reached a maximum power density of 8.1 mW cm^{-2} by supplying 8.5 M ethanol solution at room temperarure.

2.3.5 Alternative Technologies

PCB Technology

A true alternative to the use of MEMS technology in realizing micro fuel cells was proposed in 2003. O'Hayre et al. demonstrated the successful use of patterned printed-circuit board (PCB) technology to make arrays of H_2/O_2 fuel cells, including microchannel flow fields in the substrate (Figure 2.21). Performances as high as

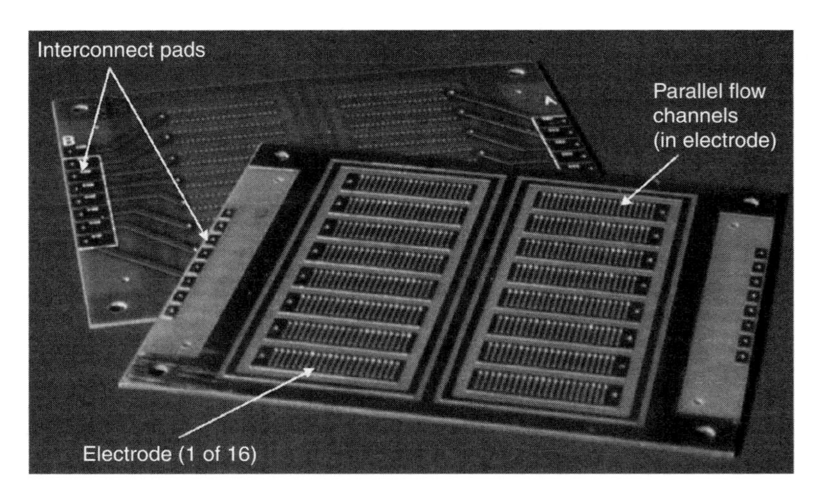

FIGURE 2.21 Details of a PCB-fuel cell prototype with array of 16-cell electrode contacts and 16 perimeter interconnection pads, *after* [61].

700 mW cm^{-2} were reached with H_2/O_2 (<200 mW cm^{-2} with H_2/air) thanks to the good compressibility of the PCB material [61].

Laminar Flow Fuel Cells

Another approach is currently under active development. Actually, there have been several papers on membraneless laminar flow-based microfluidic fuel cells [62–64], which do not require any proton exchange membrane. These micro fuel cells use a liquid-liquid interface between the fuel and the oxidant. An aqueous stream containing a liquid fuel, such as formic acid, methanol, or dissolved hydrogen, and an aqueous stream containing a liquid oxidant, such as dissolved hydrogen, permanganate, or hydrogen peroxide, are introduced into a single microfluidic channel in which the opposite sidewalls are the anode and cathode [67]. Thanks to the dominant laminar convective transport, fuel crossover can be reduced, and the anode dry-out and cathode flooding can be avoided (Figure 2.22). Furthermore, these fuel cells are media flexible, i.e., the composition of the fuel and oxidant streams (e.g., pH) can be chosen independently, allowing improvement in reaction kinetics and open cell potentials. For instance, using a new dual electrolyte H_2/O_2 fuel cell system, Cohen et al. exhibited thus an OCV of 1.4 V and 1.5 mW generated from a single planar device with Si microchannels [65, 66]. With the integration of a gas-diffusion electrode-enabling oxygen delivery directly from air, the laminar

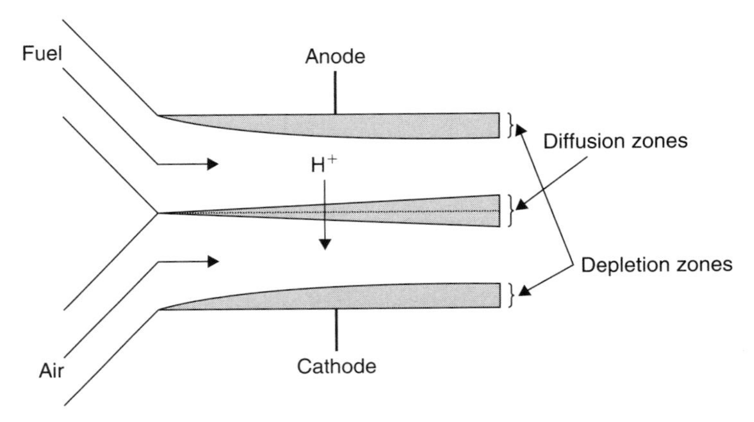

FIGURE 2.22 Illustration of the laminar flow fuel cell principle.

flow-based fuel cell proposed by Jayashree et al. achieved performances as high as 26 mW cm^{-2} with 1 M formic acid as fuel [67]. This novel technology is already competing with proton exchange membrane fuel cells and allows the consideration of further miniaturization of fuel cells.

Solid-Oxide Fuel Cells

Miniature SOFCs (for solid-oxide fuel cells) functioning at temperatures as low as 500°C (low temperature for SOFCs usually functioning at 800°C or more) have also been reported [68–72]. They are based on thin films technology such as those employed for MEMS-based PEMFC and DMFC (photolithography, etching, physical vapor deposition). However decreasing the operating temperature involves the decrease of performances of the fuel cell. Besides, these temperatures remain inadequate for portable devices because they imply major heat transfer and thus require an additional cooling system and/or very efficient insulation, which involves an increase of the overall volume of the fuel cell.

2.3.6 Summary

This brief overview shows that it is not only the fuel or the substrate material that really matters in achieving high performances, but mostly the ways the fluids are managed (GDL/flow fields) and the stack is sealed (electrode-electrolyte interface). Even though Si is still the material most commonly used, the Fraunhöfer Institute with stainless steel foils, for instance, or the promising performances reached with PMMA by Chan et al. [31], have shown a way that true alternatives to silicon can also lead to functional FCs. One of the most exciting solutions may come from the use of microscale effects such as the laminar flow for membraneless FCs, though their performances are not currently equivalent to the classical membrane FCs. Other interesting and recent reviews on micro fuel cells have to be reported, especially the ones by Kundu et al. [46] and Morse [73], that may bring further information and details.

Tables 2.1 and 2.2 summarize the main characteristics of the previously cited fuel cells.

As the preceding review has shown, the role of the electrolyte for hydrogen-fed FCs and DMFCs is almost always played by

TABLE 2.1 Main characteristics of the reviewed micromachined hydrogen fed fuel cells, with Pmax in mW cm^{-2}, OCV the open circuit voltage in V, RT for room temperature and T for working temperature in °C, and "air" is for passive air breathing. NR is for non-reported information.

Ref.	First Author	Year	Substrate	PEM	Catalyst	Pmax	OCV	Oxidant	T
[31]	Chan	2005	PMMA	Nafion® 1135	Pt (ink)	315	0.35	O_2	RT
						82	NR	air	RT
[26]	Chen	2006	silicon	Nafion® 1135	Pt (carbon paper)	144	0.93	air	RT
[28]	Hahn	2004	stainless steel	commercial MEA	NR	80	NR	air	60
[32, 33]	Hsieh	2004	PMMA	Nafion® 117	Pt (sputtering)	25	0.78	O_2	RT
		2005	SU-8	Nafion® 117	Pt (sputtering)	30	NR	air	RT
[75]	Jankowski	2002	silicon	NR	NR	37	0.9	air	40
[20]	Lee	2002	silicon	Nafion® 115	NR	40	NR	O_2	RT
[77]	Marsacq	2005	silicon	Nafion®	Pt (inkjet)	300	NR	air	RT
[18, 19]	Myers	2002	silicon	Nafion®	NR	60	NR	NR	RT
[21, 22]	Min	2006	silicon	Flemion®	Pt	0.8	NR	air	RT
[29]	Müller	2003	metal	Gore MEA	NR	50	0.87	O_2	RT
[61]	O'Hayre	2003	printed circuit	Nafion® 112	Pt/C (ink)	713	0.95	O_2	RT
[37]	Park	2006	pyrolyzed Cirlex	Nafion® 115	Pt (carbon paper)	1.21	NR	O_2	RT
[85, 86]	Pichonat	2006	silicon	porous silicon	Pt (carbon cloth)	58	0.68	air	RT

(Continued)

TABLE 2.1 *Continued*

Ref.	First Author	Year	Substrate	PEM	Catalyst	Pmax	OCV	Oxidant	T
[34, 35]	Shah	2004	PDMS	Nafion® 112	Pt (sputtering)	0.8	0.79	air	60
[36]	Wan	2006	PTFE	DE 2021	Pt (ink)	80	0.985	air	RT
[30]	Wan	2007	titanium	Nafion® 112	Pt (ink)	220	0.96	forced air	RT
[24, 25]	Xiao	2006	silicon	Nafion® 117	Pt (sputtering)	13.7	NR	O_2	RT
[58]	Yeom	2005	silicon	Nafion® 112	Pt black (electrodeposited)	35	NR	O_2	RT
[23]	Yu	2003	silicon	Nafion® 112	NR	193	NR	O_2	RT
[27]	Zhang	2007	silicon	Nafion® 112	Pt	104	0.9	air	RT

TABLE 2.2 Main characteristics of the other reviewed micromachined fuel cells, with Pmax in mW cm^{-2}, OCV the open circuit voltage in V, FA for formic acid, SA for sulfuric acid, MeOH for methanol, EtOH for ethanol, RT for room temperature and T for working temperature in °C. NR is for non-reported information.

Ref.	First Author	Year	Substrate	PEM	Catalyst	Pmax	OCV	Oxidant	T
[44]	Cha	2004	SU-8	Nafion® 115	Pt black (cathode) Pt-Ru (anode) (spray)	8	0.45	2M MeOH/O_2	RT
[42, 43]	Liu	2006	silicon	modified Nafion®	NR	4.9	NR	MeOH/air	RT
[38, 39]	Lu	2004	silicon	Nafion® 112	NR	16/50	NR	2M MeOH/forced air	RT/60
[40]	Lu	2005	stainless steel	Nafion® 112	NR	34/100	NR	2M MeOH/forced air	RT/60
[56, 57]	Chu	2006	silicon	porous silicon	Pt black (cathode) Pd-black (anode) (ink)	94	NR	5M FA + 0.5 SA/air	RT
[16, 17]	Ha	2006	metal	Nafion® 117	Pt black (cathode) Pt-Ru black (anode) (ink)	177	0.53	10M FA/air	RT
[59]	Jayashree	2005	silicon	Nafion® 112	Pd (anode) (electrodeposited)	28	NR	FA/O_2	RT
[60]	Yeom	2006	silicon	Nafion®	Pd (anode) (electrodeposited)	12.3	0.3	10M passive FA/air	RT
[15]	Aravamudhan	2005	silicon	Nafion® 115	Pt	8.1	NR	8.5M etOH/air	RT

perfluorinated carboxylic acid polymer membranes such as Nafion® from DuPont, Flemion® from Asahi Glass, or Gore membrane. Few alternatives were proposed, such as the use of porous silicon. Yet these proton-conducting polymers are not really suitable for microfabrication: they cannot be patterned with standard photolithography and their volumetric changes with humidification are a major problem to deal with at a micrometer scale.

In the following section, another approach consisting of using standard MEMS techniques, porous silicon and chemical grafting to create a proton-conducting membrane will be described as an example of a miniature fuel cell manufactured using microfabrication technology.

2.3.7 Toward a Commercial Mass Production of Fuel Cells?

Concerning forthcoming commercial applications, only public and internet website announcements allow us to know a few details on FCs about to come on the market. Toshiba, Hitachi, Casio, NEC, Motorola, Samsung, Neah Power Systems, to quote only some of the best-known companies, have presented FC prototypes, small or not, supplying portable equipment. They regularly announce forthcoming commercialization but, as far as we know, none of the main actors has yet made their products really available for the masses. That is only an additional proof of the great difficulties to overcome to propose a finished product capable of satisfying the average customer. However, a few companies have begun to develop products for specific applications, such as UltraCell for military purposes. Only a few details have filtered out about the technologies employed, whether MEMS-based or not. Nevertheless, in order to give a clear overview on the trends of the major protagonists of the domain, main characteristics of their prototypes will be reported, again according to the fuel feeding system chosen.

Hydrogen-Fed Micro Fuel Cells

Hydrogen-fed micro FCs give the best performances, provided that hydrogen can be stored safely and efficiently or produced conveniently from water [46]. Four major companies, namely Angstrom Power Inc., Casio, Hitachi Maxell, and NTT DoCoMo, were currently leading in this direction and have already proposed prototypes including hydrogen storage. The first two used metal hydride ($LaNi_5$ for Casio) for hydrogen storage. Angstom Power Inc. proposed a prototype of $160\,cm^3/350\,g$ developing 2W to be used as a charger for cell phones, PDAs, and digital cameras [74]. The claimed maximum operation time is 6h without recharging and they also elaborated a recharging station for refueling the metal with hydrogen. As for the Casio prototype, it has a total volume of $35\,cm^3$ allowing a power of 2.1W.

The possibility of using water for hydrogen storage is, of course, very attractive. Actually, pure hydrogen can be obtained from water through metal powders (Al, Mg), such as through Al oxidation, theoretically with a very good efficiency (1g Al could produce 1.3L hydrogen), and generally requiring sodium hydroxide as a catalyst:

$$Al + 3H_2O \xrightarrow{NaOH} Al(OH)_3 + \frac{3}{2}H_2 \qquad (2.2)$$

Hitachi Maxell has succeeded in producing hydrogen from water and Al powder without the help of a catalyst. The complex procedure is based on the activation of the Al particles by sequential thermal shocks, and the particle size is clearly the primary factor for getting hydrogen. NTT DoCoMo is also involved in developing water-based FC systems, for the purpose of charging Li-ion batteries.

More information is forthcoming from laboratories connected to the industry, often referred to as "thin film technology." For instance, for many years now, Morse et al. (Lawrence Livermore National Laboratory) have been working on SOFCs and PEMFCs [75, 76]. They combined thin film and MEMS technologies to fabricate miniature FCs. A silicon modular design served as the platform for FCs based on either PEM or SO membranes. In 2002, the PEMFC yielded a computed peak power of $37\,mW\,cm^{-2}$ at 0.45V and 40°C, whereas the SOFC reached $145\,mW\,cm^{-2}$ at 0.35V and 600°C. Since 2002, UltraCell has an exclusive license with Lawrence Livermore National Laboratory to exploit their technology to develop commercial FCs.

The French Nuclear Research Center CEA has also announced the successful fabrication of high performance prototypes (Figure. 2.23)

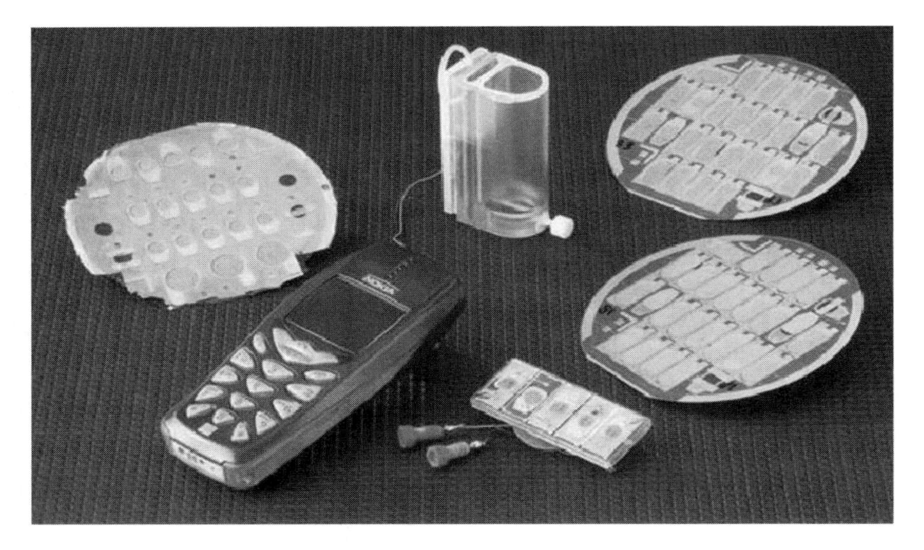

FIGURE 2.23 From the wafer to the phone, fuel cells from the CEA, *after* [77].

fueled by H_2 and based on thin film type structures on Si substrate obtained by microelectronics fabrication techniques (RIE for fuel microchannels, PVD for anode collector, CVD, serigraphy, inkjet for Pt catalyst, lithography) with a Nafion® membrane [77]. A power density of $300\,mW\ cm^{-2}$ with a stabilization around $150\,mW\ cm^{-2}$ during one hundred hours was reported.

μ-*RHFC*

Four companies were notably involved in the development of micro RHFCs: UltraCell (with the Lawrence Livermore National Laboratory), Casio, and Battelle with methanol steam reformers, and Seiko, betting on $NaBH_4$ (previously cited in an earlier section). Since 2002, Casio Computer has been producing and improving prototypes for notebook charging, first using Si as the basic material, then glass since 2004. The fuel processor produced hydrogen from methanol heated to 280°C in only 6 s and the package surface temperature was about 40°C, thanks to glass substrates with vacuum insulation and an inside thin Au layer to minimize radiant heat. Battelle has also developed a methanol reformer for 15 W (military purposes) and 400-mW (portable electronics) fuel

FIGURE 2.24 Prototype of methanol steam RHFC, *after* [78].

cells. Here the reforming was performed at about 350°C with a Pd/ZnO catalyst. The weight of the 15 W prototype was 1 kg (100 g for the fuel processor), and only 1 g (0.3 cm^3) for the methanol reformer of the 400 mW prototype. UltraCell [78] has recently proposed two products (one for military purpose and the other for portable electronics) with a maximum power of 25 W (Figure 2.24). The methanol reforming conceived by the Lawrence Livermore National Laboratory was obtained at 270°C with a Cu/ZnO-based catalyst deposited on Si-based microchannels by sputter coating.

DMFC

Various companies have bet on DMFCs for supplying mobile phones and laptop computers. While Toshiba is still recognized as the industry leader in DMFC development [46] and has presented different prototypes over the last years, especially for powering MP3 players, Samsung, Motorola, NEC, Fujitsu Smart Fuel Cell GmbH, and Antig Technology have already proposed prototypes for portable electronics. Samsung (SDI and AIT) seems currently one of the most advanced competitors, with a DMFC capable of sustaining a laptop for 8 hours daily during an entire month and rapid advancements concerning the size of the fuel cell (Figure 2.25).

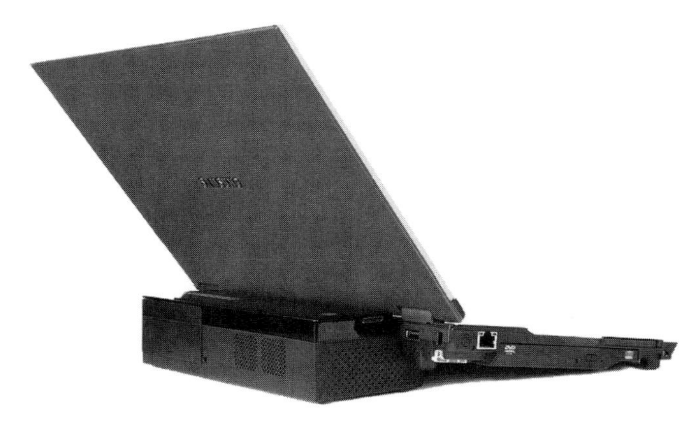

FIGURE 2.25 Samsung DMFC prototype for laptop computer, presented in August 2007.

2.4 EXAMPLE: GRAFTED POROUS SILICON-BASED MINIATURE FUEL CELLS

This section focuses on the fabrication of a particular proton exchange membrane (PEM) for small FCs using microtechnology. The solution has been developed in FEMTO-ST (Besançon, France) as an alternative to classical ionomeric membranes like Nafion® and is based on a porous silicon (PS) membrane on which is grafted a proton-conducting silane. Like previously cited Chu et al. [55–57], previous works have reported the relevance of using PS directly as the membrane of the FC, either as the support for a Nafion® solution [79–82] or with appropriate proton-conducting molecules grafted on the pore walls [80, 83–85]. This second solution involves the chemical grafting of silane molecules containing sulfonate (SO_3^-) or carboxylate (COO^-) groups on the pores' walls in order to mimic the structure of an ionomer.

2.4.1 Manufacturing Process

Membranes

The complete process for the fabrication of the proton-conducting membranes, previously reported in [79, 83], can be described as follows. A 4-inch 520 μm-thick n+ (100)-oriented double-side polished silicon wafer is first thermally oxidized in an oven at

1000°C under O_2 and water steam flows to obtain a 2 μm-thick SiO_2 layer on both sides of the substrate.

These layers will allow the electrical insulation between the two parts of the future fuel cell. Then these previous layers are covered with sputtered Cr-Au layers on both sides. The Cr layers are used as adherence layers for the Au layers and are relatively thin (30 nm). The Au layers are 800 nm thick and will serve as current collector layers for the fuel cells. They are also useful as masking layers for the next different etchings, since Au is not etched either by KOH solution or by HF solution, the two wet etchants used in the next steps.

This is followed by classical photolithography (with adapted etch for the metal layers) and chemical wet etching with a wet KOH solution to produce 50 μm-thick double-sided membranes. These membranes are then made porous by anodization in an ethanoic HF bath with an average pore diameter of 10 nm (size controlled with the input current density). In order to be sure that all the pores are open on the rear side of the membranes, a short reactive ion etching (RIE) is added to the process with a classical silicon etching process using SF_6 and O_2 gases. Figure 2.26 sums up the different steps of the microfabrication process, and Figures 2.27 and 2.28 show the result of the process.

Proton Conduction and Membrane Electrode Assembly

The previous manufactured membranes are not usable as they are, since PS is not naturally proton-conducting. In order to make the membranes proton-conducting, a solution consists of grafting silane molecules containing proton-conducting groups on the pores walls. A commercially available silane salt from United Chemical Technologies Inc (UCT) containing three carboxylate groups has been used for the first investigations. The first step consists of creating silanol functions (Si-OH) at the surface of PS. A soft process involving a UV-ozone cleaner has been successfully implemented. The grafting of silane molecules is then realized by immersing the hydrophilic porous membranes into a 1% silane solution in ethanol for 1 h at room temperature and ambient air. In order to replace -Na endings from the silane salt by -H endings to get the real carboxylic behavior for the grafted function, membranes are immersed for 12 hours in a 20% solution of sulfuric

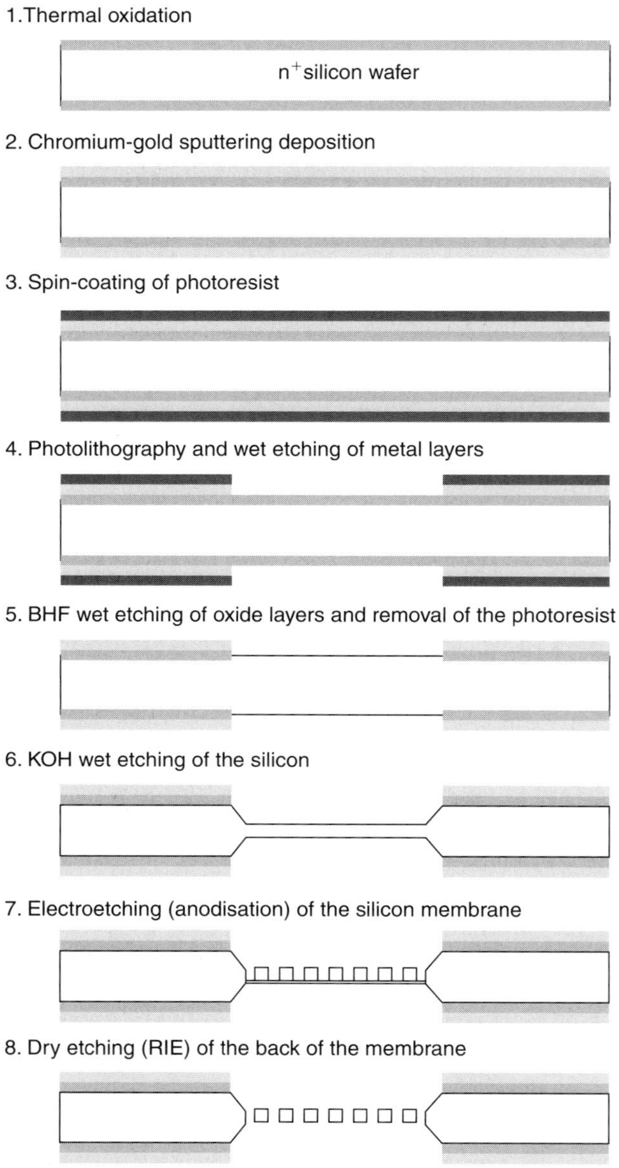

1.Thermal oxidation

n^+silicon wafer

2. Chromium-gold sputtering deposition

3. Spin-coating of photoresist

4. Photolithography and wet etching of metal layers

5. BHF wet etching of oxide layers and removal of the photoresist

6. KOH wet etching of the silicon

7. Electroetching (anodisation) of the silicon membrane

8. Dry etching (RIE) of the back of the membrane

FIGURE 2.26 Sketch of a complete process for manufacturing porous silicon membranes for miniature fuel cells.

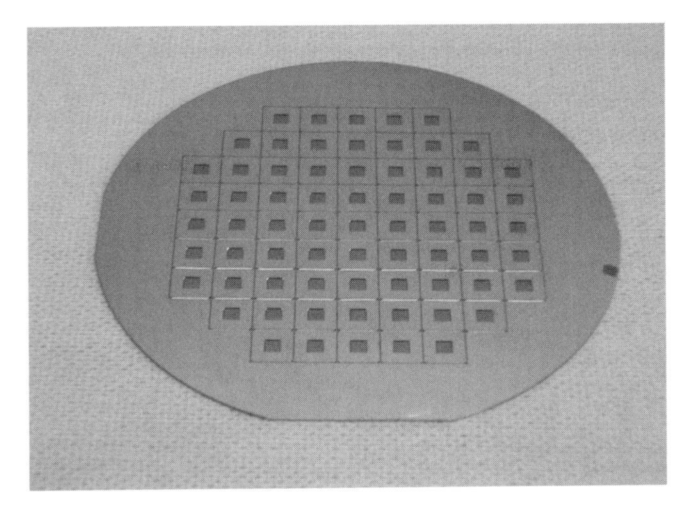

FIGURE 2.27 Photograph of the silicon wafer after the membrane etching process. sixty-nine membranes of 7 mm^2 are micromachined on a 4-inch wafer.

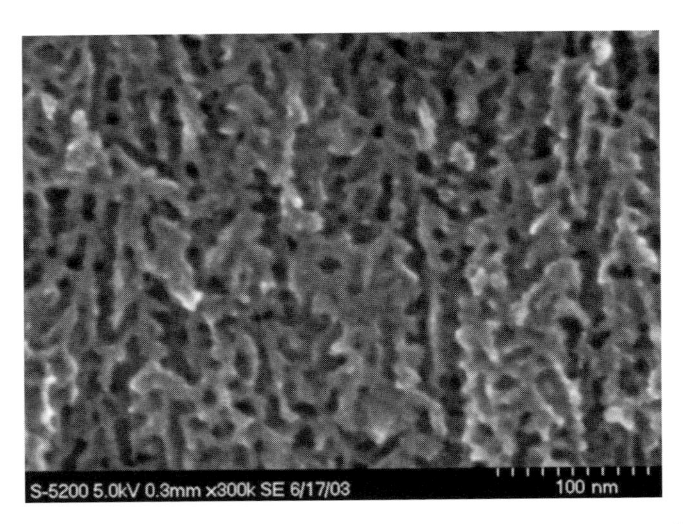

FIGURE 2.28 SEM (scanning electron microscope) cross-section view of a typical porous silicon membrane made with the process in Figure 2.26.

acid, then carefully rinsed in deionized water. To complete the FC assembly, E-TEK carbon conducting cloth electrodes filled with Pt (20% Pt on Vulcan XC-72) are added on both sides of the membrane as a H_2/O_2 catalyst. Figure 2.29 shows the final single cell.

FIGURE 2.29 Membrane-electrode assembly (8 × 8 mm), scale comparison with a 1 cent euro coin.

2.4.2 Results

Figure 2.29 shows (top view) a typical 8 mm by 8 mm FC realized with an active area (in black on the figure) of 7 mm^2.

Measurements were carried out at room temperature. H_2 feeding was provided by a 20% NaOH solution electrolysis while passive ambient air was used at the cathode. In order to bring the gas to the membrane, a homemade test cell was used in which up to four single cells can be tested separately or together with a serial or parallel connection. A maximum power density of 58 mW cm^{-2} at 0.34 V was reached with an open circuit voltage (OCV) of 0.68 V (Figure 2.30). The relative low voltage is mostly due to the crossover of H_2 through the pores.

In order to reduce this crossover and then to increase the OCV, the pore diameter should be reduced to at least as small as that of assumed pores in Nafion® (3 to 5 nm). Moreover, to increase the power density, the grafting density should be controlled to be sure that all the internal surface of the pores is grafted. The use of a catalyst ink instead of a Pt-loaded carbon cloth should create a more intimate link between the membrane and the catalyst and should further increase the performances. A study on the thickness of current-collecting metal layers [26] had also recently shown that

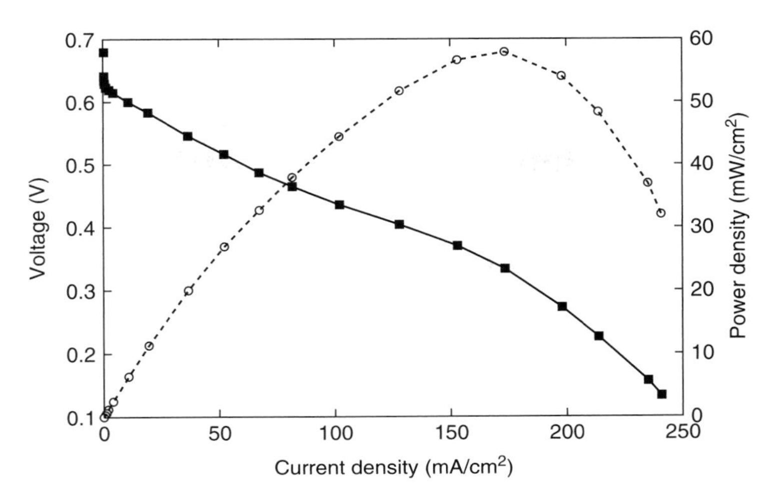

FIGURE 2.30 Performances of the miniature fuel cell obtained with a mesoporous silicon membrane grafted with proton-conducting silane (voltage vs. current density in continuous line, power density vs. current density in dashed line).

the thickness of the Au layer was probably too thin to provide the lowest electrical resistance.

2.5 CONCLUSION

Though no commercial product is available yet, microfabrication techniques have now proven themselves to be very useful tools for the development of miniature fuel cells. These techniques enable the miniaturization and the mass fabrication of almost every component of a fuel cell: flow channels for the proper circulation of fuel and waste with photolithography and reactive ion etching or micromolding, catalyst support by etching and deposition, current collector by metal deposition and proton-conducting membrane with insulator deposition, photolithography, wet etching, electroetching, and post-fabrication chemical treatment.

Several examples of miniature fuel cells using microfabrication techniques have been presented. Among the numerous solutions developed today, the basic structure of fuel cells remains the same: thin film planar stack (generally silicon, foils, polymer or glass) with commercial ionomer, most often Nafion®, the reported layers being micromachined (microchannels or porous media) for gas/liquid management and coated with gold for current collecting. Performances range from a tenth of μW cm^{-2} to several hundreds of mW cm^{-2} according principally to fluids management and sealing. As the basic material for MEMS technologies, silicon remains the most employed material for MEMS-based fuel cells, but foils and polymers have shown interesting potential for future commercial development. Thus, in most of the cases, this is often the simple application of microtechnology to conventional fuel cell structures. Yet new ways of conceiving fuel cells using real microscale effects, such as membraneless laminar flow-based fuel cells, are beginning to appear. They show that new technological breakthroughs have to be expected, especially to get rid of ionomer membranes that have disadvantages, most notably their change in size with humidification and their incompatibility with microtechnology.

As a detailed example of a micromachined fuel cell, an alternative solution which does not use an ionomer for the proton exchange membrane has also been reported. It consists of a porous silicon membrane with a proton-conducting silane grafted on the pore walls. With this membrane, performances as high as 58 mW cm^{-2} have been achieved. This promising membrane can still be improved. Future work should focus on the reduction of the pore diameter to decrease the gas crossover, the control of the grafting density into the porous silicon, the replacement of the electrode carbon cloth by an ink, and the use of a more proton-conducting silane (with SO_3^- terminations).

ACKNOWLEDGMENTS

All previously published figures have been reproduced with kind permission from the editors and/or the authors.

References

[1] C.K. Dyer, Sci. Am. 281 (1) (1999) 88–93.
[2] C.K. Dyer, IEEE Symposium on VLSI Circuits, Digest of Technical Papers, 2004, pp. 124–127.
[3] C.K. Dyer, J. Power Sources 106 (2002) 31–34.
[4] V. Boisard, Etats Unis Microélectronique 22 (2000) 4–16.

[5] F. Davis, S.P.J. Higson, Biosens. Bioelectron. 22 (2007) 1224–1235.

[6] C. Hebling, A. Heinzel, Fuel Cells Bull. 7 (2002) 8–16.

[7] F. Chollet, H.B. Liu, A (not so) Short introduction to micro electromechanical systems, version 2.0, August 2006, <http://memscyclopedia.org/IntroMEMS.html>.

[8] O. Millet, P. Bernardoni, S. Régnier, P. Bidaud, E. Tsitsiris, D. Collard, L. Buchaillot, Sens. Actuators, A 114 (2004) 371–378.

[9] B. Legrand, D. Collard, L. Buchaillot, Actuators and microsystems transducers' 05, in: Proceedings of the 13th IEEE International Conference On Solid-State Sensors, Korea, 2005, pp. 57–60.

[10] T. Pichonat, L.O. Vasquez (Ed.), Fuel Cell Research Trends, Nova Publishers, New York, 2007.

[11] S.M. Sze, VLSI Technology, 2nd edition., McGraw-Hill, New York, 1988.

[12] S.D. Senturia, Microsystem Design, Springer Science, New York, 2001.

[13] M.J. Madou, Fundamentals of Microfabrication: The Science of Miniaturization, 2nd edition., CRC PRESS, New York, 2002.

[14] N.-T. Nguyen, S.H. Chan, J. Micromech. Microeng. 16 (2006) R1–R12.

[15] S. Aravamudhan, A.R.A. Rahman, S. Bhansali, Sens. Actuators, A 123-124 (2005) 497–504.

[16] S. Ha, B. Adams, R.I. Masel, J. Power Sources 128 (2004) 119–124.

[17] S. Ha, Z. Dunbar, R.I. Masel, J. Power Sources 158 (2006) 129–136.

[18] H.L. Maynard, J.P. Meyers, in: Proceedings of the 2nd Annual Small Fuel Cells and Battery Technologies for Portable Power Applications International Symposium, New Orleans, LA, U.S.A., 2000.

[19] J.P. Meyers, H.L. Maynard, J. Power Sources 109 (2002) 76–88.

[20] S.J. Lee, A. Chang-Chien, S.W. Cha, R. O'Hayre, Y.I. Park, Y. Saito, F.B. Prinz, J. Power Sources 112 (2002) 410–418.

[21] K.B. Min, S. Tanaka, M. Esashi, in: Proceedings of the 16th International Conference on MEMS, Kyoto, Japan, 2003, pp. 379–382.

[22] K.B. Min, S. Tanaka, M. Esashi, J. Micromech. Microeng. 16 (2006) 505–511.

[23] J. Yu, P. Cheng, Z. Ma, B. Yi, J. Power Sources 124 (2003) 40–46.

[24] Z. Xiao, G. Yan, C. Feng, P.C.H. Chan, I-M. Hsing, Actuators and microsystems transducers' 05, in: Proceedings of the 13th International Conference on Solid-State Sensors, Seoul, Korea, 2005, pp. 1856–1859.

[25] Z. Xiao, G. Yan, C. Feng, P.C.H. Chan, I-M. Hsing, J. Micromech. Microeng. 16 (2006) 2014–2020.

[26] C. Chen, X. Li, T. Wang, X. Zhang, J. Li, P. Dong, D. Zheng, B. Xia, J. Microelectromech. Syst. 15 (5) (2006) 1088–1097.

[27] X. Zhang, D. Zheng, T. Wang, C. Chen, J. Cao, J. Yan, W. Wang, J. Liu, H. Liu, J. Tian, X. Li, H. Yang, B. Xia, J. Power Sources 166 (2007) 441–444.

[28] R. Hahn, S. Wagner, A. Schmitz, H. Reichl, J. Power Sources 131 (2004) 73–78.

[29] M. Müller, C. Müller, F. Gromball, M. Wölfe, W. Menz, Microsyst. Technol. 9 (2003) 159–162.

[30] N. Wan, C. Wang, Z. Mao, Electrochem. Commun. 9 (2007) 511–516.

[31] S.H. Chan, N.-T. Nguyen, Z. Xia, Z. Wu, J. Micromech. Microeng. 15 (2005) 231–236.

[32] S.-S. Hsieh, J.-K. Kuo, C.-F. Hwang, H.-H. Tsai, Microsyst. Technol. 10 (2004) 121–126.
[33] S.-S. Hsieh, C.-F. Hwang, J.-K. Kuo, H.-H. Tsai, J. Solid State Electrochem. 9 (2005) 121–131.
[34] K. Shah, W.C. Shin, R.S. Besser, J. Power Sources 123 (2003) 172–181.
[35] K. Shah, W.C. Shin, R.S. Besser, Sens. Actuators, B 97 (2004) 157–167.
[36] W. Wan, G. Wang, J. Power Sources 159 (2006) 951–955.
[37] B.Y. Park, M.J. Madou, J. Power Sources 162 (2006) 369–379.
[38] T.J. Yen, N. Fang, X. Zhang, G.Q. Lu, C.Y. Wang, Appl. Phys. Lett. 83 (2003) 4056–4058.
[39] G.Q. Lu, C.Y. Wang, T.J. Yen, X. Zhang, Electrochim. Acta 49 (2004) 821–828.
[40] G.Q. Lu, C.Y. Wang, J. Power Sources 144 (2005) 141–145.
[41] S.C. Yao, X. Tang, C.-C. Hsieh, Y. Alyousef, M. Vladimer, G.K. Fedder, C.H. Amon, Energy 31 (2006) 636–649.
[42] X. Liu, C. Suo, Y. Zhang, H. Zhang, W. Chen, W. Lu, in: Proceedings of the 8th Symposium on Design, Test, Integration and Packaging of MEMS/ MOEMS (DTIP '06), Stresa, Italy, 2006, pp. 360–364.
[43] X. Liu, C. Suo, Z. Zhang, H. Zhang, X. Wang, C. Sun, L. Li, L. Zhang, J. Micromech. Microeng. 16 (2006) S226–S232.
[44] H.Y. Cha, H.G. Choi, J.D. Nam, Y. Lee, S.M. Cho, E.S. Lee, J.K. Lee, C.H. Chung, Electrochim. Acta 50 (2004) 795–799.
[45] A. Kundu, D.H. Kim, Y.G. Shul, T.S. Zhao, K.-D. Kreuer, T. Nguyen (Eds.), Advances in Fuel Cells Book Series, vol.1, Elsevier Publications, 2006, pp. 417–470.
[46] A. Kundu, J.H. Jang, C.R. Jung, H.R. Lee, S.-H. Kim, B. Ku, Y.S. Oh, J. Power Sources 170 (2007) 67–78.
[47] A.V. Pattekar, M.V. Kothare, J. Microelectromech. Syst. 13 (1) (2004) 7–18.
[48] A. Kundu, J.H. Jang, H.R. Lee, J.H. Gil, S.-H. Kim, C.R. Jung, H.Y. Cha, B. Ku, C.M. Miesse, K.S. Chae, Y.S. Oh, ECS Trans. 5 (2006) 655–663.
[49] A. Kundu, J.M. Park, J.E. Ahn, S.S. Park, Y.G. Shul, H.S. Han, Fuel 86 (2007) 331–1336.
[50] Y. Kawamura, N. Ogura, T. Yamamoto, A. Igarashi, Chem. Eng. Sci. 61 (2006) 1092–1101.
[51] Z.T. Xia, S.H. Chan, J. Power Sources 152 (2005) 46–49.
[52] J.-H. Wee, J. Power Sources 155 (2006) 329–339.
[53] J.-H. Wee, K.-Y. Lee, S.H. Kim, Fuel Process. Technol. 87 (2006) 811–819.
[54] F. Iwasaki, in: Proceedings of the 3rd International Hydrogen & Fuel Cell Expo (FC-11), 2007, pp. 41–79.
[55] S. Gold, K.-L. Chu, C. Lu, M.A. Shannon, R.I. Masel, J. Power Sources 135 (2004) 198–203.
[56] K.-L. Chu, S. Gold, V. Subramanian, C. Lu, M.A. Shannon, R.I. Masel, J. Microelectromech. Syst. 15 (2006) 671–677.
[57] K.-L. Chu, M.A. Shannon, R.I. Masel, J. Electrochem. Soc. 153 (2006) A1562–A1567.
[58] J. Yeom, G.Z. Mozsgai, B.R. Flachsbart, E.R. Choban, A. Asthana, M.A. Shannon, P.J.A. Kenis, Sens. Actuators, B 107 (2005) 882–891.
[59] R.S. Jayashree, J.S. Spendelow, J. Yeom, C. Rastogi, M.A. Shannon, P.J.A. Kenis, Electrochim. Acta 50 (2005) 4674–4682.

[60] J. Yeom, R.S. Jayashree, C. Rastogi, M.A. Shannon, P.J.A. Kenis, J. Power Sources 160 (2006) 1058–1064.
[61] R. O'Hayre, D. Braithwaite, W. Hermann, S. Lee, T. Fabian, S. Cha, Y. Saito, F. Prinz, J. Power Sources 124 (2003) 459–472.
[62] R. Ferrigno, A.D. Stroock, T.D. Clark, M. Mayer, G.M. Whitesides, J. Am. Chem. Soc. 124 (2002) 12930–12931.
[63] E.R. Choban, L.J. Markoski, A. Wieckowski, P.J.A. Kenis, J. Power Sources 128 (2004) 54–60.
[64] E.R. Choban, J.S. Spendelow, L. Gancs, A. Wieckowski, P.J.A. Kenis, Electrochim. Acta 50 (2005) 5390–5398.
[65] J.L. Cohen, D.A. Westly, A. Pechenik, H.D. Abruna, J. Power Sources 139 (2005) 96–105.
[66] J.L. Cohen, D.J. Volpe, D.A. Westly, A. Pechenik, H.D. Abruna, Langmuir 21 (2005) 3544–3550.
[67] R.S. Jayashree, L. Gancs, E.R. Choban, A. Primak, D. Natarajian, L.J. Markoski, P.J.A. Kenis, J. Am. Chem. Soc. 127 (2005) 16758–16759.
[68] J.D. Morse, R.T. Graff, J.P. Hayes, A.F. Jankowski, in: Proc. Mat. Res. Soc. Symp. 575 (2002) 321–324.
[69] V.T. Srikar, K.T. Turner, T.Y.A. Le, S.M. Spearing, J. Power Sources 125 (2004) 62–69.
[70] C.D. Baertsch, K.F. Jensen, J.L. Hertz, H.L. Tuller, S.K. Vengallatore, S.M. Spearing, M.A. Schmidt, J. Mat. Res. 19 (9) (2004) 2604–2615.
[71] Y. Tang, K. Stanley, J. Wu, D. Ghosh, J. Zhang, J. Micromech. Microeng. 15 (2005) S185–S192.
[72] X. Chen, N.J. Wu, L. Smith, A. Ignatiev, Appl. Phys. Lett. 84 (14) (2004) 2700–2702.
[73] J.D. Morse, Int. J. Energy Res. 31 (2007) 576–602.
[74] http://www.angstrompower.com.
[75] A.F. Jankowski, J.P. Hayes, R.T. Graff, J.D. Morse, in: Proc. Mat. Res. Soc. Symp., San Francisco, CA, 730 (2002) 93–98.
[76] http://www.llnl.gov.
[77] D. Marsacq, Clefs CEA 50-51 (2004).
[78] http://www.ultracellpower.com.
[79] T. Pichonat, B. Gauthier-Manuel, D. Hauden, Chem. Eng. J. 101 (1–3) (2004) 107–111.
[80] T. Pichonat, Study, conception and realization of a miniature fuel cell for portable applications, Ph.D. Thesis, 2004. Available from: <http://www.tel.ccsd.cnrs.fr/tel-00010710>.
[81] T. Pichonat, B. Gauthier-Manuel, J. Membr. Sci. 280 (2006) 494–500.
[82] T. Pichonat, B. Gauthier-Manuel, Microsyst. Technol. 12 (4) (2006) 330–334.
[83] T. Pichonat, B. Gauthier-Manuel, J. Micromech. Microeng. 15 (2005) S179–S184.
[84] T. Pichonat, B. Gauthier-Manuel, J. Power Sources 154 (2006) 198–201.
[85] T. Pichonat, B. Gauthier-Manuel, in: Proceedings of the 8th Symposium on Design, Test, Integration and Packaging of MEMS/MOEMS (DTIP' 06), Stresa, Italy, 2006, pp. 354–359.
[86] T. Pichonat, B. Gauthier-Manuel, Microsyst. Technol. 13 (11-12) (2007) 1671–1678.

CHAPTER

3

Advances in Microfluidic Fuel Cells

Erik Kjeang, Nedjib Djilali, and David Sinton

Department of Mechanical Engineering, and Institute for Integrated Energy Systems (IESVic), and University of Victoria, British Columbia, Canada

Since its invention in 2002, microfluidic fuel cell technology has developed rapidly. This entry provides background and reviews recent advances in microfluidic fuel cells. In microfluidic fuel cells, all components and functions including fluid delivery, reaction sites, and electrodes are confined to a microfluidic channel. Microfluidic fuel cells typically operate in a co-laminar flow configuration without a physical barrier, such as a membrane, to separate the anode and the cathode. The background covered here includes the theory, fabrication, unit cell development, performance achievements, design considerations, and scale-up options. Advances in this area will be described with particular emphasis on microfluidic fuel cell architectures—an area where significant gains have been made to date. In addition, relevant microfluidic biofuel cell developments are described, particularly as they present opportunities for future work in this multi-disciplinary field. Looking ahead, the main challenges associated with the technology are described along with suggested directions for further research and development.

3.1 INTRODUCTION

Small fuel cells are widely considered as miniaturized power sources for the next generation consumer electronics [1]. Foremost, the accelerating power demands of portable devices are already compromised by the size, weight, and reliability of existing battery technology. Small, integrated fuel cells enable higher overall energy density than battery systems, and the cost tolerance associated with portable electronics is in favor of new, conceptually redesigned power packages with extended runtime. There are numerous technical challenges, however, related to the development of miniaturized fuel cell technologies. Hydrogen PEM fuel cells require hydrogen storage and/or reformer units that are generally too large for consumer electronics. Direct liquid fuel cells have compact fuel storage solutions, but the performance is restricted by slow electrochemical kinetics and fuel crossover from anode to cathode.

Microfluidic fuel cells, membraneless fuel cells, and laminar flow-based fuel cells are all part of a recent classification of fuel cell technology capable of operation within the framework of a microfluidic chip. In microfluidic fuel cells, all functions and components related to fluid delivery and removal, reaction sites, and electrode structures are confined to a microfluidic channel. Microfluidic fuel cells typically operate in a co-laminar flow configuration without a physical barrier, such as a membrane, to separate the anode and the cathode.

The microfluidic fuel cell concept was invented in 2002 [2, 3]. A schematic of an early microfluidic fuel cell is shown in Figure 3.1. Fuel and oxidant (in this case both are vanadium redox species) are introduced to a single microfluidic channel with electrodes patterned on the bottom walls.

Building on this concept, many different microfluidic fuel cell architectures have been developed since 2002. These developments are marked by many journal publications, and a recent review [4]. The field of microfluidic fuel cells has developed as a subset of microstructured fuel cells, or micro fuel cells. For a more general overview of micro fuel cell technology the reader is referred to recent review articles [1, 5]. Figure 3.2 shows schematically the main microfluidic fuel cell architectures presented to date. The first architectures involve two streams combined horizontally in a T- or Y-channel with: electrodes on bottom (Figure 3.2a); electrodes on sides (Figure 3.2b); and porous electrodes on bottom (Figure 3.2c). Alternatively, with an F-channel configuration, streams may be combined vertically via sheath-flow with electrodes

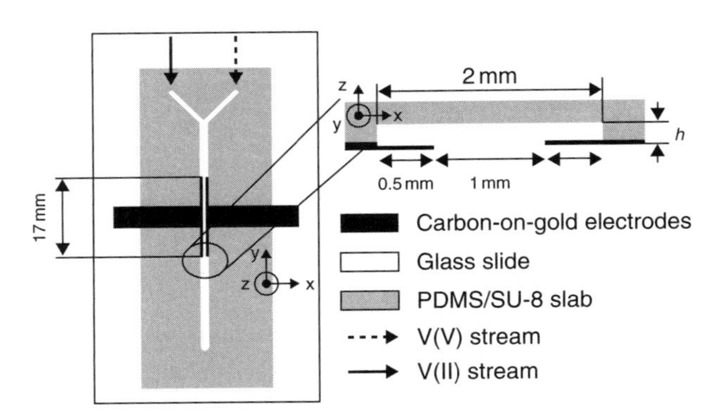

FIGURE 3.1 Schematic of an early microfluidic fuel cell. The fuel and oxidant are introduced into a single microchannel pre-patterned with electrodes. Vanadium redox species were used here for both fuel and oxidant, and reacted on carbon-on-gold electrodes as shown. Reprinted with permission from Ferrigno et al. [3]. Copyright 2005 American Chemical Society.

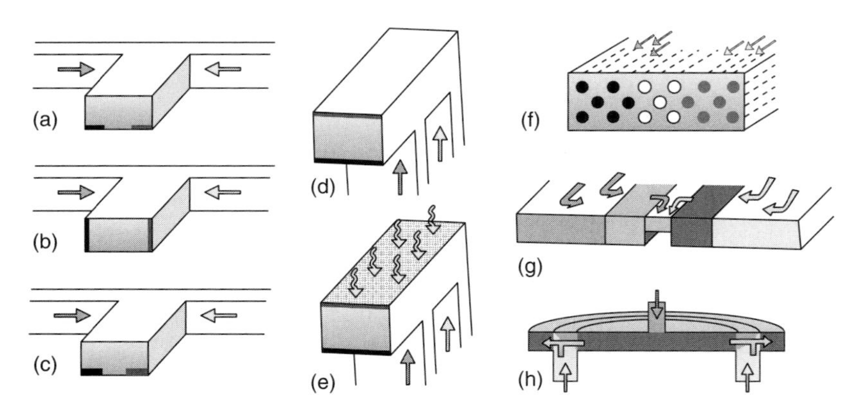

FIGURE 3.2 Co-laminar microfluidic fuel cell architectures developed since 2002. As shown, laminar streaming facilitates the separation of fuel and oxidant in the absence of a physical membrane. Fuel is shown in dark gray, oxidant in light gray, cathodes in black, anodes in gray, and porous electrodes are shown textured. Two streams are combined horizontally in a T- or Y-channel with electrodes on bottom (a), electrodes on sides (b), and porous electrodes on bottom (c). An F-channel configuration (d), and with the addition of a porous electrode to facilitate air-breathing (e). An electrode array microfluidic fuel cell (f). A flow through porous electrode microfluidic fuel cell (g). A radial porous electrode fuel cell architecture (h).

on top and bottom (Figure 3.2d). The F-channel configuration permits the use of a porous electrode on top as well. This configuration may be employed to facilitate air-breathing (Figure 3.2e). The microfluidic array fuel cell employs microfluidic flow between fixed, free standing electrodes in a hexagonal pattern (Figure 3.2f). With the flow through porous electrode microfluidic fuel cell (Figure 3.2g), reactants flow through and react within porous electrodes prior to combining in a co-laminar waste stream. The radial porous electrode fuel cell (Figure 3.2h) architecture uses sequential flow of fuel and oxidant in a concentric fashion.

The most common configurations of microfluidic fuel cells utilize the laminar flow characteristic of microfluidic flows to delay convective mixing of fuel and oxidant. Flow in this regime is characterized by low Reynolds numbers, $Re = \rho U D_h/\mu$, where ρ is the fluid density, U is the average velocity, D_h is the hydraulic diameter, and μ is the dynamic viscosity. At low Reynolds numbers, both aqueous streams (one containing the fuel and one containing the oxidant) will flow in parallel down a single microfluidic channel, as shown in Figure 3.2a. The anolyte (fuel) and catholyte (oxidant) also contain supporting electrolyte that facilitates ionic transport within the streams, thereby eliminating the need for a separate electrolyte. Mixing of the two streams is limited to diffusive effects only, and is thus restricted to an interfacial width at the center of the channel. The resulting mixing width can be controlled by modification of channel dimensions and flow rate. In general, the electrodes are integrated on one or more walls of the manifold with sufficient separation from the co-laminar interdiffusion zone in order to prevent fuel crossover.

Some of the issues encountered in PEM-based fuel cells, for example humidification, membrane degradation, and fuel crossover, are limited or eliminated with microfluidic fuel cells. It is also possible to select the composition of the anolyte and catholyte streams independently, thus providing an opportunity to improve reaction rates and cell voltage. In addition to compactness, miniaturization of fuel cells provides a further advantage. Specifically, miniaturization is accompanied by an increase in surface-to-volume ratio, which scales as the inverse of the characteristic length. Since electrochemical reactions are surface-based, the performance of the fuel cell benefits directly from miniaturization of the device. Perhaps the most prominent benefit associated with microfluidic fuel cells, however, is cost. Microfluidic fuel cells can be manufactured by inexpensive, well-established micromachining and microfabrication methods and the cost associated with the membrane, which is significant for most other fuel cells, is eliminated. While catalyst is still required in microfluidic fuel cells, a variety of catalyst-free cells have been developed with electrodes based on carbon that are orders of magnitude less expensive than platinum required in traditional hydrogen fuel cells. In addition, microfluidic fuel cells generally do not require auxiliary humidification, water management, or cooling considerations and may be operated at room temperature. The challenges that lie ahead for microfluidic fuel cells, however, are in the area of energy density and fuel utilization. While significant gains have been made in this area, many of which will be highlighted in this chapter, these challenges remain to a large degree.

In this chapter, the fundamentals behind microfluidic fuel cell technology are described first. Next, the development of microfluidic fuel cells to date is presented with several examples. Consideration is given to choice of reactants, electrochemical reactions, transport characteristics and, particularly here, cell architectures. Microfluidic fuel cell architecture is an area that has seen particularly rapid development, as illustrated in Figure 3.2. Biofuel cell technology is discussed here as well, in the context of microfluidic fuel cells, and identified as one of the exciting research avenues ahead.

3.2 MICROFLUIDIC FUEL CELL FUNDAMENTALS

Microfluidics has been described as both a science and a technology [6, 7]. It is defined as the study and application of fluid flow and transport phenomena in microstructures with at least one characteristic dimension in the range of 1–1000 μm [6, 7]. This multidisciplinary field involves engineering, chemistry, and biology and serves a wide range of applications including biomedical diagnostics, lab-on-chip technologies, drug discovery, proteomics, and energy conversion. Squires and Quake [8] and Gad-El-Hak [9] provide in-depth reviews of the physics of microfluidics. Erickson [10] provides an overview of numerical modeling as applied to microfluidic transport phenomena. Fluid flow in microscale devices can be expected to be laminar under most conditions. Microfluidics is thus characterized by low Reynolds number where viscous effects dominate over inertial effects and surface forces play a dominant role over body forces.

Due to the laminar nature of microfluidics, the velocity field $\overline{u}$ for incompressible Newtonian fluids can be determined directly from the Navier-Stokes equations for momentum conservation in 3D. The use of the Navier-Stokes equations presupposes that the fluid may be treated as a continuum; however, this assumption is generally valid in microscale liquid flows [7–9], and may be applied with reasonable accuracy into the nanofluidic range. At very low Reynolds numbers, the nonlinear convective terms in the Navier-Stokes equations may be safely neglected, resulting in linear and predictable Stokes flow:

$$\rho\frac{\partial\overline{u}}{\partial t} = -\nabla p + \mu\nabla^2\overline{u} + \overline{f} \tag{3.1}$$

where p represents pressure and $\overline{f}$ summarizes the body forces per unit volume. Furthermore, mass conservation for fluid flow obeys the continuity equation:

$$\frac{\partial \rho}{\partial t} + \nabla \cdot (\rho \overline{u}) = 0 \tag{3.2}$$

For a fluid with constant density this equation simplifies to the incompressibility condition $\nabla \cdot \overline{u} = 0$. For a simple geometry such as parallel plates, or a cylindrical tube, Eqs. (3.1) and (3.2) lead to the familiar parabolic pressure-driven velocity profile.

The surface area to volume ratio, which is inversely proportional to the characteristic length, is high in microfluidic devices. A high surface to volume ratio is favorable for surface-based (i.e., heterogeneous) chemical reactions such as the electrochemical reactions occurring in fuel cells. Reducing the size of the channel does, however, increase the frictional parasitic load required to drive the flow. Thus, when discussing microfluidic fuel cells it is important to consider the role of pressure drop due to friction. The pressure drop required to generate a pressure-driven laminar flow with mean velocity U in a channel of length L and hydraulic diameter D_h may be expressed as [11]:

$$\nabla p = \frac{32 \mu L U}{D_h^2} \tag{3.3}$$

The pumping power W required may be obtained by multiplying the pressure drop by the flow rate Q:

$$W = \nabla p Q = \frac{32 \mu L U Q}{D_h^2} \tag{3.4}$$

It is important to note that the pumping power calculated from Eq. (3.4) provides an estimate based on fully developed flow in a straight channel. The contributions from inlet and outlet feed tubes and any minor losses are not included. In typical microfluidic fuel cell designs, however, relatively long thin channels are used and losses of the type indicated in Eq. (3.4) typically dominate.

The laminar nature of microfluidic flows enables a great deal of control over fluid-fluid interfaces [12] and provides unique functionality. Most important for microfluidic fuel cells is co-laminar streaming. Specifically, when two liquid streams of similar fluids (in terms of viscosity and density) are brought into a single microfluidic channel, a parallel co-laminar flow is developed. The resulting fluid-fluid interface may be used to observe chemical reactions in real-time, or serve as a lens, or—as in the case of microfluidic fuel cells—separate reactants. Species transport within microscale flows can result from convection, diffusion, and electromigration. In the absence of electromigration, mixing between two co-laminar streams occurs by transverse diffusion alone. Microscale devices generally experience high Péclet numbers, $Pe = UD_h/D$. High Péclet numbers indicate that the rate of mass transfer via transverse diffusion is much lower than the streamwise convective velocity, and diffusive mixing is therefore restricted to a thin interfacial width in the center of the channel. The interdiffusion zone has an hourglass shape with maximum width (δ_x) at the channel walls that varies according to the following scaling law [13] for pressure-driven laminar flow of two aqueous solutions:

$$\delta_x \infty \left(\frac{DHz}{U} \right)^{1/3} \tag{3.5}$$

where D represents the diffusion coefficient, z is the downstream position, and H is the channel height. Eq. (3.5) is limited, however, to liquids of similar density. When two liquids of different densities are employed, a gravity-induced reorientation of the co-laminar liquid-liquid interface can occur [14]. The physics of co-laminar flow is the core mechanism in several microfluidic devices such as the T-sensor [15], Y-mixer [16], and H-filters [17] that have applications in lab-on-chip diagnostic technologies. Multi-stream laminar flow can also be used to selectively pattern microfluidic systems [18].

In most microfluidic fuel cell architectures, one laminar stream contains the fuel, and a second laminar stream contains the oxidant. For the purposes of this discussion, consider the common microfluidic fuel cell geometry shown in Figure 3.1 and also schematically in Figure 3.2a. As the fuel and oxidant streams flow in a co-laminar fashion in a single microchannel, the liquid-liquid interface serves as a virtual separator without the need for a membrane.

The position of the electrodes on the channel walls is thus constrained by the width of the co-laminar interdiffusion zone. To prevent fuel and oxidant crossover effects, the electrodes must have sufficient separation from the liquid-liquid interface throughout the channel. The position and orientation of the electrodes also influences fuel utilization, as will be shown later in this chapter. In order to provide ionic charge transport between the electrodes and to close the electrical circuit, both co-laminar streams must have a relatively high ionic conductivity. High conductivity is normally facilitated by the addition of a supporting electrolyte that contains ions with high mobility such as hydronium and hydroxide ions. The supporting electrolyte also stabilizes the co-laminar flow with respect to electromigration of fuel and oxidant species, since it is these highly mobile constituents that redistribute and shield the effects of the electric field in the channel. The ohmic resistance for ionic transport in the channel can be expressed in terms of the average charge-transfer distance between the electrodes (d_{ct}), the cross-sectional area for charge transfer (A_{ct}), and the ionic conductivity (σ), as follows:

$$R_f = \frac{d_{ct}}{\sigma A_{ct}} \tag{3.6}$$

Eq. (3.6) indicates that a strong supporting electrolyte with high ionic conductivity and a high microchannel with closely spaced electrodes are therefore desired. This design is in conflict, to some degree, with that required for efficient separation of fuel and oxidant; specifically the interdiffusion width according to Eq. (3.5) places a lower limit on the electrode spacing. The trade-off between species transport and ionic conductivity is a common one in microfluidic fuel cells. Ohmic resistance in these cells is generally higher than in membrane-based fuel cells due to a larger separation of the electrodes. One route to reducing ohmic resistance is increasing the concentration of the supporting electrolyte. Ultimately, however, the choice of supporting electrolyte should be made with consideration of optimum reaction kinetics as well. The co-laminar configuration does allow the composition of the two streams to be chosen independently, thus providing an opportunity to improve reaction rates and cell voltage. Likewise, the open-circuit voltage of the cell can be increased by tweaking the reversible half-cell potentials by pH modification of the individual streams.

Reactant transport from the bulk flow to the surface of the electrode takes place primarily by convection and diffusion, and in the absence of significant electrical migration. In this case, species conservation takes the form:

$$\nabla \cdot (C_i \overline{u}) = -\nabla \cdot \overline{J}_i + R_i \tag{3.7}$$

where C_i is the local concentration of species i and R_i is a source term that describes the net rate of generation or consumption of species i via homogeneous chemical reactions. Under the infinite dilution assumption, the diffusive flux of species i is given by Fick's law:

$$\overline{J}_i = -D_i \nabla C_i \tag{3.8}$$

where D_i is the diffusion coefficient of the species i in the appropriate medium. Heterogeneous electrochemical reactions at the fuel cell electrodes are the boundary conditions of Eq. (3.8). During fuel cell operation, a concentration boundary layer develops over the electrode starting at the leading edge. Assuming the electrochemical reactions are rapid, the maximum current density of a microfluidic fuel cell is determined by the rate of the convective/diffusive mass transport from the bulk to the surface of the electrode. In this transport limited case, the reactant concentration is zero at the entire surface of the electrode. Kjeang et al. [19] provided scaling laws for microfluidic fuel cell operation in the transport controlled regime based on pseudo-3D flow over a flat plate. More generally, there is not one dominating limiting factor and current density of a microfluidic fuel cell is controlled by a combination of mass transport, electrochemical kinetics, and ohmic resistance. This trio of potential limiting factors must be considered when designing a microfluidic fuel cell.

3.3 CHANNEL FABRICATION, ELECTRODE PATTERNING, AND INTEGRATION

Fabrication methods developed originally for electronics, and later microfluidics, have been applied to good effect in the area of microfluidic fuel cells. Most commonly, these devices consisted of a microchannel, a pair of electrodes, and a liquid-tight support structure as shown schematically in Figure 3.2a–c. Microchannels have generally been fabricated by rapid prototyping, using standard photolithography and soft lithography protocols [20–22]. The channel structures were commonly molded in poly(dimethylsiloxane) (PDMS) and subsequently sealed to a solid substrate that was prepatterned with the desired electrode pattern. Alignment of the channel structure with the electrode pattern could be done with a suitably modified mask aligner, if available, or by hand. Detailed microfluidic fuel cell fabrication procedures are available in several reports, for example Kjeang et al. [23].

The common, soft lithography-based procedure for forming the channel structure is summarized as follows: a pretreated substrate such as a microscope glass slide or a silicon wafer is coated with a thin layer of photoresist by spin-coating. The coated substrate is then baked on a hot plate to stabilize the photoresist. The substrate is then exposed to UV light through a photomask that defines the desired channel structure (typically drafted in CAD software and printed on a transparency by a high-resolution image setter). After an additional bake, the unexposed parts of the photoresist may be removed by immersion of the substrate in developer liquid. The result is a master with a positive pattern defined by the remaining polymerized photoresist ridges. This master may be reused many times, depending on the quality of the original coating and the intricacy and aspect ratio of the features. The channel structure is fabricated by pouring a liquid polymer (often PDMS) over the master, followed by degassing in a vacuum, and subsequent curing on a hot plate. The polymer part containing a negative imprint of the channel structure is then cut from the mold and removed from the master. The obtained channel structure is then sealed to a substrate, typically glass or PDMS, either reversibly as is, or irreversibly following plasma-treating of both parts. Plasma-treating also renders the PDMS channel walls hydrophilic, which also promotes wetting and reduces pressure drop in the channel. PDMS generally has benign properties for fuel cell applications; it is relatively inert and compatible with most solvents and electrolytes [24].

Microchannel structures for microfluidic fuel cells can also be fabricated by photolithographic techniques directly in the photoresist material [3, 25]. In this case, the channel is made in negative relief. Alternatively, four separate parts, each contributing one channel wall, may be assembled to form the microfluidic channel. This method was employed by Choban et al. [26, 27] to incorporate electrodes on the side walls of a microfluidic channel with side-by-side streaming (Figure 3.2b). In that case, two graphite plates were aligned using separators (spacers) and sealed on the top and bottom with PDMS films and polycarbonate capping layers. The number of required parts can be reduced to three using a channel stencil, which has an open area defining the channel along with its two side walls. The stencil approach has been found particularly useful for microfluidic fuel cells employing vertically layered streaming, given that top and bottom walls are provided by electrode substrates. The channel height is fixed by the thickness of the channel stencil, which is in turn restricted by the geometry and the stiffness of the material. PDMS channel stencils of ~1 mm height have suitable mechanical strength for handling, and can be fabricated in PDMS by pressing a solid plate against the ridge pattern.

Another common material employed in microfluidic fuel cell fabrication is poly(methylmethacrylate), PMMA. Jayashree et al. [28, 29] fabricated their air-breathing cells by joining a 1 mm high PMMA stencil to a top graphite plate anode and Toray carbon paper-based gas diffusion cathode on the bottom. Alternatively, preformed sheets of PMMA may be cut directly with a CO_2 laser cutting system, as done by Li et al. [30]. Laser micromachining enables precise control of the geometry by adjusting the speed and power of the laser beam, and is generally very rapid (~ few seconds). The PMMA stencil of Li et al. [30] was bound between two top and bottom PMMA parts using adhesive gaskets, also cut to size by the laser. Prior to assembly, the top and bottom PMMA parts were mechanically roughened with fine sandpaper to promote adhesion and the gold electrodes were sputtered onto the top and bottom surfaces.

Microfluidic fuel cells have also been produced from mechanical stencils formed by silicon etching [31, 32]. A silicon-based microfluidic fuel cell was developed by Cohen et al. [32]. In brief, a silicon wafer was coated with a layer of positive photoresist. The channel pattern was removed from the photoresist layer via photolithography. A negative relief was obtained in the silicon wafer by potassium hydroxide etching at 90°C. In this case, the etching was continued until the entire thickness of the silicon wafer was penetrated. The obtained silicon stencil was coated with an insulator material to prevent electrical short-circuiting, and subsequently sealed between two flexible polyamide electrode films.

Some of the fuel cells discussed above incorporated electrode materials as channel walls. Most microfluidic fuel cell devices, however, have employed patterned electrodes positioned in parallel on the bottom wall of the channel. The bottom substrate was typically a glass plate with electrodes patterned by the photolithographic approaches of either lift-off or etch. The lift-off approach uses a photoresist layer on the substrate with an imprint of the desired electrode pattern in negative. The photoresist-patterned substrate is then coated with a conductive material such as graphite, gold, or platinum over an adhesive layer (often chromium or titanium) by standard evaporation or sputtering techniques. The photoresist layer is then removed, hence the name lift-off, leaving only the desired electrode pattern. The etch approach utilizes a similar strategy, where the substrate is first uniformly coated with a conductive layer, and then a photoresist layer pattern, in positive. An etching step removes the conductive material, except for the electrode pattern as protected by the photoresist. The remainder of the photoresist may then be removed revealing the desired pattern. It is also possible to start this method with commercial pre-fabricated gold slides as an alternative to the initial evaporation or sputtering processes [23]. Alternatively, rigid, self-contained electrode structures such as graphite rods [19] and carbon paper strips [33–35] have also been employed as both electrodes and current collectors in cells with side-by-side streaming. Graphite rods have been additionally employed as structural elements as well as current collectors and electrodes [28].

The constituents of the electrode surfaces, and the total surface area of the electrodes have been tuned by electrodeposition of a first or additional layer of conductive material and/or catalyst [3, 23, 25, 36–38]. A similar result was achieved by deposition of nanoparticle-supported catalyst suspended in a Nafion-based ink solution applied by pipette or sprayed on the electrode [26–29]. Other catalyst deposition techniques used in microfluidic fuel cell electrodes include electron-beam evaporation [25, 31, 32, 37, 38], sputtering [39, 40], and micromolding [41]. The surface roughness and chemical composition of electrode surfaces are critical to the performance of electrochemical devices and microfluidic fuel cells are no exception in this regard.

Sealing is a common concern in microfluidic systems. The most common way to accomplish a liquid-tight seal during operation of a microfluidic fuel cell is to physically clamp the assembled parts together using Plexiglas or aluminum plates. This approach works well when gaskets are used or the channel or membrane material is smooth elastomeric, as in the case of PDMS channels. The advantage of a mechanically enforced seal is that it may be easily undone following testing to salvage different components. The use of a clamping device, however, requires additional parts and space. Alternatively, an irreversible seal may be achieved betwen PDMS/PDMS or PDMS/glass by plasma-treating of both parts prior to assembly [23, 33, 34, 42]. Fuel and oxidant are generally injected to the microfluidic fuel cell via a syringe pump interfaced with polyethylene tubing. To accommodate the tubes, access holes for inlets and outlets were commonly punched, drilled, or machined into the microfluidic chip.

3.4 TECHNICAL ADVANCES IN MICROFLUIDIC FUEL CELLS

As of 2008, the research advances in the area of microfluidic fuel cells have included more than 30 scientific publications, and the technology is currently being developed for commercial and military applications by INI Power Systems (Morrisville, NC), with intellectual property licensed from the University of Illinois at Urbana-Champaign. Prototype microfluidic fuel cell devices have been demonstrated based on hydrogen [31, 37, 38], methanol [25–28], formic acid [23, 29, 30, 32, 33, 36, 42, 43], hydrogen peroxide [40], and vanadium redox species [3, 19, 34, 35] as fuel. Several microfluidic fuel cell concepts have also been demonstrated for biofuel cells, including microfluidic bioanodes based on ethanol [41] and glucose [39, 44] fuel. Oxygen [26–32, 36–39, 41, 44] in aqueous or gaseous form is the most commonly used oxidant, followed by hydrogen peroxide [23, 25, 30, 40], vanadium redox species [3, 19, 34, 35], potassium permanganate [36, 42, 43], and sodium hypochlorite [33]; all of which were employed in aqueous media. Most devices incorporated a supporting electrolyte within the fuel and oxidant streams to promote the ionic charge transport mechanism and reduce the ohmic resistance. This was accomplished by adding a strong acid or base with highly mobile and soluble ionic components, such as sulfuric acid or potassium hydroxide. The supporting electrolyte was generally passive and did not get consumed during cell operation. The co-laminar flow was implemented using horizontal streaming with vertical liquid-liquid interface or vertical sheath flow with horizontal liquid-liquid interface. In the case of horizontal streaming, the fuel cells had T-, Y-, Ψ-, or H-shaped microchannel designs with electrodes positioned either in parallel on the bottom wall (Figures 3.2a and 3.2c), or on opposite side walls (Figure 3.2b). For cells using vertical sheath flow, the channels were F-shaped with electrodes positioned on the top and bottom walls (Figure 3.2d) or in a concentric arrangement (Figure 3.2h). In contrast, some microfluidic fuel cells with selective catalysts and stable reactants do not require co-laminar flow to prevent mixing and related crossover effects. Consequently, the fuel and oxidant species can be mixed in a single I-shaped stream [25, 39, 41].

The invention of the co-laminar microfluidic fuel cell in 2002 was followed by two early demonstrations that provided the foundation for future technology advances [3, 36]. The proof-of-concept cell presented by Choban et al. [36] had a Y-shaped microchannel with platinum-coated electrodes positioned on the side walls, housing an aqueous formic acid fuel stream and an aqueous oxygen-saturated oxidant stream. The power density of that early cell was, however, significantly restricted by the rate of mass transport to the active sites, primarily in the cathodic half-cell, and the overall system performance suffered from low fuel utilization. The cathodic transport limitation was confirmed by switching to potassium permanganate oxidant. Due to the significantly higher solubility of this oxidant in aqueous media, an order of magnitude higher power density was achieved [36]. The high-conductivity liquid electrolyte employed enables the use of an external reference electrode to characterize individual half-cells and measure ohmic resistance *in situ* during fuel cell operation. With this experimental approach, the overall cathodic mass transport limitation of dissolved oxygen-based systems was verified [27]. For early devices using formic acid in the anodic stream and dissolved oxygen in the cathodic stream, the measured power density levels were in the range of $0.2\,mW\,cm^{-2}$ [32, 36], and were primarily constrained by the low solubility (1–4 mM) and diffusivity ($2 \cdot 10^{-5}\,cm^2\,s^{-1}$) of oxygen in the aqueous electrolyte (and by CO-poisoning of the Pt catalyst used for formic acid oxidation as well). Implementing bimetallic Pt/Ru nanoparticles for methanol oxidation that are less susceptible to CO-poisoning, improved power densities to up to $2.8\,mW\,cm^{-2}$ [27]. An additional benefit of the nanoparticle-based catalysts is a very high electrocatalytic surface area (roughness factor of ~500) that further promoted the electrochemical kinetics of both electrodes. Alternatively, adatoms of Bi can be adsorbed on Pt to reduce CO-poisoning, as shown by Cohen et al. [32]. This approach was demonstrated in a formic acid/dissolved oxygen co-laminar fuel cell, resulting in drastically improved performance and durability. The fuel cell had a unique F-shaped microchannel design, as depicted in Figure 3.2d. The co-laminar streams were vertical layers and relatively large electrodes were used on the top and bottom walls, which is advantageous in terms of overall power output and space-efficient stacking of multiple cells [32].

Poly(dimethylsiloxane) (PDMS) is quite permeable to gases [45], as has been previously exploited in the application of microfluidics to biological systems that require oxygen. This property also presents interesting opportunities for microfluidic fuel cells. It is possible, for instance, to provide gaseous reactant through thin layers of PDMS to a pair of electrodes separated by an electrolyte channel [37, 38]. The power densities of the hydrogen/oxygen [37] and hydrogen/air [38] fuel cells that were based on this concept were in the range of $0.7\,mW\,cm^{-2}$. In these cases, quite severe crossover effects in the absence of electrolyte flow, and the permeation rate of hydrogen through the polymer limited the performance of the cells.

3.4.1 Improved Performance through Mixed Media Operation

The co-laminar microfluidic fuel cell format enables mixed media operation, in contrast to traditional types of fuel cells operating under all-acidic or all-alkaline conditions. This feature allows independent adjustment of half-cell conditions for optimum reaction kinetics and enhanced cell potential. Cohen et al. [31] demonstrated that the open circuit potential of a hydrogen/oxygen fuel cell can be raised well beyond the standard cell potential of 1.23 V using mixed media. In that case an alkaline dissolved hydrogen stream and an acidic dissolved oxygen stream were implemented in a co-laminar microfluidic fuel cell of the previously reported F-shaped architecture [32]. The increased cell potential results from the negative shift of the hydrogen oxidation potential in the alkaline environment. In that case, the power produced in the dual electrolyte configuration was more than twice that for the corresponding single electrolyte systems. These results were observed despite the mitigating effect of relatively slow kinetics of the hydrogen oxidation reaction in alkaline media. The media flexibility was also studied by Choban et al. [26] by operating a methanol/oxygen fuel cell under all-acidic, all-alkaline, and mixed media conditions. This microfluidic fuel cell design eliminated the issue of membrane clogging by carbonate products formed in alkaline media (an issue with many conventional direct methanol fuel cells). Alkaline conditions had positive effects on the reaction kinetics at both electrodes. In addition, the cell potential was increased under mixed media conditions (alkaline anolyte and acidic catholyte) up to an impressive 1.4 V. A peak power density of 5 mW cm^{-2} at 1.0 V cell voltage was achieved with the methanol/oxygen fuel cell under mixed media conditions. This value is over double that of 2.4 and 2.0 mW cm^{-2} for all-acidic and all-alkaline conditions, respectively. Additionally, it was observed that at cell voltages below 0.8 V, the oxygen reduction reaction at the cathode was complemented by proton reduction to hydrogen, thereby providing an additional contribution to the cathodic mass transport controlled cell current. This result is a direct consequence of the mixed media configuration, and a significantly enhanced fuel cell performance with a peak power density of 12 mW cm^{-2} was achieved. Interestingly, Hasegawa et al. [40] also used the mixed media approach to operate a microfluidic fuel cell in which hydrogen peroxide served as both fuel and oxidant, in alkaline and acidic media, respectively. The direct hydrogen peroxide fuel cell produced relatively high power densities up to 23 mW cm^{-2}. This cell was limited primarily by the rate of spontaneous hydrogen peroxide decomposition on the cathode and associated oxygen gas evolution that could perturb the co-laminar flow interface.

Operation under mixed media conditions, however, does have some general disadvantages. For instance, mixed media operation causes exothermic neutralization of OH^- and H^+ at the co-laminar flow interface resulting in the formation of a liquid junction potential and locally reduced ionic strength that can reduce cell potential and increase ohmic cell resistance. Furthermore, the overall cell reaction included net consumption of supporting electrolyte in each of the three cases described above—that is, protons consumed at the cathode and hydroxide ions consumed at the anode. The supporting electrolyte must therefore be taken into account when considering the overall energy density of the fuel cell system.

3.4.2 Gas-Permeable Cathodes

The oxygen solubility limitation, common to many microfluidic fuel cells discussed thus far, may be addressed by incorporating cathodes that access the surrounding air. Ambient air has four orders of magnitude higher diffusivity ($0.2\,cm^2\,s^{-1}$) and several times higher concentration (10 mM) than dissolved oxygen in aqueous media [29]. Hydrophobic porous gas diffusion electrodes are key components for PEM-based fuel cells that allow gaseous reactants to pass, while limiting liquid transport. Jayashree et al. [29] introduced the first microfluidic fuel cell with an integrated air-breathing cathode, using a graphite plate anode covered with Pd black nanoparticles and a porous carbon paper cathode covered with Pt black nanoparticles. A schematic of the fuel cell is provided in Figure 3.3 (also Figure 3.2e). To facilitate ionic transport to the cathodic reaction sites and sufficient separation between the interdiffusion zone and the cathode, the air-breathing cell architecture requires a blank cathodic electrolyte stream. With this cell a peak power density of $26\,mW\,cm^{-2}$ was achieved using 1 M formic acid in 0.5 M sulfuric acid anolyte and a blank 0.5 M sulfuric acid catholyte flowing at $0.3\,mL\,min^{-1}$ per stream. The air-breathing cell architecture was also evaluated using methanol [28], which enables higher overall energy density than formic acid. Relatively modest power densities were obtained with 1 M methanol fuel ($17\,mW\,cm^{-2}$), however, improved reaction kinetics resulted in an increase in the open-circuit cell voltage from 0.93 V to 1.05 V. The air-breathing cells also enabled significantly higher coulombic fuel

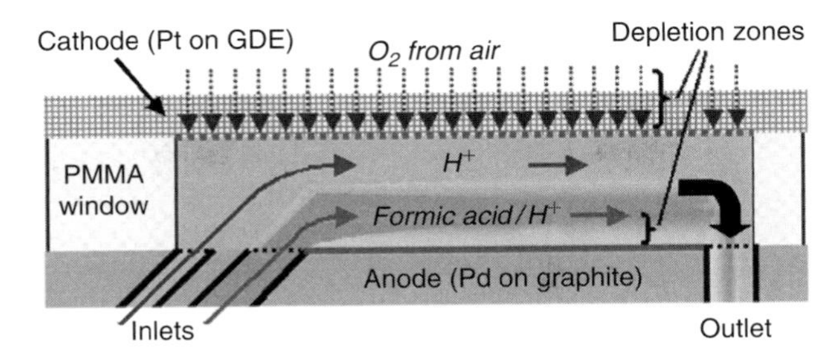

FIGURE 3.3 Schematic of an air-breathing microfluidic fuel cell. This cell captures oxygen from the surrounding air via gas diffusion through the dry side of the porous cathode structure. The opposite side is in contact with a blank electrolyte stream, establishing a three-phase interface between the gas, electrolyte, and catalyst/solid electrode phases. Reprinted with permission from Jayashree et al. [29]. Copyright 2005 American Chemical Society.

utilization than the cells based on dissolved oxygen, up to a maximum of 33% [29].

The scale-up and integration of multiple air-breathing fuel cells, while maintaining sufficient oxidant access, can be complicated. INI Power Systems (Morrisville, NC) is developing direct methanol microfluidic fuel cells with integrated gas diffusion cathode for commercial applications. Notably, recent improvements of electrodes and catalysts, optimization of methanol concentration and flow rates, and the addition of a gaseous flow field on the cathode side have resulted in impressive power densities on the order of 100 mW cm^{-2} [46]. As compared to other direct methanol fuel cells, these microfluidic fuel cells are competitive.

Very recently, Tominaka et al. [47] developed a monolithic microfluidic fuel cell with air-breathing capabilities. The architecture of their silicon-based device is shown schematically in Figure 3.4. In this case a microchannel is employed that is open to the atmosphere on one side. The configuration provides air-breathing access to oxidant at the porous cathode. The liquid fuel is contained in the microchannel by capillary forces; however, there is some potential for fuel evaporation.

3.4.3 Liquid Oxidants

Another avenue toward improved performance of typically mass transfer-limited microfluidic fuel cells is the use of oxidants

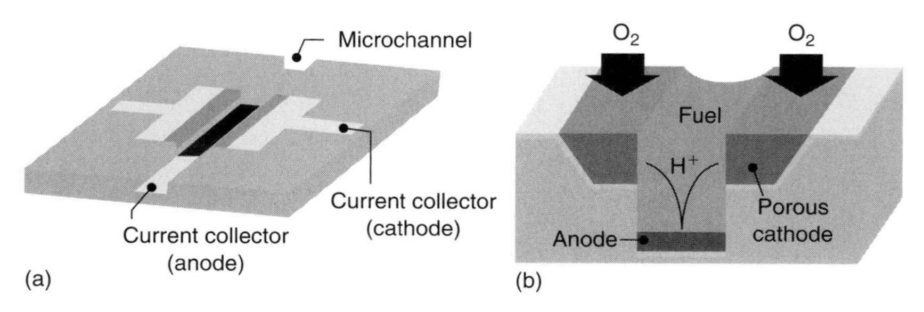

FIGURE 3.4 An air-breathing microfluidic fuel cell with porous cathode. Schematics indicate current collector layout (a), and cross-sectional view of the device (b). This architecture uses an open microchannel that provides air-breathing access to oxidant at the cathode. The liquid fuel is contained by capillary forces; however, there is some potential for fuel evaporation. Reproduced with permission from Tominaka et al. [47]. Copyright 2008 American Chemical Society.

soluble at higher concentrations than dissolved oxygen. There are many liquid fuels available with relatively high specific energy density, including sodium borohydride, methanol, formic acid, and other liquid hydrocarbons [5, 48, 49]. Normally these conventional liquid fuels are paired with oxygen or air cathodes. With the exception of hydrogen peroxide (which was previously employed as an oxidant in various direct sodium borohydride/hydrogen peroxide fuel cells [50–52]), liquid oxidants are less common.

Formic acid fuel with an acidic hydrogen peroxide oxidant solution was employed in a laser-micromachined microfluidic fuel cell device [30] in an F-shaped co-laminar format. This cell produced power densities up to 2 mW cm^{-2} while operating at flow rates of 0.4–1.6 mL min^{-1} per stream. The performance of this cell was primarily restricted by the low hydrogen peroxide concentration used (10 mM). Unfortunately, direct hydrogen peroxide reduction on common catalysts such as Pt and Pd is accompanied by vigorous oxygen gas evolution. This results from parasitic oxidation (decomposition) that must be accommodated by strategic cell design without compromising the stability of the co-laminar flow. Several methods to stabilize the co-laminar liquid-liquid interface have recently been proposed. These stabilization strategies include magnetically separated streams [53], integration of a third electrolyte stream [42], and utilization of a grooved microchannel geometry that serves as a guide for the gaseous products [23]. The magnetic field-induced approach was develop by Aogaki et al. [53].

Their magnetic virtual wall separated a paramagnetic oxidant solution from a diamagnetic fuel solution at a liquid-liquid interface. A zinc/copper flow cell was employed and the magnetic field was provided by a permanent magnet. This approach enables rapid removal of bubbles and solid particles, and precise crossover control. The common Y- or T-shaped microfluidic fuel cell was revised into a Ψ-shape by Sun et al. [42]. The third electrolyte stream, in the center of the co-laminar flow, contained blank electrolyte. Its purpose was to promote the separation of the fuel and oxidant streams and prevent interfacial reaction. The fuel cell employed formic acid and potassium permanganate, and during operation both cell voltage and current density could be optimized via precise flow rate control of the central electrolyte stream. Kjeang et al. [23] recently reported a microfluidic fuel cell based on formic acid and hydrogen peroxide, exploiting a grooved microchannel design. Steady operation without crossover issues was demonstrated at flow rates as low as $3\,\mu L\ min^{-1}$, and practical power densities up to $30\,mW\ cm^{-2}$ were achieved. In this design, the microchannel is grooved over each electrode so as to effectively capture all gas bubbles formed by the electrochemical reactions. This architecture prevented the otherwise destabilizing effect that bubble formation has on co-laminar flow.

The rate of gas evolution in hydrogen peroxide reduction is dependent on the catalyst. Gold has been identified as an alternative catalyst for hydrogen peroxide reduction that minimizes gas evolution while still providing high current density [54]. Gold may also be used as a catalyst for direct borohydride oxidation with limited hydrogen evolution [54]. The end product in that case, sodium metaborate, is highly soluble in aqueous media. In conventional direct borohydride/hydrogen peroxide fuel cells, the anolyte and catholyte are physically separated by a membrane. In a microfluidic fuel cell, however, the feasibility of coupling sodium borohydride and hydrogen peroxide is restricted by chemical stability issues. Specifically, sodium borohydride is only stable in alkaline solutions, and hydrogen peroxide requires an acidic environment to prevent fast decomposition. Co-laminar mixing of an alkaline borohydride solution with an acidic peroxide solution leads to vigorous gas formation and heating. A microfluidic fuel cell that couples these two components in a single channel is therefore not feasible. Individual half-cells based on either alkaline borohydride or acidic peroxide might, however, find some application in a cell architecture that accommodates low rates of gas evolution, such as the previously described grooved microchannel design.

An alternative strategy to circumvent the stability problem associated with hydrogen peroxide oxidant is to employ selective catalysts on both electrodes. In this case, crossover is not a concern at all, and the fuel and oxidant solutions can be mixed in a single laminar stream. The mixed reactant approach is, however, restricted to fuel and oxidant pairs that do not react spontaneously upon mixing and have sufficiently high kinetics in a singular electrolyte. Sung and Choi [25] demonstrated a single-stream microfluidic fuel cell in alkaline solution. A nickel hydroxide anode was employed for methanol oxidation and a silver oxide cathode was used for hydrogen peroxide reduction. These catalysts are not entirely selective and the result was a very low open-circuit voltage (0.12 V) and low power density (0.03 mW cm^{-2}).

Vanadium redox battery technology utilizes soluble vanadium redox couples in both half-cells for regenerative electrochemical energy storage units [55]. The combination of aqueous redox pairs in vanadium redox cells, V^{2+}/V^{3+} and VO^{2+}/VO_2^+ as fuel and oxidant, respectively, provides many benefits for microfluidic fuel cell operation: they have high solubility and enable relatively high redox concentrations up to 5.4 M [56]; they feature well-balanced electrochemical half-cells (in terms of both transport characteristics and reaction rates); they provide a high open-circuit voltage (up to ~1.7 V at uniform pH) as a result of the large difference in formal redox potentials; and the electrochemical reactions are facilitated by plain carbon electrodes without expensive catalysts. Accordingly, the first journal publication in the emerging field of microfluidic fuel cells was an all-vanadium microfluidic redox fuel cell introduced by Ferrigno et al. in 2002 [3]. The proof-of-concept cell featured a Y-shaped microchannel with planar graphite-covered gold electrodes patterned on the bottom wall, and produced comparatively high power densities up to 38 mW cm^{-2}. These power levels however required a high flow rate of 1.5 mL min^{-1} per stream and the fuel utilization was limited to ~0.1%. Kjeang et al. [19] developed a conceptually similar vanadium redox fuel cell based on graphite rod electrodes. Graphite rods, of the same type as those employed in mechanical pencils, are inexpensive and provide combined electrodes and current collectors in basic self-contained structures with high electrical conductivity. The prototype graphite rod fuel cell delivered useful power density levels at high flow rates. In addition, the high-aspect ratio (width/height) cross-sectional geometry of the microchannel enabled fuel cell operation at low flow rates with unprecedently high levels of fuel utilization up to 63% per single pass. A cell voltage breakdown analysis confirmed that the performance was mainly controlled by convective/diffusive species transport from the bulk fluid. The cell design was further developed by replacing the graphite rods with integrated porous carbon electrodes (Figure 3.2c) [35]. Porous electrodes have the capability to delay otherwise commonly encountered transport limitations by increasing the active surface area and enhancing the convective transport characteristics via a small parallel flow inside the top portion of the porous medium. A peak power density of 70 mW cm^{-2} was achieved with the microfluidic vanadium redox fuel cell with porous electrodes.

3.4.4 Architectures for Improved Reactant Transport

Microfluidic fuel cells rely on cross-stream diffusion to transport reactants to the active sites. Diffusion on the microscale is often, however, relatively slow and this leads to the common transport limitations discussed earlier. Approaches to improve reactant transport are discussed in this section.

While the inner structure of porous electrodes provides increased surface area and aids diffusive species transport, conventional microfluidic fuel cells with electrodes replaced by porous electrodes fail to take full advantage of these structures. Kjeang et al. [34] modified the vanadium redox fuel cell architecture by sealing the porous electrodes between the top and bottom substrates such that reactant crossed directly through the electrodes. The flow-through porous electrode cell is shown schematically in cross-section in Figure 3.2g, in profile in Figure 3.5a, and in operation in Figure 3.5b and c. As shown, the two reactant streams meet in an orthogonally arranged central channel where they are directed toward the outlet in a co-laminar format. Due to the disparity in flow resistance between the channel and the electrode, the flow distribution was very uniform

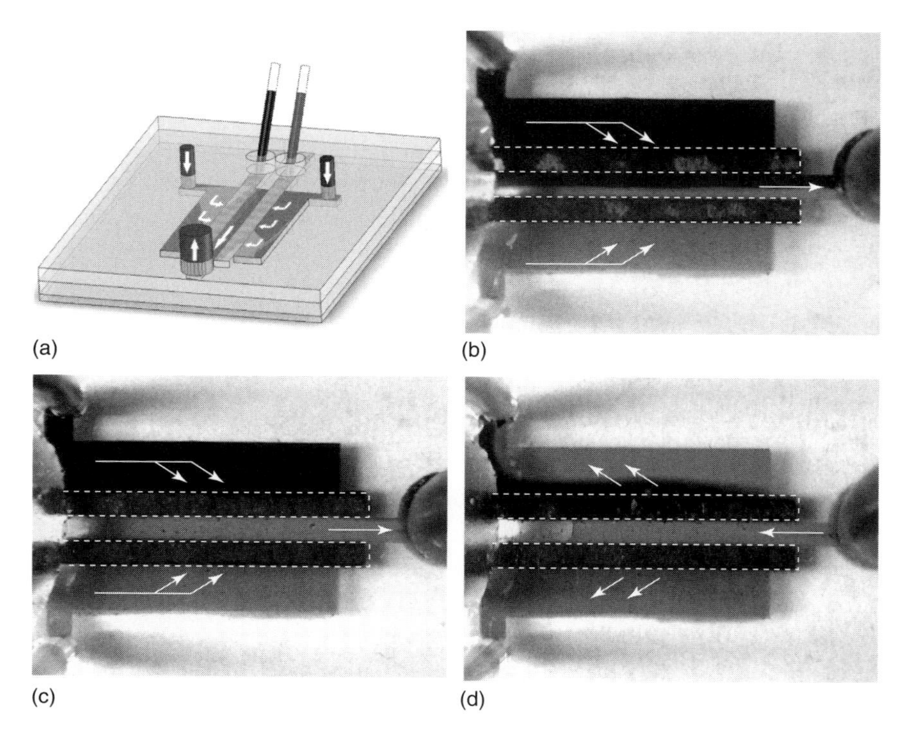

FIGURE 3.5 Microfluidic fuel cell with flow-through porous electrodes: (a) Schematic; (b–d) Annotated images of the flow through porous electrode microfluidic fuel cell in operation at (b) open-circuit and (c) 0.8 V cell voltage. The vanadium electrolytes contain V^{2+} (purple) and V^{3+} (light green) at the anode (solution appears here as gray), and VO_2^+ (black) and VO^{2+} (turquoise) at the cathode (solution appears here as black). The flow rate was 1 μL min^{-1} per stream. (d) Image of the fuel cell operating in reverse, demonstrating the *in situ* regeneration capability. Fully mixed waste solution (~50/50 V^{3+}/VO^{2+}) is flowing in the reverse direction from right to left at 1 μL min^{-1} per stream, with an applied cell voltage of 1.5 V. Dashed lines indicate the extent of the porous electrodes, and arrows indicate flow directions. Reproduced and adapted with permission from Kjeang et al. [34]. Copyright 2008 American Chemical Society.

through the entire porous electrode. This aspect enabled utilization of the full depth of the porous medium and associated active area, and provided enhanced species transport from the bulk to the active sites. The various colors inherent to different vanadium solutions enable convenient visualization of the fuel cell operation. At open circuit (Figure 3.5b), the waste stream shows a co-laminar flow of fuel and oxidant. In operation at 0.8 V (Figure 3.5b), fuel and oxidant are largely consumed in the porous electrode structure prior to reaching the co-laminar waste channel as indicated by the uniform, light blue color. The flow-through porous electrode architecture enabled class-leading performance levels at room temperature, including steady state power densities up to 131 mW cm^{-2} and near complete fuel utilization. The fuel cell also had the capability to combine high fuel utilization with high cell voltages. As an example, at 1 μL min^{-1} flow rate an active fuel utilization of 94% per single pass was achieved at 0.8 V. This level of fuel utilization is equivalent to an overall energy conversion efficiency of 60%.

An advantage of an all-vanadium fuel cell is that the mixed products may be regenerated directly for re-use. With previous microfluidic fuel cell designs this regeneration would have to occur off-chip. However, the structure of the flow-through cell enables *in situ* regeneration. Specifically, proof-of-concept *in situ* regeneration of the initial fuel and oxidant species was established by operating the fuel cell in reverse mode. Figure 3.5d shows an image of the fuel cell in regeneration mode. The regeneration of reactants in reverse mode is evidenced by the color changes in the supply channels and the boundary layer-like flow indicated.

While transport and active area aspects significantly improved with flow-through porous electrode designs, the main factor limiting further energy density improvements on this design is the use of vanadium redox couples and their limited solubility. One approach to address this issue was to operate a flow-through porous electrode cell in alkaline mode using formate fuel and hypochlorite oxidant [33]. Reactant solutions of formic acid and sodium hypochlorite are both available and stable as highly concentrated liquids. Formate oxidation and hypochlorite reduction in alkaline media on porous Pd and Au electrodes were shown to have rapid kinetics at low overpotentials while preventing gaseous CO_2 formation by carbonate absorption. The prototype formate/hypochlorite fuel cell with flow-through porous electrode architecture exhibited a peak power density of 52 mW cm^{-2}. This performance was primarily constrained by ohmic resistance and it is noteworthy that concentrations well below solubility limits were used. Specifically, the hypochlorite solution employed was that of commercially available household bleach.

A promising fuel cell architecture with flow-through porous electrodes was developed by Salloum et al. [43]. As shown in

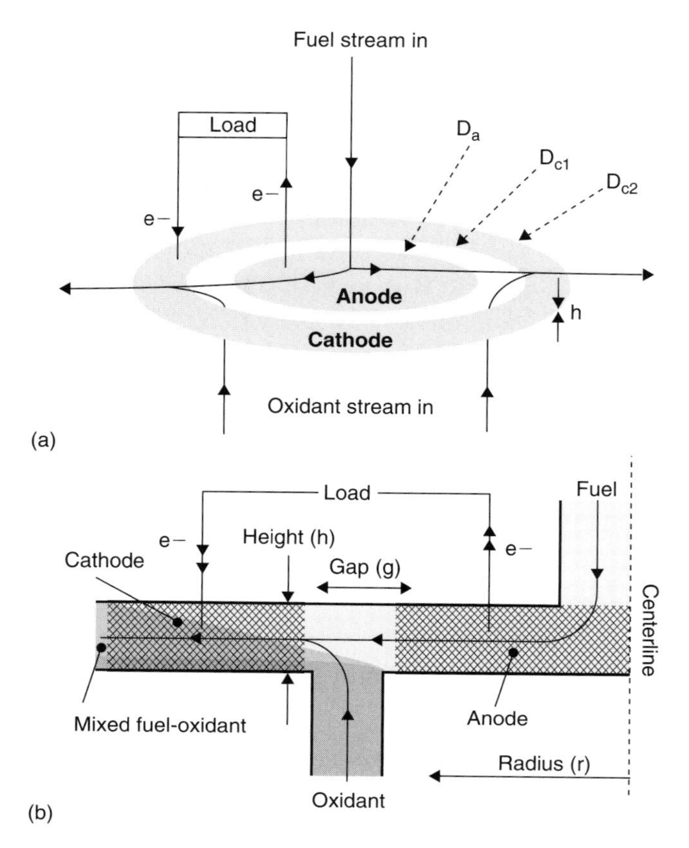

FIGURE 3.6 A radial membraneless fuel cell with porous electrodes. As shown schematically in isometric projection (a) and cross-sectional view (b), the fuel enters via the center of the anode disc and oxidizes as it flows radially outward. The stream then receives oxidant prior to flowing through a ring-shaped porous cathode. Reproduced with permission from Salloum et al. [19]. Copyright Elsevier (2008).

Figure 3.2h, and in more detail in Figure 3.6, this cell employs sequential radial flow through concentric porous electrodes. The anolyte flow enters through the center of a disc-shaped anode and flows radially toward a ring-shaped cathode. The partially consumed anolyte is then blended with a catholyte stream prior to entering the porous cathode. The concentric cell design enables independent control over the fuel and oxidant flow rate, although the impact of the fuel crossover to the cathode must be considered.

Another approach to improving reactant transport is active boundary layer control. The concentration boundary layers that develop in operating microfluidic fuel cells can be replenished via strategic design modifications in order to enhance the overall fuel cell performance. Yoon et al. [57] proposed three different strategies for active control of concentration boundary layers: removing consumed species through multiple periodically placed outlets; adding reactants through multiple periodically placed inlets; and generating a secondary transverse flow by topographical herringbone patterns on the channel walls. The benefits of these approaches were evaluated through simulations and experiments using the ferricyanide/ferrocyanide redox couple as a model system. A chronoamperometric study revealed that by adding two extra inlets, the transport-controlled current could be enhanced by ~30%, without increasing the overall flow rate.

Microfluidic mixing has been studied extensively in an effort to increase cross-stream mixing of species for lab-on-chip applications. Toward this goal Stroock et al. [58] employed topographical herringbone ridge patterns. Later analyzed theoretically by Kirtland et al. [59] for microreactor systems, these patterns induce a secondary spiralling flow that enhances the rate of cross-stream transport in microchannels, depending on the exact geometry of the pattern. In this case, a chronoamperometric demonstration showed a 10–40% increase in current density with the addition of herringbone ridges to one wall. For microfluidic fuel cells, this approach is promising for improving fuel utilization; however, it must be used cautiously to avoid increased fuel and oxidant mixing at the co-laminar flow interface. Additional disadvantages include increased fabrication complexity and/or increased parasitic pumping power required to drive the flow. Lim and Palmore [60] suggested passive boundary layer control achieved by splitting the electrodes into smaller units separated by a gap. They fabricated a prototype microfluidic redox fuel cell with five sets of consecutive electrodes. The local current density at the second set of electrodes was enhanced by the passive replenishment of the concentration boundary layer in the gap section. A relationship between the measured current density and the distance between consecutive electrodes was developed; however, since the geometrical surface area required for the gap did not contribute any net current, the overall current density of the fuel cell was not improved. This work was followed by a numerical optimization study by Lee et al. [61], recommending arrays of miniaturized electrodes, i.e., nanoelectrodes, along the same lines.

3.4.5 Advances from Computational Fluid Dynamics

Computational fluid dynamics (CFD) is an essential tool in the development of microfluidic and nanofluidic processes. In contrast to macroscale fluid mechanics where there are challenges in modeling turbulence, the main challenges in CFD modeling of micro- and nanoflows are in the application of appropriate boundary conditions and in modelling species transport. In the context of microfluidic fuel cells, this area was first investigated by Bazylak et al. [62]. A computational model was employed to analyze a T-shaped formic acid/dissolved oxygen microfluidic fuel cell with side-by-side streaming. Different cross-sectional channel geometries and electrode configurations were studied computationally, targeting enhanced fuel utilization while minimizing fuel/oxidant mixing. The multidimensional nature of the flow required a 3D solution using a computational fluid dynamics framework coupled with convective/diffusive mass transport (infinite dilution) and electrochemical reaction rate models for both anode and cathode. A high aspect ratio (width/height) channel geometry with electrodes placed on the top and bottom walls was found to enable significantly improved fuel utilization and reduced interdiffusional mixing width. As shown in Figure 3.7, the numerical study also suggested the implementation of a tapered electrode design that accommodates the growth of the co-laminar mixing zone in the downstream direction.

Chang et al. [63, 64] provided an extended model with Butler-Volmer electrochemical reaction kinetics and the capability of predicting complete polarization curves. The results obtained for Y-shaped [63] and F-shaped [64] formic acid/dissolved oxygen-based cells were in good agreement with previous experimental studies [32, 36], and confirmed the cathodic activity and mass transport limitation of these cells. Consequently, the predicted cell performance was essentially independent of anodic formic acid concentration. These numerical results also recommended

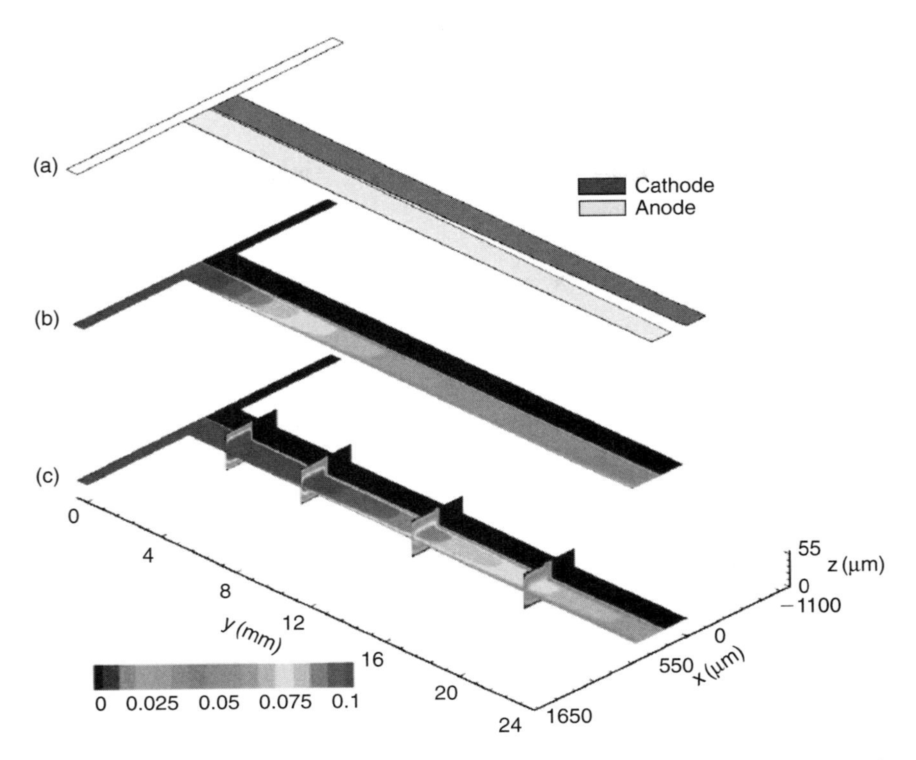

FIGURE 3.7 Computational modelling results of a microfluidic fuel cell with electrodes on top and bottom surfaces. Shown at top (a) is the tapered electrode geometry, designed to increase fuel utilization while accommodating downstream growth of the inter-diffusion zone in the center of the channel. Shown below are computationally predicted fuel concentration contours in the center plane (b), and with vertical projections (c). Reproduced with permission from Bazylak et al. [62]. Copyright Elsevier (2005).

high aspect ratio channel geometry, high Péclet number, and high oxygen concentration as well as a thick cathode catalyst layer to improve the performance. This work was later extended by a 2D theoretical model of the cathode kinetics under co-laminar flow [65]. A Butler-Volmer model [66] was also developed for the previously discussed microfluidic fuel cell using hydrogen peroxide as both fuel and oxidant in mixed media [40], and applied to study the effects of species transport and geometrical design. The simulated fuel cell performance results were invariant at flow rates above 0.1 mL min^{-1}, indicating the absence of the commonly encountered cathodic transport limitation. Two-phase flow and transport effects related to oxygen gas evolution from hydrogen peroxide decomposition were not considered in that study. However, it was found that increasing the surface area and thickness of the catalyst layers can enhance current density.

3.4.6 Microfluidic Fuel Cells with Biocatalysts

Biocatalysts are a promising alternative to traditional catalysts, particularly in the context of microfluidic fuel cells. So-called biofuel cells [67, 68] utilize biological molecules such as enzymes and microbes to catalyze the chemical reactions, thereby replacing traditional electrocatalysts. The name biofuel cell is somewhat of a misnomer as these cells are not restricted to biofuels, the usage of this name is, however, prevalent.

In conventional biofuel cells, biocatalytic entities are placed in a two-compartment electrochemical cell containing buffered solution with concentrated fuel and oxidant in the anolyte and catholyte compartments, respectively. In most configurations, these compartments are separated by an ion-exchange membrane or a salt bridge [67]. Each compartment also includes a redox couple acting as a diffusional electron mediator (or cofactor), which is necessary for efficient catalyst utilization. The rate of electron transfer is, however, generally confined by the rate of diffusion of these cofactors and the ion permeability of the membrane that separates the two compartments [68]. Moreover, enzymes in solution are generally only stable for a few days. Recently, several novel methodologies have been developed for the functionalization of electrode surfaces and immobilization of active enzymes in order to improve electron transfer characteristics and stability. These approaches include covalent polymer tethering of cofactor units to multilayered enzyme array assemblies, cross-linking of affinity complexes formed between redox enzymes and immobilized cofactors on functionalized conductive supports, and noncovalent coupling by hydrophobic/hydrophilic or affinity interactions [69].

Biofuel cells with non-selective electrochemistry (i.e., cells using diffusional redox mediators) can utilize the established co-laminar microfluidic fuel cell design, which also enables the tailoring of independent anolyte and catholyte compositions for optimum enzymatic activity and stability. Alternatively, immobilized enzymes and localized cofactors may be employed in a microfluidic fuel cell format. Full selectivity of both anodic and cathodic half-cells with co-immobilized redox relays allows microfluidic biofuel cell operation in a single microchannel without the need for co-laminar flow. In this configuration, initially mixed fuel and oxidant would flow together in a single channel with species-specific oxidation and reduction occurring at the biocatalyst electrodes.

Relatively few microfluidic biofuel cell works have been presented to date. The area was pioneered by Moore et al. [41]. They presented a microchip-based bioanode with NAD-dependent alcohol dehydrogenase enzymes immobilized in a tetrabutylammonium bromide-treated Nafion membrane. The bioanode was assembled on a micromolded carbon electrode integrated on a glass substrate. A PDMS microchannel was used to deliver the fuel solution containing ethanol and NAD^+ in phosphate buffer to the bioanode. The microfluidic bioanode produced an open-circuit voltage of 0.34 V and a maximum current density of 53 $\mu A\ cm^{-2}$, when operated versus an external platinum cathode. The performance of the cell was expected to be limited by the rate of diffusion of NADH within the membrane. Later, an integrated microfluidic fuel cell based on the unique modified Nafion-based enzyme immobilization technique was developed [70, 71]. The technology is currently licensed to Akermin Inc. of St Louis, MO, under the trademark "stabilized enzyme biofuel cells." In terms of microfluidic biofuel cell power density and stability, this technology is considered state-of-the-art.

Much work to date has focused on bioanode development. A microfluidic bioanode based on vitamin K_3–mediated glucose oxidation by the glucose dehydrogenase enzyme was developed by Togo et al. [39]. The bioanode was immobilized inside a fluidic chip containing a PDMS-coated conventional Pt cathode with an integrated Ag/AgCl reference electrode. The bioanode was positioned downstream of the Pt cathode to minimize contamination. The flow cell produced 32 $\mu W\ cm^{-2}$ at 0.29 V running on 1 mL min^{-1} of air-saturated pH 7–buffered fuel solution containing 5 mM glucose and 1 mM NAD^+. The current density of the proof-of-concept cell declined by 50% over 18 hours of continuous operation, and this was attributed to swelling effects.

Figure 3.8 shows a complete microfluidic biofuel cell developed by Togo et al. [44]. In this work they replaced their previous

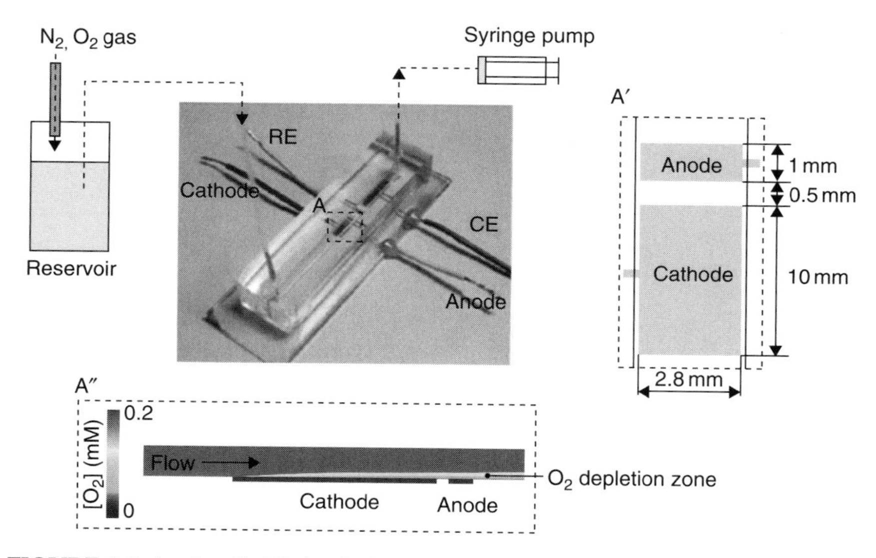

FIGURE 3.8 A microfluidic biofuel cell with an upstream biocathode and a downstream bioanode integrated on the bottom channel wall. The biofuel cell consumed a mixed reactant feed of oxygen-saturated glucose solution. The reaction zone (A) is shown magnified to illustrate the electrode configuration (A′) and the growth of the oxygen concentration boundary layer formed on the cathode (A″). Reproduced with permission from Togo et al. [44]. Copyright Elsevier (2008).

Pt cathode with a bilirubin oxidase-adsorbed biocathode, and the power output of the biofuel cell was comparable to the previous device. A parametric study of flow rate, channel height, and electrode geometry demonstrated restricted access of dissolved oxygen to the biocathode. The present cell design mitigates this limitation by enlarging the cathode area to ten times the anode size.

A significant opportunity for biofuel cell technologies, particularly with respect to fuel utilization, is current extraction from a series of consecutive biocatalyzed reactions. To harness consecutive reactions, strategic patterning of multiple enzyme electrodes is required. This opportunity was investigated by Kjeang et al. [72], using a generic computational model of species transport in microchannels with heterogeneous chemical reactions and Michaelis-Menten enzyme kinetics as boundary conditions. This first computational study of microfluidic biofuel cell technology provided guidelines for the design and fabrication of microfluidic biofuel cells exploiting consecutive reactions. Separated and mixed enzyme patterns in different proportions were analyzed for various Péclet numbers. The mixed transport regime, at medium Pe, was shown to be particularly attractive while current densities were maintained close to maximum levels. Mixed enzyme patterning tailored with respect to individual turnover rates was found to enable high current densities combined with nearly complete fuel utilization and provide the best overall performance.

3.4.7 Scale-Up of Microfluidic Fuel Cells

As the field is only a few years old, the majority of microfluidic fuel cell devices reported to date have been proof-of-concept unit cells. The voltage and overall power output of these unit cells were generally less than 1 V and 10 mW, respectively. While promising advances have been made, on their own these fuel cells are insufficient for most portable power applications. Thus, scale-up or integration of multiplexed microfluidic fuel cells is critical to the application of microfluidic fuel cell advances. Microfluidic fuel cell technology can be scaled up based on various methodologies. Ferrigno et al. [3] demonstrated a planar array of three cells with separate inlets and outlets. The unit cells were based on the familiar Y-shaped channel geometry with side-by-side streaming and electrodes positioned on the bottom wall. The array produced roughly three times the power of the unit cell, and when connected in series, provided useful operational voltages up to 2.4 V. Although planar arrays of this type are convenient from a fabrication perspective, they require substantial "overhead" volume of passive materials that restrict the volumetric power density of the device. Cohen et al. [31, 32] reported a more compact planar expansion methodology for parallel microchannels with combined inlets and outlets. The prototype five-microchannel array employed five parallel 1 mm wide and 5 cm long microchannels of the F-shaped design (Figure 3.2d), and was operated on aqueous formic acid or hydrogen fuel together with dissolved oxygen in the cathode stream. The flow and distribution of reactants was sufficiently uniform to achieve a power output that scaled linearly with the results obtained for a single microchannel [32]. This planar design facilitates custom fabricated channels with dimensions tuned for specific power requirements. The expansion of a single silicon-machined microchannel was demonstrated up to 5 mm in width for a 5 cm long channel, with power output comparable to the five-microchannel array (i.e., linear scaling again). The potential for vertical stacking was also evaluated. Two 1 mm wide microchannels, placed on top of each other and separated by electrodes, produced twice the power of a single channel without increasing the total volume of the fuel cell device. While these results are promising, it is important to note that the overall power output of these devices was only about 1 mW or less, and they were still restricted by transport and solubility of dissolved oxygen, and the fuel utilization was poor in general.

In contrast to film-deposited electrodes, the geometry and mechanical properties of rod-shaped electrodes enable unique three-dimensional microfluidic fuel cell architectures. An array architecture fuel cell was developed by Kjeang et al. [19] based on a hexagonal array of graphite rods mounted in a single cavity. The developed cell is shown Figure 3.9. The array cell consisted of 29 graphite rods of 0.5 mm diameter, spaced an average of 0.2 mm apart, and the flow area in between the rods exhibited microfluidic flow characteristics similar to those of a planar unit cell. The array cell had 12 anodes and 12 cathodes, and the five rods in the center were electrically insulated to compensate for the co-laminar interdiffusion zone. When operated on the all-vanadium redox system, the array cell produced an order of magnitude more powerful than a planar unit cell given the same total flow rate. Specifically, power and current levels of 28 mW and 86 mA were demonstrated, and the fuel utilization was significantly higher than that for the planar unit cell at any given flow rate. The array cell configuration may be readily expanded in both vertical (preferable) and horizontal directions to increase capacity. Scale-up requires only an enlarged cavity, in contrast to the volumetric costs of stacking of planar cells. However, the solubility and concentration of the vanadium redox species ultimately limit the overall energy density of this array fuel cell system.

A prototype microfluidic fuel cell system based on the direct methanol laminar flow fuel cell technology [46] has been reported by INI Power Systems. A combination of planar and vertical stacking was employed to scale the system. With respect to fuel utilization,

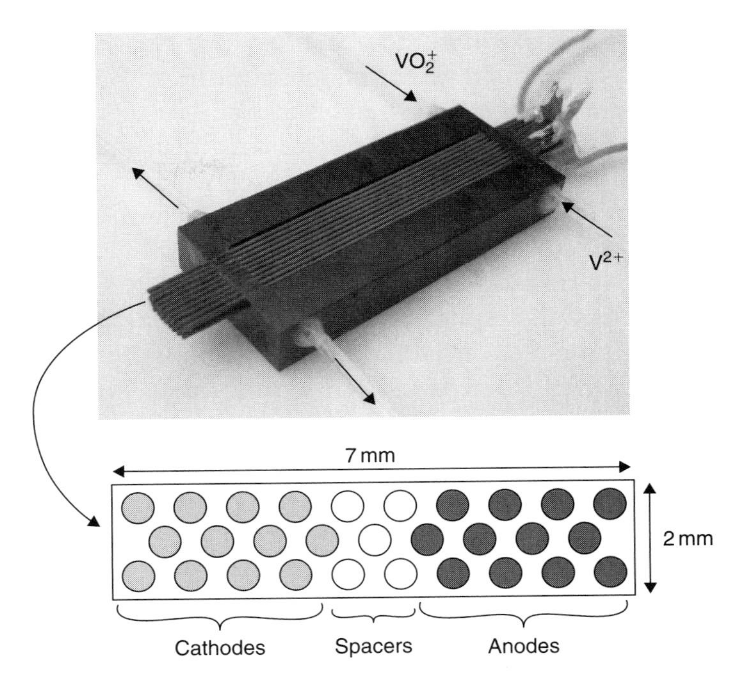

FIGURE 3.9 A microfluidic fuel cell expanded in three dimensions via the array cell architecture. Graphite rods are mounted in a single CNC-machined cavity, as shown in the image (top) and the cross-sectional schematic view (bottom). The cell comprises 12 anodes and 12 cathodes that are essentially independent, and could be connected externally in response to load requirements. The five rods in the center of the device are exposed to co-laminar inter-diffusion of anolyte and catholyte and are not electrically connected. This architecture provides scalability in both vertical and horizontal directions, although vertically is preferred from the perspective of ohmic losses. Reproduced with permission from Kjeang et al. [19]. Copyright Elsevier (2007).

a fuel and electrolyte separation and recirculation system was proposed at the cost of added complexity and reduced energy density of the complete fuel cell system.

3.5 CONCLUSION AND CHALLENGES AHEAD

Considering the invention of the microfluidic fuel cell is relatively recent, the number of advances in this field is impressive. Devices have been developed based on various fuels and oxidants, with competitive power densities and cell voltages obtained at room temperature. The levels of fuel utilization have been raised from below 1% to nearly 100% per single pass in some cases. Many of these advances, as discussed in this chapter, have stemmed from improving transport through microfluidic fuel cell architecture and running conditions. Several scale-up methodologies have also been demonstrated recently. These scale-up efforts show promise in translating advances made in unit cells to highly functioning integrated devices.

While developments in the field of microfluidic fuel cells have been rapid, much further work is required to facilitate a major commercial break-through. Ideally, the microfluidic fuel cell system, including auxiliary equipment and fluid storage, would output in the 1–20 W range in a compact integrated package with simple connections to established external infrastructure. This is a tall order. We end this chapter with some of the main challenges, and opportunities, that face the field.

The most pressing challenge for current microfluidic fuel cells is to increase the relatively low energy density (defined as energy output per system volume or mass). The physics of the typical co-laminar flow configuration require that both streams are liquid and contain an electrolyte. Although reactants may be added to the system at high concentration, the energy density of all devices presented to date has been low compared to other microstructured fuel cells. The low energy density is due primarily to the impractical single-pass use of liquid electrolyte without any form of recirculation or recycling. The implementation of a recirculation system for the electrolyte is a challenging task due to the space constraints, mixing/contamination issues, and the overhead associated with increased complexity. While fuel utilization data up to 100% per single pass have been presented, the fuel utilization at practical flow rates and useful cell voltages has generally been much lower, and frequently below 10%.

A number of integration issues remain. Specifically, there is a lack of engineering solutions for important functions such as integration of fuel and oxidant storage, waste handling, and low-power microfluidics-based fluid delivery (normally driven by a syringe pump via external tubing) using integrated micropumps and microvalves. It is hoped that microfluidic fuel cells can, at least to some extent, exploit engineering solutions already developed for other small scale fuel cell technologies.

Improvements in microfluidic fuel cell architecture, as described in this chapter, have addressed to some extent the mass transport limitation of dissolved oxygen that plagued many early cells. With these advances, the performance of current microfluidic fuel cell technologies is typically restricted by a complex combination of mass transport, electrochemical kinetics, and ohmic resistance. Firstly, with respect to mass transport, further challenges include increasing diffusive transport by further reducing the average cross-stream distance that a reactant molecule has to travel to reach the active site, or using reactant species with higher diffusivity. Secondly, electrochemical limitations at the fuel cell electrodes are caused by activation overpotentials and slow reaction rates. Choosing appropriate electrocatalysts and increasing the overall active surface area of the electrodes are key to reducing these limitations. The choice of electrolyte media and pH may also influence the electrochemical kinetics. Alternatively, the kinetics could be increased by raising the operational temperature. Increased temperature may be a beneficial side-effect of increased integration or scale-up. Thirdly, the combined ohmic resistance of a microfluidic fuel cell includes ionic charge transfer resistance in the electrolyte and electrical resistance in electrodes, contacts, and wires. All of these resistances need to be minimized for an efficient fuel cell system. The detrimental effects of ohmic resistance become increasingly important for devices with high power density, and thus ohmic resistance is expected to be a major challenge as other advances push power density to higher levels.

The cost of microfluidic fuel cell systems is an important consideration. The technology shows potential for very cost-effective units because of inexpensive fabrication techniques that have been employed, the elimination of the membrane and associated issues, and the range of (often inexpensive) catalysts that may be employed. Strategic cost reductions on the unit cell level are possible in several areas, including materials, manufacturing procedures, and the choice (and source) of fuel, oxidant, electrolyte, and low-cost alternative catalysts. It is noteworthy that each type of microfluidic fuel cell has features that will influence eventual device and implementation cost. For instance, the vanadium cell with flow-through porous electrodes, *in situ* fuel/oxidant regeneration can relieve the cost, hazards, and inertia associated with establishing fuelling infrastructure.

A number of opportunities exist for optimizing current microfluidic fuel cell technologies. New fuel and oxidant combinations are still possible, and the prospects of finding new electrolytes with advanced functionality are high. Examples include the use of ionic liquids and nonaqueous solvents, as well as the use of a phase boundary to separate fuel and oxidant. The opportunities for two-phase flows are numerous, with the fuel carried in one phase (likely liquid) and the oxidant in the other (likely gas). There is also ample opportunity to improve the micro- and nano-structure of the electrodes employed in microfluidic fuel cells. An ideal electrode would incorporate a wide range of characteristic length scales from nanometers to millimeters and have a very large active surface area. Lastly, there is much room for improvement in the area of scale-up. The expansion methodologies reported to date have confirmed that microfluidic fuel cell technology and the associated manufacturing processes are well-suited for scale-up. The volumetric energy density of the prototype devices was, however, insufficient for actual applications and the total power output was less than 0.1 W.

Biofuel cells represent an important growth area for microfluidic fuel cell technology. There are opportunities for microfluidic fuel cells to improve current biofuel cells in several ways: (i) enhancing convective mass transport, enabling higher enzyme loading while maintaining the enzymatic turnover rates at peak; (ii) harnessing the high surface to volume ratio inherent to microstructured devices to promote reactions catalyzed by the immobilized enzymes; and (iii) providing scale-up opportunities for practical devices with automated reactant supply and on-chip integration.

In summary, research to date has resulted in operational devices with promising room temperature performance in terms of power density and operational cell voltage. More must be done, however, to realize practical, efficient, and competitive devices with high energy density and high fuel utilization. While there have been many advances in microfluidic fuel cells to date, it is likely many more are ahead.

ACKNOWLEDGMENTS

This research was funded by the Natural Sciences and Engineering Research Council of Canada (NSERC) and Angstrom Power Inc. Infrastructure funding from Canada Foundation for Innovation (CFI) and the British Columbia Knowledge Development Fund (BCKDF) is highly appreciated.

References

[1] C.K. Dyer, J. Power Sources 106 (2002) 31–34.
[2] E.R. Choban, L.J. Markoski, J. Stoltzfus, J.S. Moore, P.A. Kenis, Power Sources, vol. 40, Cherry Hill, NJ, 2002, pp. 317–320.
[3] R. Ferrigno, A.D. Stroock, T.D. Clark, M. Mayer, G.M. Whitesides, J. Am. Chem. Soc. 124 (2002) 12930–12931.
[4] E. Kjeang, N. Djilali, D. Sinton, J. Power Sources 186 (2009) 353–369.
[5] A. Kundu, J.H. Jang, J.H. Gil, C.R. Jung, H.R. Lee, S.H. Kim, B. Ku, Y.S. Oh, J. Power Sources 170 (2007) 67–78.
[6] G.M. Whitesides, Nature 442 (2006) 368–373.
[7] N.T. Nguyen, S.T. Wereley, Fundamentals and Applications of Microfluidics, Artech House, Boston, MA, 2002.
[8] T.M. Squires, S.R. Quake, Rev. Mod. Phys. 77 (2005) 977–1026.
[9] M. Gad-el-Hak, Phys. Fluids 17 (2005).
[10] D. Erickson, Microfluid. Nanofluid. 1 (2005) 301–318.
[11] C.T. Crowe, D.F. Elger, J.A. Roberson, Engineering Fluid Mechanics, John Wiley & Sons Inc., New York, 2001.
[12] J. Atencia, D.J. Beebe, Nature 437 (2005) 648–655.
[13] R.F. Ismagilov, A.D. Stroock, P.J.A. Kenis, G. Whitesides, H.A. Stone, Appl. Phys. Lett. 76 (2000) 2376–2378.
[14] S.K. Yoon, M. Mitchell, E.R. Choban, P.J.A. Kenis, Lab on a Chip 5 (2005) 1259–1263.
[15] A.E. Kamholz, B.H. Weigl, B.A. Finlayson, P. Yager, Anal. Chem. 71 (1999) 5340–5347.
[16] J.B. Salmon, C. Dubrocq, P. Tabeling, S. Charier, D. Alcor, L. Jullien, F. Ferrage, Anal. Chem. 77 (2005) 3417–3424.
[17] J.P. Brody, P. Yager, Sens. Actuators, A 58 (1997) 13–18.
[18] P.J.A. Kenis, R.F. Ismagilov, G.M. Whitesides, Science 285 (1999) 83–85.
[19] E. Kjeang, J. McKechnie, D. Sinton, N. Djilali, J. Power Sources 168 (2007) 379–390.
[20] J.C. McDonald, D.C. Duffy, J.R. Anderson, D.T. Chiu, H.K. Wu, O.J.A. Schueller, G.M. Whitesides, Electrophoresis 21 (2000) 27–40.
[21] Y.N. Xia, G.M. Whitesides, Annu. Rev. Mater. Sci. 28 (1998) 153–184.
[22] D.C. Duffy, J.C. McDonald, O.J.A. Schueller, G.M. Whitesides, Anal. Chem. 70 (1998) 4974–4984.

[23] E. Kjeang, A.G. Brolo, D.A. Harrington, N. Djilali, D. Sinton, J. Electrochem. Soc. 154 (2007) B1220–B1226.
[24] J.N. Lee, C. Park, G.M. Whitesides, Anal. Chem. 75 (2003) 6544–6554.
[25] W. Sung, J.-W. Choi, J. Power Sources 172 (2007) 198–208.
[26] E.R. Choban, J.S. Spendelow, L. Gancs, A. Wieckowski, P.J.A. Kenis, Electrochim. Acta 50 (2005) 5390–5398.
[27] E.R. Choban, P. Waszczuk, P.J.A. Kenis, Electrochem. Solid-State Lett. 8 (2005) A348–A352.
[28] R.S. Jayashree, D. Egas, J.S. Spendelow, D. Natarajan, L.J. Markoski, P.J.A. Kenis, Electrochem. Solid-State Lett. 9 (2006) A252–A256.
[29] R.S. Jayashree, L. Gancs, E.R. Choban, A. Primak, D. Natarajan, L.J. Markoski, P.J.A. Kenis, J. Am. Chem. Soc. 127 (2005) 16758–16759.
[30] A. Li, S.H. Chan, N.T. Nguyen, J. Micromech. Microeng. 17 (2007) 1107–1113.
[31] J.L. Cohen, D.J. Volpe, D.A. Westly, A. Pechenik, H.D. Abruna, Langmuir 21 (2005) 3544–3550.
[32] J.L. Cohen, D.A. Westly, A. Pechenik, H.D. Abruna, J. Power Sources 139 (2005) 96–105.
[33] E. Kjeang, R. Michel, D.A. Harrington, D. Sinton, N. Djilali, Electrochim. Acta 54 (2008) 698–705.
[34] E. Kjeang, R. Michel, D.A. Harrington, N. Djilali, D. Sinton, J. Am. Chem. Soc. 130 (2008) 4000–4006.
[35] E. Kjeang, B.T. Proctor, A.G. Brolo, D.A. Harrington, N. Djilali, D. Sinton, Electrochim. Acta 52 (2007) 4942–4946.
[36] E.R. Choban, L.J. Markoski, A. Wieckowski, P.J.A. Kenis, J. Power Sources 128 (2004) 54–60.
[37] S.M. Mitrovski, L.C.C. Elliott, R.G. Nuzzo, Langmuir 20 (2004) 6974–6976.
[38] S.M. Mitrovski, R.G. Nuzzo, Lab on a Chip 6 (2006) 353–361.
[39] M. Togo, A. Takamura, T. Asai, H. Kaji, M. Nishizawa, Electrochim. Acta 52 (2007) 4669–4674.
[40] S. Hasegawa, K. Shimotani, K. Kishi, H. Watanabe, Electrochem. Solid-State Lett. 8 (2005) A119–A121.
[41] C.M. Moore, S.D. Minteer, R.S. Martin, Lab on a Chip 5 (2005) 218–225.
[42] M.H. Sun, G.V. Casquillas, S.S. Guo, J. Shi, H. Ji, Q. Ouyang, Y. Chen, Microelectron. Eng. 84 (2007) 1182–1185.
[43] K.S. Salloum, J.R. Hayes, C.A. Friesen, J.D. Posner, J. Power Sources 180 (2008) 243–252.
[44] M. Togo, A. Takamura, T. Asai, H. Kaji, M. Nishizawa, J. Power Sources 178 (2008) 53–58.
[45] T.C. Merkel, V.I. Bondar, K. Nagai, B.D. Freeman, I. Pinnau, J. Polym. Sci. Part B: Polym. Phys. 38 (2000) 415–434.
[46] L.J. Markoski, The laminar flow fuel cell: a portable power solution, in: 8th Annual International Symposium Small Fuel Cells 2006; Small Fuel Cells for Portable Applications, The Knowledge Foundation, Washington, DC, 2006.
[47] S. Tominaka, S. Ohta, H. Obata, T. Momma, T. Osaka, J. Am. Chem. Soc. 130 (2008) 10456–10457.
[48] J.D. Morse, Int. J. Energy Res. 31 (2007) 576–602.
[49] W.M. Qian, D.P. Wilkinson, J. Shen, H.J. Wang, J.J. Zhang, J. Power Sources 154 (2006) 202–213.

[50] C.P. de Leon, F.C. Walsh, A. Rose, J.B. Lakeman, D.J. Browning, R.W. Reeve, J. Power Sources 164 (2007) 441–448.
[51] G.H. Miley, N. Luo, J. Mather, R. Burton, G. Hawkins, L.F. Gu, E. Byrd, R. Gimlin, P.J. Shrestha, G. Benavides, J. Laystrom, D. Carroll, J. Power Sources 165 (2007) 509–516.
[52] N.A. Choudhury, R.K. Raman, S. Sampath, A.K. Shukla, J. Power Sources 143 (2005) 1–8.
[53] R. Aogaki, E. Ito, M. Ogata, J. Solid State Electrochem. 11 (2007) 757–762.
[54] L.F. Gu, N. Luo, G.H. Miley, J. Power Sources 173 (2007) 77–85.
[55] C.P. de Leon, A. Frias-Ferrer, J. Gonzalez-Garcia, D.A. Szanto, F.C. Walsh, J. Power Sources 160 (2006) 716–732.
[56] W. SkyllasKazacos, C. Menictas, M. Kazacos, J. Electrochem. Soc. 143 (1996) L86–L88.
[57] S.K. Yoon, G.W. Fichtl, P.J.A. Kenis, Lab on a Chip 6 (2006) 1516–1524.
[58] A.D. Stroock, S.K.W. Derringer, A. Ajdari, I. Mezic, H.A. Stone, G.M. Whitesides, Science 295 (2002) 647.
[59] J.D. Kirtland, G.J. McGraw, A.D. Stroock, Phys. Fluids 18 (2006).
[60] K.G. Lim, G.T.R. Palmore, Biosens. Bioelectron. 22 (2007) 941–947.
[61] J. Lee, K.G. Lim, G.T.R. Palmore, A. Tripathi, Anal. Chem. 79 (2007) 7301–7307.
[62] A. Bazylak, D. Sinton, N. Djilali, J. Power Sources 143 (2005) 57–66.
[63] M.H. Chang, F. Chen, N.S. Fang, J. Power Sources 159 (2006) 810–816.
[64] F.L. Chen, M.H. Chang, M.K. Lin, Electrochim. Acta 52 (2007) 2506–2514.
[65] W.Y. Chen, F.L. Chen, J. Power Sources 162 (2006) 1137–1146.
[66] F. Chen, M.-H. Chang, C.-W. Hsu, Electrochim. Acta 52 (2007) 7270–7277.
[67] R.A. Bullen, T.C. Arnot, J.B. Lakeman, F.C. Walsh, Biosens. Bioelectron. 21 (2006) 2015–2045.
[68] E. Katz, A.N. Shipway, I. Willner, Biochemical fuel cells, in: W. Vielstich, H.A. Gasteiger, A. Lamm (Eds.), Handbook of Fuel Cells – Fundamentals, Technology and Applications, vol. 1, John Wiley & Sons, Ltd., New York, 2003.
[69] I. Willner, E. Katz, Angew. Chem. Int. Ed. 39 (2000) 1180–1218.
[70] N.L. Akers, C.M. Moore, S.D. Minteer, Electrochim. Acta 50 (2005) 2521–2525.
[71] S. Topcagic, S.D. Minteer, Electrochim. Acta 51 (2006) 2168–2172.
[72] E. Kjeang, D. Sinton, D.A. Harrington, J. Power Sources 158 (2006) 1–12.

CHAPTER

4

Development of Fabrication/Integration Technology for Micro Tubular SOFCs

Toshio Suzuki
National Institute of Advanced Industrial Science and Technology, Nagoya, Japan

Yoshihiro Funahashi
Fine Ceramics Research Association, Nagoya, Japan

Toshiaki Yamaguchi, Yoshinobu Fujishiro, Masanobu Awano
National Institute of Advanced Industrial Science and Technology, Nagoya, Japan

We summarize recent development of fabrication/integration technology for micro tubular solid-oxide fuel cells (SOFCs) with 0.8–2.0 mm in diameter operable at at/under 550°C. Successful achievement will be shown from single cell to stack fabrication technology which was designed for cost-effective mass production. Optimization of the electrode microstructure and development of electrolyte coating technology realized high-performance micro tubular SOFCs and bundles, maximum power density of over 1 W cm^{-2} at 550°C operating temperature for single cell, and over 2 W cm^{-3} @ 0.7 V at 550°C for 1 cm^3 cubic bundle, respectively. The key to realize such high-performance SOFCs lies in the design of the cell/bundle/stack and novel fabrication technology that can realize optimized electrode structures.

4.1 INTRODUCTION

Solid oxide fuel cells (SOFCs) have been recognized as a keystone of the future energy economy, and the development of SOFC systems has been an important issue in recent years [1, 2]. A typical SOFC consists of doped zirconia for an electrolyte, Ni cermet for an anode and doped lanthanum manganite for a cathode, and it has shown its long-term stability over 20,000 h operation as well as high power output up to 2 W/cm^2 at 800°C [3–5].

Currently, many efforts focus on lowering the operating temperature of SOFCs because use of SOFCs in the intermediate temperature under 650°C can decrease material degradation and prolong stack lifetime, reducing cost by utilizing metal materials. The approaches to reducing operating temperature have widely been reported, for example:

(1) Using new electrolyte, cathode, and anode materials [6–14].
(2) Reducing the thickness of the electrolyte using traditional electrolyte materials such as Y doped ZrO_2 [4, 15].
(3) Introducing new structures for electrolytes [16–18] and electrodes.

One of the well-known obstacles for reduced temperature operation of SOFCs is the poor activity of the cathode. For this reason, a lot of new materials have been investigated for cathodes. For example, Shao and Haile have shown outstanding cell performance (peak power density = 1010 W cm^{-2} at 600°C) by introducing a (Ba, Sr)(Co, Fe)O_3 cathode using a ceria-based electrolyte on an anode supported cell. Improvements of cell performance by introducing new anode and electrolyte materials were also reported by several other research groups. For example, outstanding results of a power density of 0.8 W cm^{-2} using a Ru-Ni-ceria-based anode and ceria electrolyte, and over 1.9 W cm^{-2} using $LaGaO_3$-based electrolyte at 600°C. Introducing doped ceria to an electrolyte and electrodes appeared to be of importance to enhance the performance of SOFC in the reduced temperature operation under 600°C. However, ceria-based materials display relatively lower open circuit voltage (i.e., 0.8–0.9 V) due to high electronic conductivity at the reducing atmosphere. Despite this, the high output power of ceria-based electrolyte cells attracts the use of ceria as cell components and, so far, several studies have purported to offer solutions for these problems [16, 17]. Thus, ceria-based materials can be considered as one of the candidate electrolytes for SOFCs operated at temperatures under 600°C.

On the other hand, to increase the variety of applications such as auxiliary power units for automobile and power sources for portable devices, it is crucial to develop highly efficient small cell stacks which are robust for rapid temperature change operation. So far, tubular design SOFCs were shown to be effective in realizing such stacks since they were shown to be robust under thermal stress caused by a rapid heating cycle [19, 20]. In addition, decreasing tubular cell diameter is expected to improve the mechanical properties as well as volumetric power density of the cell stacks. Thus, it is of importance to develop fabrication technology for micro tubular SOFCs with mm to sub mm diameters, along with their bundles and stacks.

In the present chapter, development of micro tubular SOFCs will be discussed, including fabrication technology designed for mass production, as well as characterization of the cells and bundle/stack. We will begin with a brief review of the state-of-the-art micro tubular SOFC concepts, followed by introducing our new approach for fabricating micro tubular SOFCs using conventional SOFC materials. Then, cauterization of micro tubular SOFC will be discussed in detail, including current collecting loss due to the dimension of the cell. Finally, our new approach for fabricating micro tubular SOFC bundle/stack development will be presented.

4.2 FABRICATION AND CHARACTERIZATION OF MICRO TUBULAR SOFCs

4.2.1 Micro Tubular SOFC Concepts

Shape design of the SOFC is an important factor in improving the performance of SOFCs. Tubular design was first introduced for commercialized SOFCs by Siemens Westinghouse using a cathode supported cell. From that point, tubular SOFCs were well investigated from fundamental to stack modulation using anode-supported tubular design up to several kW scale stacks [19–23].

Tubular design can be more beneficial when the size of the tube becomes in the range of millimeter to submillimeter, because it enables the design of SOFC stacks with high volumetric power density. Figure 4.1 shows an estimation of the volumetric power density and the pressure drop as a function of tube diameter, which were calculated at the conditions shown in Figure 4.1. As can be seen, the volumetric power density effectively increases when the tube diameter becomes in the range of less than 1 mm due to the increase of electrode area (tube surface)/cell volume ratio (tube volume), while the pressure drop becomes significantly large when tube diameter becomes less than 0.1 mm. Therefore, use of micro tubular SOFCs with a diameter size of mm to sub mm is attractive for realizing high power, small SOFC stacks/modules.

In addition, micro tubular SOFCs were also shown to be well-suited to accommodate repeated heat and electrical load cycling

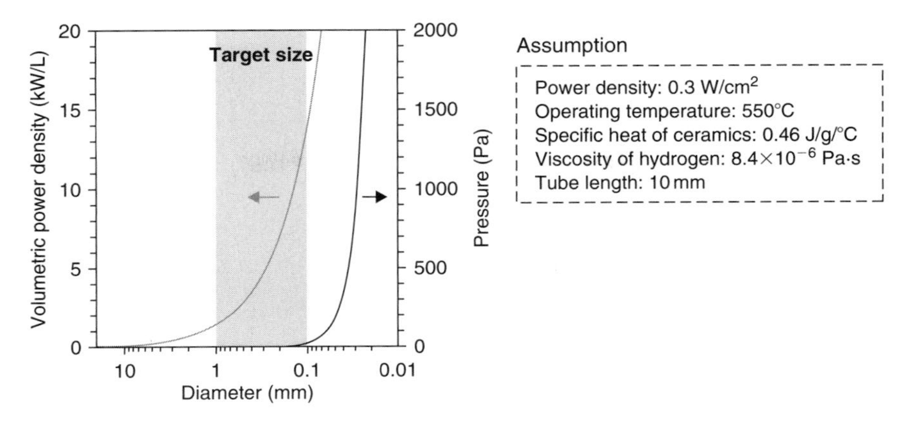

FIGURE 4.1 Volumetric power density and pressure loss of tubular SOFCs as a function of tube diameter size.

under rapid changes. Small-scale tubular SOFCs, reported by Kendall and Palin [19] and Yashiro et al. [20], could endure thermal stress caused by rapid heating up to operating temperature. Therefore, micro tubular design is promising and plays an important role for realizing highly efficient SOFC systems with quick start-up/shut-down performance.

Fabrication of Micro Tubular SOFCs

Figure 4.2 (a) shows the fabrication process of micro tubular SOFCs, and 4.2 (b) shows an image of a green anode tube, a green anode tube with an electrolyte layer, a sintered anode tube with the electrolyte, and a complete cell (1.8 mm in diameter), respectively, and 4.2 (c) shows an image of complete cells (0.8 mm in diameter). Materials used in Figures 4.2 (b) and (c) were Ni-$Gd_{0.2}Ce_{0.8}O_{2-x}$ (GDC), GDC, and $La_{0.6}Sr_{0.4}Co_{0.2}Fe_{0.8}O_{3-y}$ (LSCF)-GDC for an anode tube, an electrolyte, and a cathode, respectively. Those SOFC materials were selected from the literature as conventional materials for low temperature SOFC (under 600°C) [24]. Note that this process does not limit materials and basically any SOFC constituent materials can be chosen.

Anode tubes were made from NiO and GDC powder, binder (cellulose), and optionally, a pore former (poly methyl methacrylate beads; PMMA). The volume ratio of NiO and GDC ranged from 50:50 to 70:30. The amount of added PMMA beads was 0–40

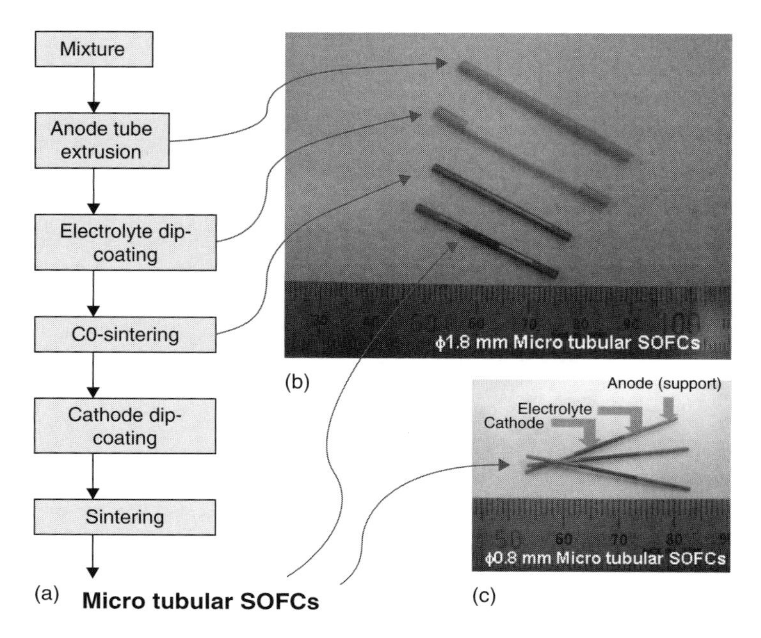

FIGURE 4.2 Fabrication procedure of the micro tubular SOFCs.

vol.% to NiO and GDC powder. Effect of PMMA on the microstructure of the anode tube will be discussed later. These ingredients were mixed for 1h by a mixer 5DMV-rr (Dalton Co., Ltd.), and, after adding the proper amount of water, it was stirred for 30min in a vacuum chamber. The mixture (clay) was then left over 15h for aging. The tubes were extruded using the clay from a metal mold set in a piston cylinder-type extruder (Ishikawa-Toki Tekko-sho Co., Ltd.). By selecting the size of the die for extrusion, micro tubular SOFCs with different diameters can be prepared.

A slurry for dip-coating the electrolyte was prepared by mixing the GDC powder, solvents (methyl ethyl ketone and ethanol), binder (poly vinyl butyral), dispersant (polymer of an amine system) and plasticizer (dioctyl phthalate) for 24h. The anode tubes were dipped in the slurry and coated at the pulling rate of 1 to 3mm/sec. During the coating process, the edge of the tube was sealed to protect inside the anode tube. After drying, they were sintered at 1100–1400°C for 1h in air. The anode tubes with electrolyte were, again, dip-coated with the cathode slurry, which was prepared in the same manner using LSCF and GDC powder and organic ingredients. The weight ratio of LSCF and GDC powder

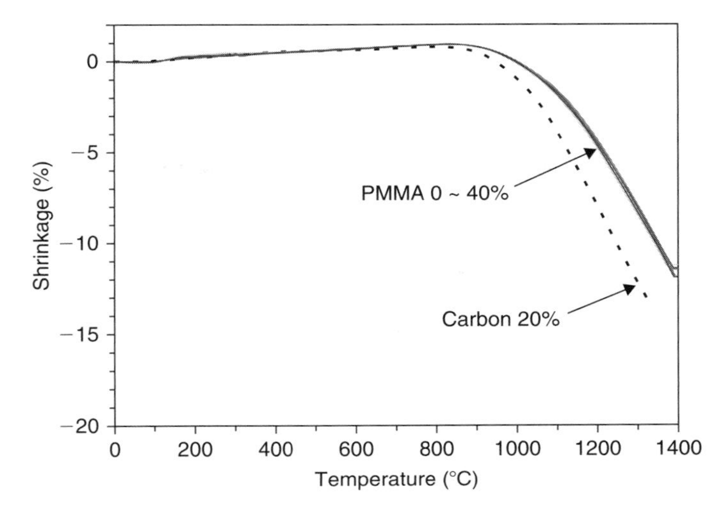

FIGURE 4.3 Shrinkage behaviors of the anode tubes as a function of temperature. Data were obtained for the samples with different PMMA contents.

was 70:30. After dip-coating, the tubes were dried and sintered at 1050°C for 1 h in air to complete a cell [25].

4.2.2 Optimization of Anode Microstructure

To increase the performance of SOFCs, it is important to optimize the microstructure of electrodes where reactant gases are diffused and electrochemical reactions take place. In the fabrication process, attempts were made to control and optimize anode microstructure by varying the sintering temperature and the amount of pore former added in the clay.

Figure 4.3 shows the shrinkage of the anode tube as a function of sintering temperature. As can be seen, the amount of PMMA beads did not change the sintering behavior of the anode tube, while the anode tube with carbon powder showed different behavior. This seemed to be caused by the fact that the carbon powder was bulkier than PMMA beads, which leads to the difference in the packing density of the powders. Since the amount of PMMA beads does not influence the sintering behavior of the anode tube, the use of PMMA beads as a pore former is preferable in this process. A change in the sintering behavior due to the addition of a pore former may cause a deficiency in the electrolyte layer during the co-sintering process, which should be avoided.

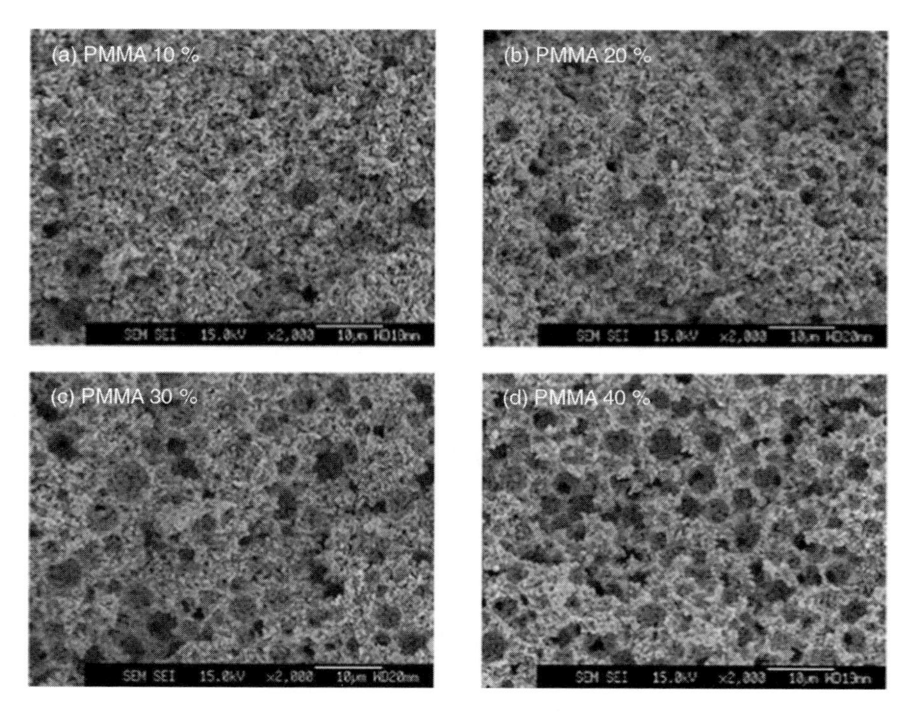

FIGURE 4.4 Microstructure of the anode tubes after reduction: (a) PMMA 10 vol.%; (b) PMMA 20 vol.%; (c) PMMA 30 vol.%; (d) PMMA 40 vol.%.

Figure 4.4 shows the microstructure of the anode tubes prepared from different amounts of PMMA beads: (a) 10, (b) 20, (c) 30, and (d) 40 vol.%, respectively. As can be seen, the porosity of the anode tubes can be controlled by the amount of the PMMA beads, which were 36, 39, 41, 47, and 51% after reduction for samples with 0, 10, 20, 30, and 40 vol.% PMMA, respectively. As can be seen, a well-developed 3D network structure was realized, especially for the sample with 40 vol.% PMMA.

Figure 4.5 shows the distribution of the pore diameter in the anode tubes sintered at 1400°C that was measured by a mercury porosimeter. The main peak of the pore diameter distribution shifted from 0.2 μm to 1 μm as the amount of PMMA increased. These results were consistent with the SEM observation as shown in Figure 4.4.

Figure 4.6 shows the cumulative pore distribution of the anode tube as a function of sintering temperature for the sample with 40 vol.% PMMA. Up to 1200°C, the total pore volume simply decreases

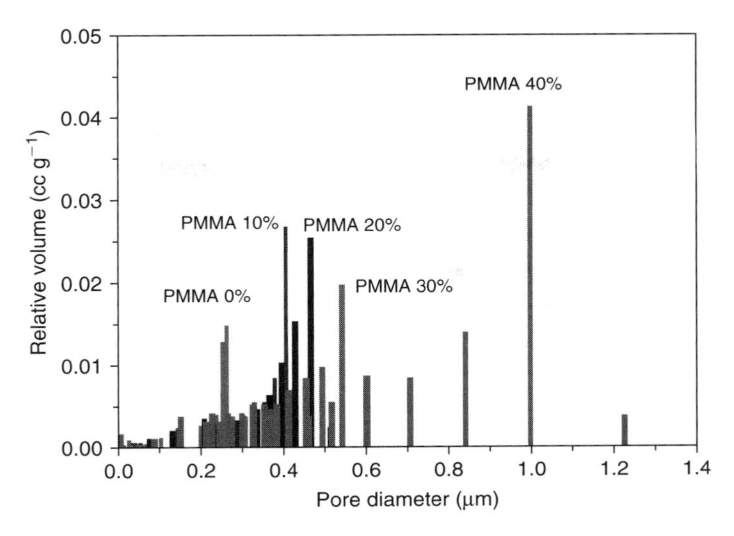

FIGURE 4.5 The pore diameter distribution of the anode tubes with different PMMA contents, sintered at 1400°C

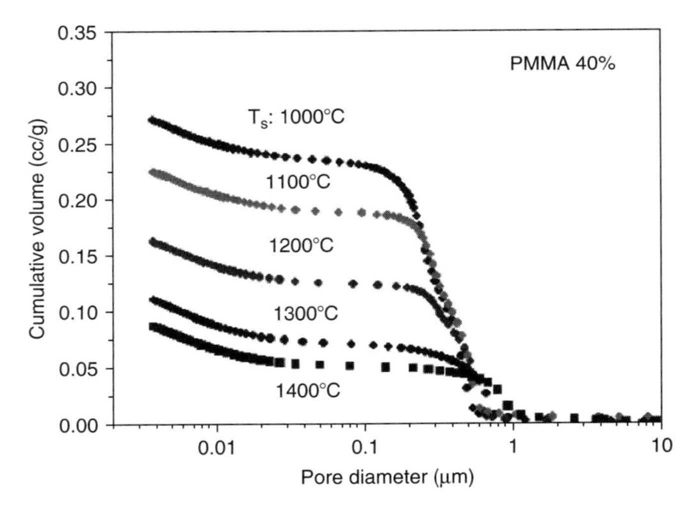

FIGURE 4.6 The cumulative pore distribution of the anode tubes prepared at different sintering temperatures.

as the sintering temperature increases due to densification of the anode tube; however, an increase of pore volume for the 0.5–1 μm range with increasing sintering temperature was observed. SEM observation of these samples is shown in Figure 4.7, which displays

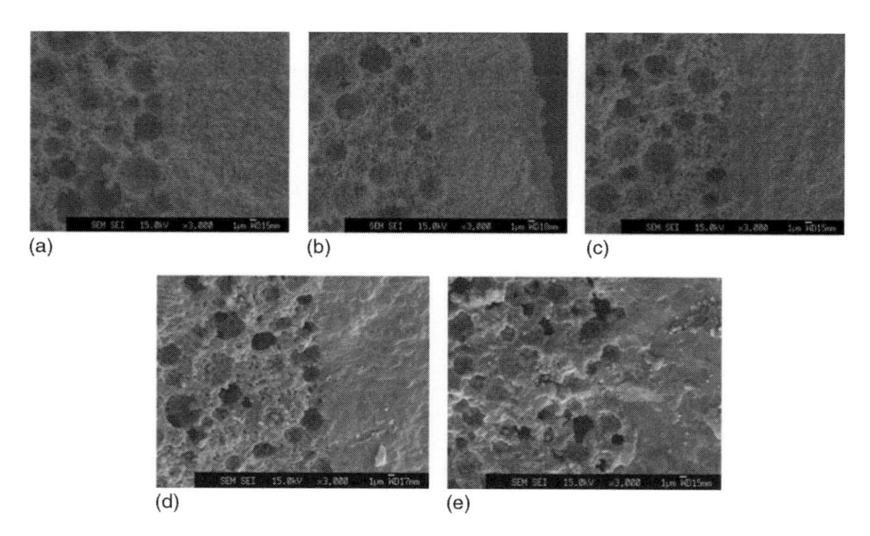

FIGURE 4.7 Microstructure of the anode tubes with the electrolyte layer (before reduction) prepared at the sintering temperature of (a) 1000°C; (b) 1100°C; (c) 1200°C; (d) 1300°C; (e) 1400°C.

cross-sectional images of the anode tube/electrolyte interface. As can be seen, the size of large pores from PMMA beads does not change up to a 1200°C sintering temperature. Then, above 1300°C, the large pore starts to shrink. Figure 4.8 summarizes the relationship between the porosity of the anode tubes and sintering temperature. Finally, two stages of densification process were observed: first, at lower temperatures, small pores between powders start to densify and then, large pores created by the pore former begin to shrink. Thus, microstructure can also be effectively controlled by changing sintering temperature. Of course, sintering temperature affects the densification process of the electrolyte during co-sintering as well, and this will be discussed in the next section.

4.2.3 Coating Technology of Electrolyte Layer on the Micro Tubular Support

Control of Electrolyte Thickness

The thickness of the electrolyte layer on the anode tube can be controlled by three main methods: (i) changing the pull speed of a dip coater, (ii) applying multiple coating, and (iii) changing the concentration of the powder in the slurry. Once the slurry is prepared,

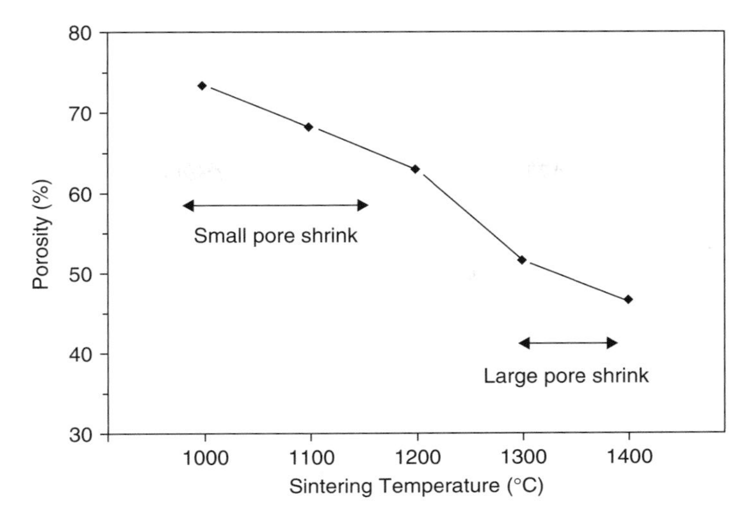

FIGURE 4.8 The porosity of the anode tube as a function of sintering temperature.

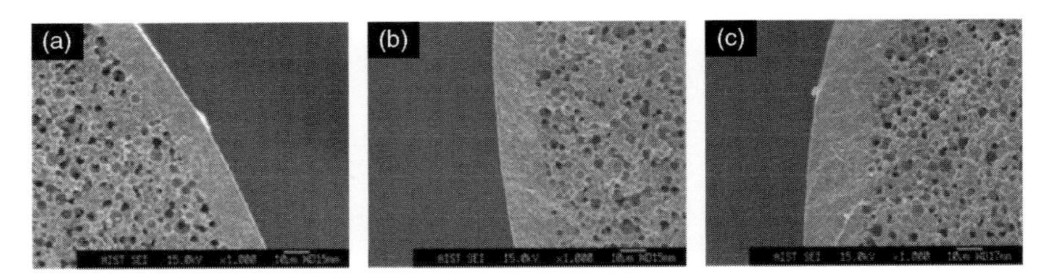

FIGURE 4.9 SEM images of the electrolyte layer and anode tube interface obtained at different coating speed (pulling speed) of (a) 1.0; (b) 2.0; and (c) 3.0 mm/sec (Sintering temperature = 1400°C). Thickness varies (a) 13.2~13.6 μm; (b) 15.6~17.5 μm; and (c) 21.6~22.0 μm, respectively.

(i) and (ii) can be more convenient for controlling the electrolyte thickness. Figure 4.9 shows the SEM images of the cross-section of the anode tube with the electrolyte (GDC) after being sintered at 1400°C, 1 h, at the pulling speed of (a) 1.0, (b) 2.0, and (c) 3.0 mm/sec. As can be seen, the thickness of 13–22 μm was obtained by changing the pulling speed of the dip coater. The thickness of the electrolyte layer also depends upon the concentration of the powder in the slurry, and this methodology allows a simple way to optimize coating conditions by combining (i)–(iii).

FIGURE 4.10 Microstructure of the electrolyte layer (surface images) prepared at the sintering temperature of (a) 1000°C; (b) 1100°C; (c) 1200°C; (d) 1300°C; (e) 1400°C.

Densification of the Electrolyte Layer

Densification behavior of the electrolyte layer on the anode-supported tube was investigated by changing sintering temperature and degree of shrinkage rate of the support. Figure 4.10 shows SEM images of the electrolyte (GDC) which initially coated the green anode tubes, and then sintered at temperatures between 1100 and 1400°C. It was observed that the microstructure of the electrolyte sintered under 1300°C included pores in the structure, and the densification of the electrolyte was completed at a sintering temperature above 1300°C.

The densification of the electrolyte layer deposited on the tubular support should also be greatly affected by the shrinkage of the support during the co-sintering process [26]. Figure 4.11 shows shrinkage behaviors of various anode tubes as a function of sintering temperature along with the result of a GDC compact prepared with a cold isostatic press. The densification of the GDC compact started at 800°C and shrank with elevating temperatures most rapidly (Figure 4.11(e)). As shown in Figure 4.11(c), the densification of the green anode tube started at 1000°C, and the shrinkage rate reached about 10% at 1400°C, while the anode tube pre-calcined

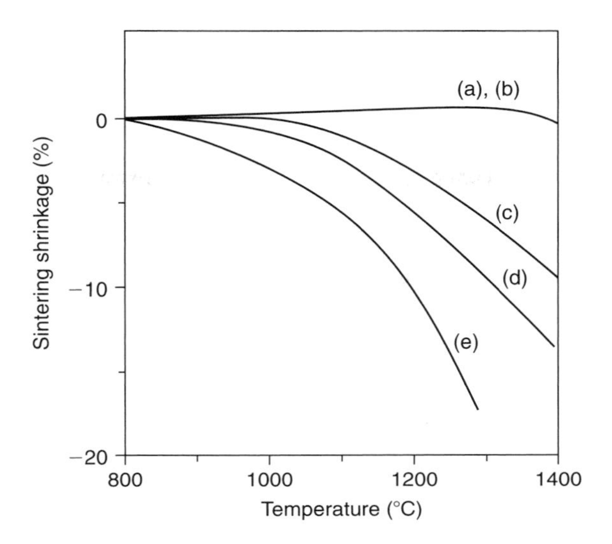

FIGURE 4.11 Shrinkage behaviors of (a) the anode tube pre-calcined at 1300°C; (b) the electrolyte coated anode tube pre-calcined at 1300°C; (c) the green anode tube; (d) the electrolyte coated green anode tube and (e) the compact GDC.

at 1300°C did not shrink at all up to 1400°C (Figure 4.11(a)). The shrinkage behaviors of the electrolyte layer coated anode tubes were also shown in Figures 4.11 (b) and (d). In the case of the electrolyte layer/green anode supported tube, the shrinkage began at about 900°C and the slope of the curve was larger than that of the noncoated green tube (Figure 4.11(c)). However, the shrinkage curve of the pre-calcined anode support coated with the electrolyte layer (Figure 4.11(b)) completely overlapped with the curve of the noncoated support (Figure 4.11(a)). These results indicate that, during the co-sintering process, the densification of the green supports was assisted by a sintering stress of the electrolyte coating layer.

Figure 4.12 summarizes the shrinkage of the anode supported tubes with the electrolyte layer (GDC) after co-sintering at 1450°C for 6h as a function of pre-calcination temperatures of the tubes, along with the microstructures of the electrolyte. As can be seen, two parts of the densification processes were clearly shown; (1) dense and thick, and (2) porous and thin, depending upon the shrinkage rate of the anode supported tubes. Dense and thick electrolyte layers were observed with the anode tube shrinkage of above

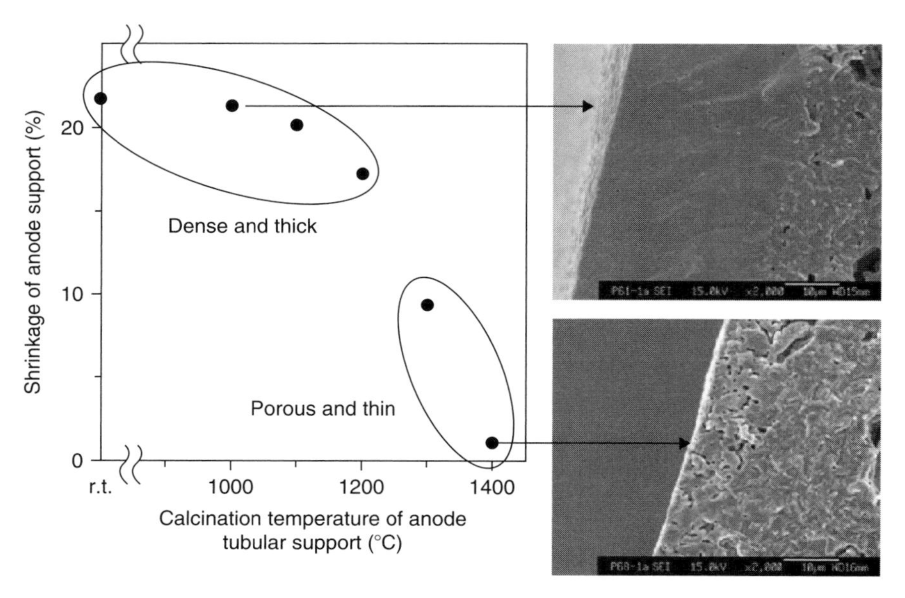

FIGURE 4.12 Relationships between the shrinkage of the anode tube with the electrolyte coating layer as a function of sintering temperature of the anode tube, and the electrolyte microstructure after co-sintering.

10% (pre-calcination temperature <1200°C), while porous and thin electrolyte layers were obtained at a shrinkage rate of less than 10% (pre-calcination temperature >1300°C). This can be explained by the fact that full densification of a close-packed compact with a packing density of 0.74 through the sintering process requires a linear shrinkage of about 10%, which is also required for the anode supported tubes to realize a dense electrolyte layer. This can also explain the fact that densification of the electrolyte was completed at 1300°C (Figure 4.10) at which point the shrinkage rate reached over 10% in Figure 4.11 (d).

As a result, a crack-free electrolyte layer without delaminating from the anode tube was realized by utilizing the densification process of the supported tubes during the co-sintering process, which turned out to be preferable for mass production. Thickness of the electrolyte and microstructure of the anode tube were also easily controlled by the fabrication process; therefore, desirable micro tubular SOFCs can be offered through the newly developed fabrication technology.

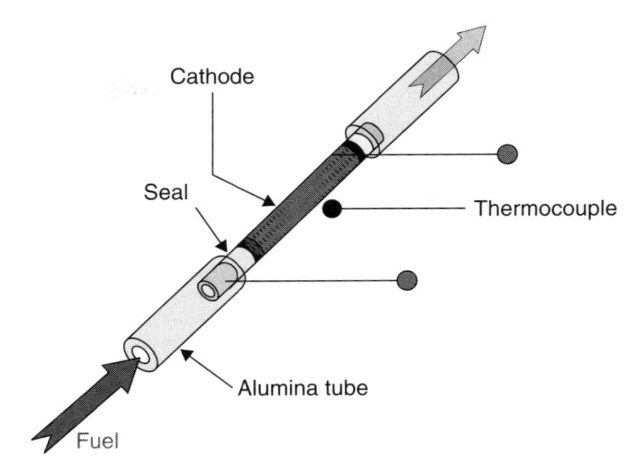

FIGURE 4.13 Experimental apparatus of the single tubular cell measurement.

4.3 CHARACTERIZATION OF MICRO TUBULAR SOFCs

4.3.1 Cell Performance

Test Setup for Single Cell Measurement

Figure 4.13 shows the experimental setup for the single micro tubular SOFC measurement. The tubular SOFC was fixed in aluminum tubes using a sealing paste after applying current collectors. As can be seen, current collections were made from the edge of the anode tube and the surface of the cathode by using Ag wire and paste. This setup was placed in the furnace, placing a thermocouple close to the surface of the micro tubular SOFC, which was used for furnace temperature control and monitoring.

The cell performance was investigated by using a potentiostat (Solartron 1296). Typical cell size for single cell measurement was 15 mm in length with cathode length of 5 mm, avoiding excess heat generation during the experiment. Hydrogen (humidified by bubbling water at room temperature) was flowed inside of the tubular cell at the rate of 5–20 mL min^{-1} diluted by adding nitrogen flow of 0–30 mL min^{-1}. The cathode side was open to the air without flowing gas.

Performance of 1.6 mm Diameter Tubular SOFC

First, a 1.6 mm diameter tubular SOFC was examined in a cell performance test, which consists of Ni-$Gd_{0.2}Ce_{0.8}O_{2-x}$ (GDC) 70:30

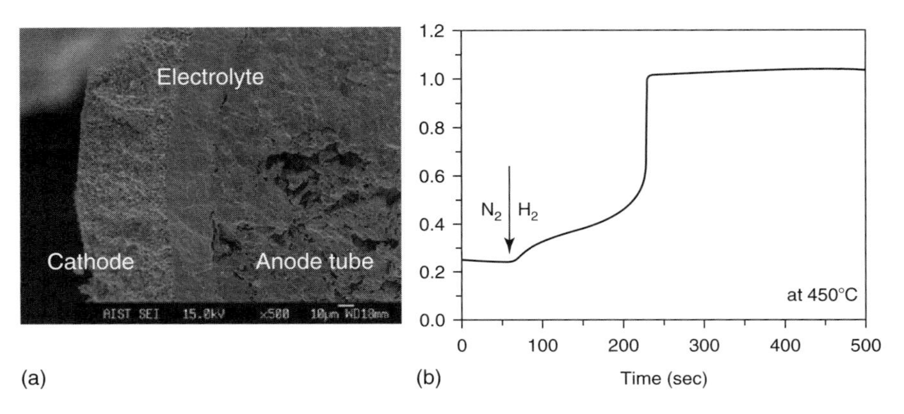

FIGURE 4.14 (a) A cross-sectional SEM image of the 1.6 mm diameter tubular SOFC prepared without pore former; (b) Initial start-up behavior of the micro tubular SOFC.

weight ratio, GDC, and $La_{0.6}Sr_{0.4}Co_{0.2}Fe_{0.8}O_{3-y}$ (LSCF)-GDC for an anode tube, an electrolyte, and a cathode, respectively. Figure 4.14 shows a cross-sectional SEM image of the tubular cell. As can be seen, the anode supported tube was prepared without using a pore former (PMMA); however, a porosity of about 30% before reduction was achieved. A crack-free dense electrolyte with a thickness of about 20 μm has been successfully prepared on the anode-supported tube by co-sintering techniques. The thickness of the anode tube was about 0.4 mm.

Initial start-up behavior of a fresh tubular cell sample is shown in Figure 4.14 (b). First, the initial sample, which was not reduced, was set in the sample holder placed in the furnace. While N_2 gas was flowing inside the tube, the sample was heated up to 450°C and then N_2 gas was switched to fuel gas. As indicated in Figure 4.14 (b), it only took about 2 min to reduce the anode and activate the cell at 450°C, where the open circuit voltage reached as high as 1.03 V.

The performance of the micro tubular SOFC was then investigated, and the results are shown in Figure 4.15. As can be seen, the open circuit voltages were dropped from 1.03 to 0.84 V as furnace temperature increased from 450 to 570°C, which is usually explained by an increase of electronic conductivity in the ceria electrolyte (this will be discussed in the next section). The power density of the cell was estimated from the area of the cathode (0.35 cm^2). The peak power density of 203, 400, 857, and 1000 mW cm^{-2} was obtained at 450, 500, 550, and 570°C, respectively. Fuel utilization for each point

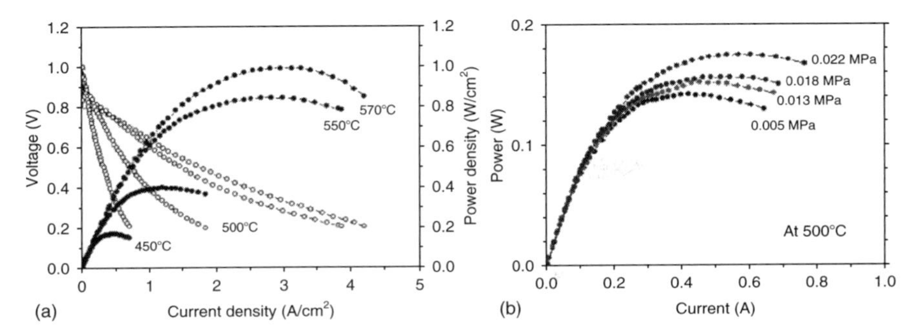

FIGURE 4.15 The performance of the micro tubular SOFC; (a) cell voltage and power density as functions of current and temperature; (b) Power as functions of current and backpressure of the fuel gas.

where the peak power was obtained was estimated to be 7.5, 31, 56, 62% for 450, 500, 550 and 570°C furnace temperatures, respectively.

Since cell components were prepared from typical materials, these outstanding performances seem to result from the anode microstructure favorable for gas diffusion and electrochemical reactions. To observe the effect of the fuel gas pressure, the backpressure of the fuel gas was varied while keeping the same gas flow rate. Figure 4.15 (b) shows the cell power as a function of current for various backpressures of the anode gas. Note that results in Figure 4.15 (a) were obtained at the backpressure of 0.005 MPa. As shown in Figure 4.15 (b), the cell power (actual cell power from a single cell with a cathode area of $0.35\,cm^2$) at 500°C improved about 30%, from 140 to 180 mW when the backpressure increased from 0.005 to 0.022 MPa. This result indicates that the performance of the cell strongly depends upon the anode side of the cell [27].

High Performance 0.8 mm Diameter Micro Tubular SOFC

Figure 4.16 shows a cross-sectional SEM image of a 0.8 mm diameter micro tubular SOFC. The anode supported tube was prepared using a pore former (40 vol.% PMMA) for further improvement of the anode microstructure. As can be seen, a crack-free dense electrolyte with a thickness of about 10 μm has been successfully prepared on the anode supported tube with uniform porous structure (before reduction). Reducing the size of the cell allows effective reduction of the cell weight and anode thickness, which decreases the gas diffusion path. As shown in Figure 4.16, the thickness of the tube was about 175 μm with a porosity of about 46% (before

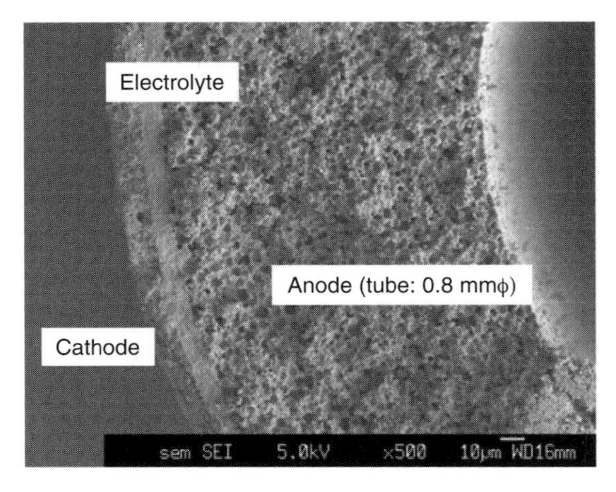

FIGURE 4.16 A cross-sectional SEM image of the 0.8 mm diameter tubular SOFC prepared with 40% pore former (PMMA beads): consists of three layers: cathode, electrolyte, and anode supported tube.

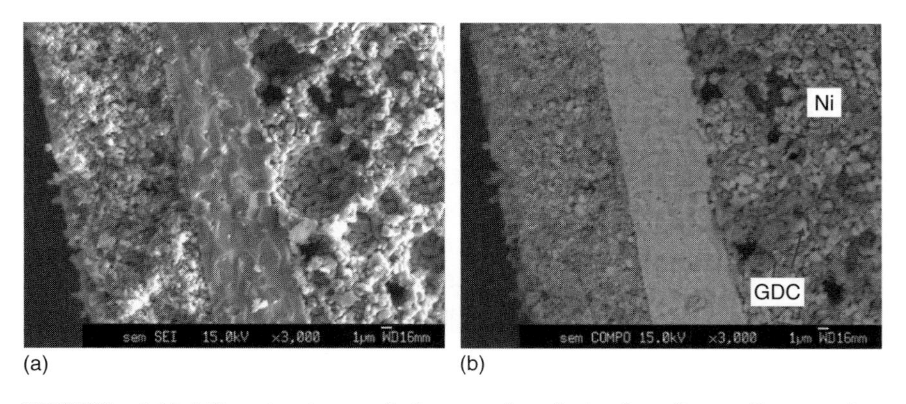

FIGURE 4.17 Microstructure of the anode electrode after cell operation (reduced structure) (a) A SEM image; (b) A BS image of the same spot, Ni: dark, GDC: light.

reduction). As a result, the weight of a single tube became 0.015 g per 1 cm tube length (0.06 g per unit electrode area, over 16 cm^2 electrode area/g).

Figure 4.17 shows that a net-shaped structure with high porosity was realized after reduction. It appears that the larger pores provide paths for gas diffusion, with micro pores between particles being a triple phase boundary, where electrochemical reactions

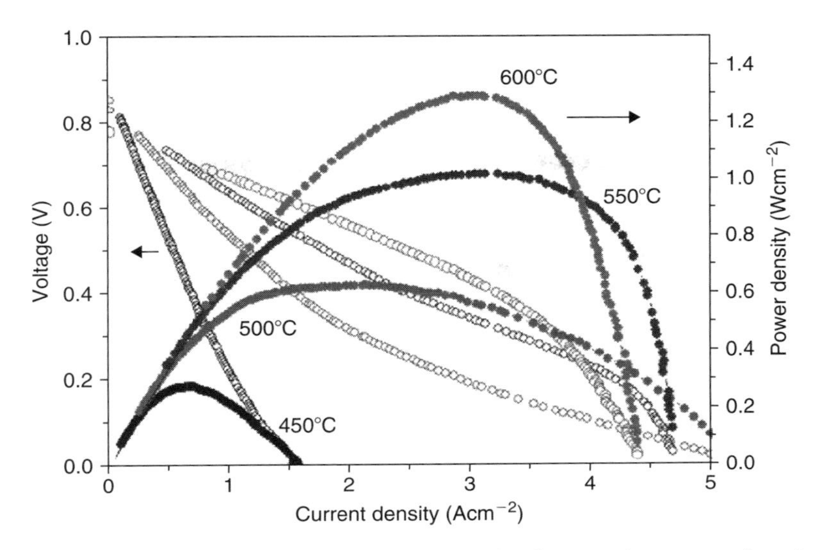

FIGURE 4.18 The performance of the SOFC; cell voltage and power as functions of current and temperature.

take place. Also well-distributed Ni and GDC particles appeared to organize a net-shape structure, which allows sufficient gas diffusion and electrochemical reactions. It is important to note that use of doped ceria to an electrolyte and electrodes was proven to enhance the performance of the SOFC in reduced temperature operation under 600°C [28, 29].

The I-V characteristic and power density of the cell was estimated from the area of the cathode and was shown in Figure 4.18 [30]. The peak power densities of 273, 628, 1017, and 1294 mW cm^{-2} were obtained at 450, 500, 550, and 600°C operating temperature, respectively. The open circuit voltages were dropped from 0.89 to 0.80 V as the temperature increased from 450 to 600°C. Compared to the previous results obtained for the cell with an electrolyte thickness of about 20 μm, the voltage drop was shown to be more serious. Again, lower OCV of the cell results from the use of the ceria-based electrolyte [31, 32], as well as use of a thinner electrolyte. Generally, use of thinner ceria-based electrolyte causes a severe voltage drop due to the reduction of the electrolyte. Several efforts were made to overcome this problem, such as use of an interlayer inside ceria electrolyte to block the leak current [33]. On the other hand, there is a report that the leak current can be canceled out during cell operation [34]. This means that the efficiency drop due to use of the ceria-based electrolyte can be minimized by optimizing the design as well as the operating conditions of an SOFC system.

The results of impedance analysis are shown in Figure 4.19. Ohmic resistances at 450, 500, and 550°C obtained from impedance analysis were 0.29, 0.15 and 0.09 Ωcm^2, respectively. These results are slightly higher compared to the literature data that were obtained in air atmosphere [28]. Thus, two possible explanations can be made: (1) the electrolyte was partially reduced, which could lower OCV on the level of 0.1–0.15 V by increasing electronic conductivity, (2) cell temperature increased during operation. However, the temperature change seemed to be small and did not cause overestimation of the cell performance due to joule and/or reaction heats. Use of a small effective cell area (~0.13 cm^2) seemed to help cell temperature to remain as stable as possible.

The results also showed that the micro tubular cell had relatively small electrode overpotential resistances, which corresponded to the size of the semicircle in Figure 4.19. Thus, the cell performance can be further improved by reducing the thickness of the electrolyte (which may be difficult due to the severe

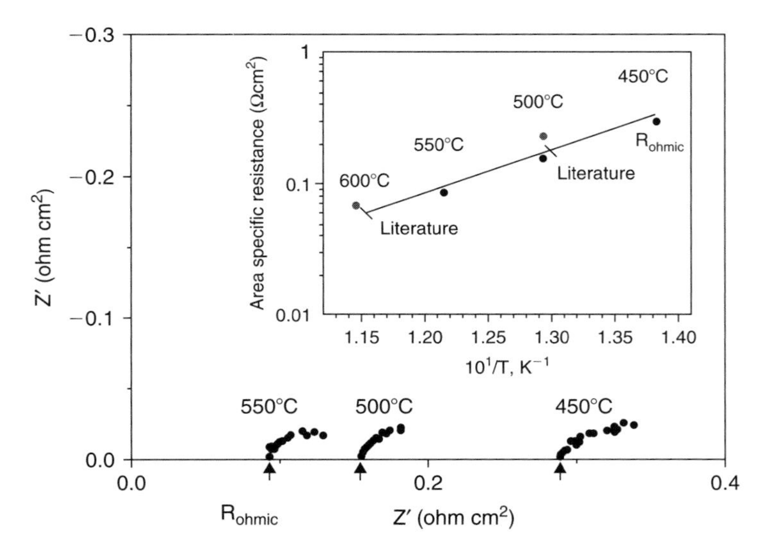

FIGURE 4.19 Impedance spectra of the tubular cell obtained at various temperatures. In box: the area specific resistance (ohmic) as a function of temperature. Literature data were calculated from the conductivity of GDC 20.

voltage drop, however) and/or use of materials with higher ionic conductivity.

The performance of the cell per weight turned out to be 4.55, 10.5, 16.9, and 21.8 W g^{-1} at 450, 500, 550, and 600°C, respectively, and improvement of the anode microstructure appeared to be very effective in improving the cell performance.

4.3.2 Estimation of Current Collecting Loss

Models for Single Micro Tubular SOFC

Because the structure of the micro tubular SOFC is so unique, the anode supported tube must also be used as a current collector from the anode side (the current is collected from the edge of the tube). Thus, the resistance of the anode, typically determined from the length, diameter, wall thickness, and porosity, becomes crucial and may cause serious performance loss. This likely happens most often when the cell resistance becomes significantly low (at higher temperature). Therefore, it is significantly important to understand the impact of the anode resistance on the cell performance and to develop a simulation model which allows further improvement of the cell/bundle/stack.

To estimate the performance loss due to current collection methods from the anode tube, the two following models proposed in this study are shown in Figures 4.20 and 4.21, which were considered from single and double terminal current collecting methods, respectively.

(a) Single terminal (ST) model

Figure 4.20 (a) shows the schematic image of a single terminal (ST) current collecting method. Current collection from the anode side is made from single terminal of the anode tube, while the whole area of the cathode was used for current collection from the cathode side. Then, the anode tube (length, L; diameter, d; tube thickness, t) was divided into N and each of them was assigned to an equivalent circuit (highlighted in Figure 4.20 (b)). R_1, R_2, …, R_N shown in Figure 4.20 were given by the following equations:

$$\begin{aligned} R_1 &= \Delta R_{cell} + \Delta R_{collector} \\ R_2 &= \Delta R_{cell} R_1/(\Delta R_{cell} + R_1) + \Delta R_{collector} \\ &\vdots \\ R_{N-1} &= \Delta R_{cell} R_{N-2}/(\Delta R_{cell} + R_{N-2}) + \Delta R_{collector} \\ R_N &= \Delta R_{cell} R_{N-1}/(\Delta R_{cell} + R_{N-1}) + \Delta R_{collector} \end{aligned} \tag{4.1}$$

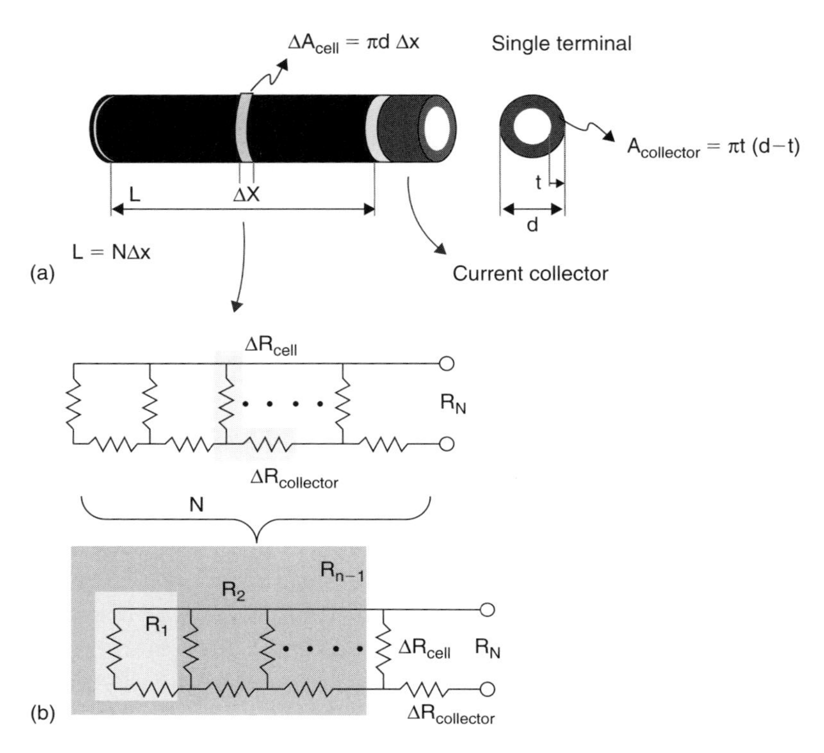

FIGURE 4.20 (a) Schematic image of the single terminal current collecting method; (b) A model equivalent circuit for the single terminal current collection.

where ΔR_{cell} and $R_{collector}$ are given as

$$\Delta R_{cell} = ASR / \Delta A_{cell} \tag{4.2}$$

and

$$\Delta R_{collector} = \Delta x / (A_{collector} \sigma), \tag{4.3}$$

where Δx, ASR, ΔA_{cell}, $A_{collector}$, and σ are the width of the sliced tube cell ($\Delta x = L/N$), the area specific resistance of the cell, the electrode area of the sliced tubular cell ($\Delta A_{\mathrm{cell}} = \pi d\ \Delta x$), the cross-section area of the anode tube ($A_{collector} = \pi t(d - t)$), and the conductivity of the reduced anode tube chosen from the literature [35], respectively. Values of ASR, area specific resistance of the

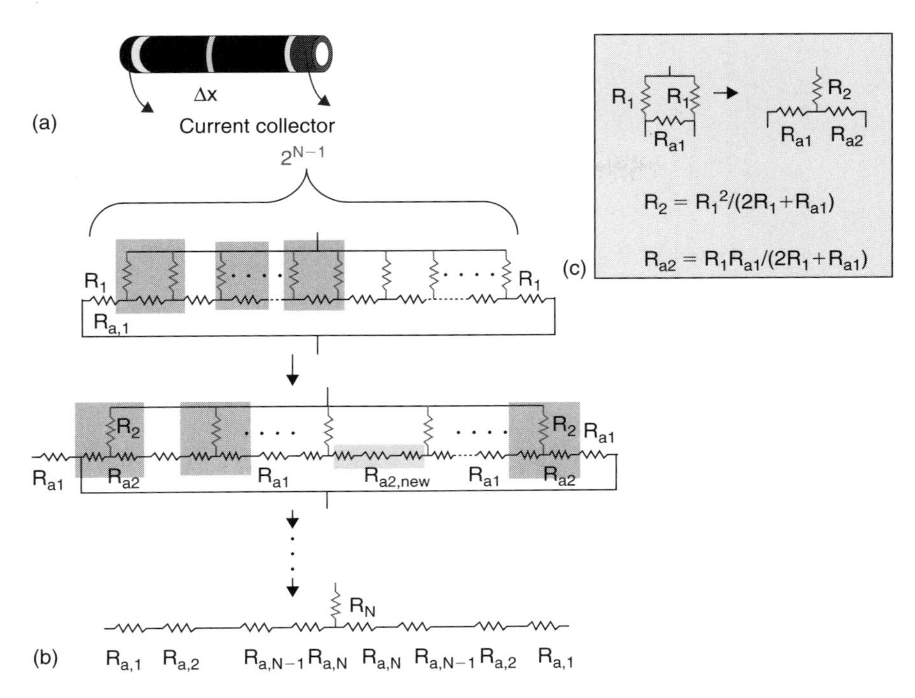

FIGURE 4.21 (a) Schematic image of the double terminal current collecting method; (b) A model equivalent circuit for the double terminal current collection; (c) The formula of Δ-Y conversion.

cell, were chosen from 0.5 Wcm^{-2} @ 0.6 V at 550°C and 0.3 W cm^{-2} @ 0.6 V at 500°C operating conditions, respectively. In this calculation, current collecting resistance of the cathode part was assumed to be negligible. Performance loss was determined using the following relation:

$$\text{performance loss} = (R_N - ASR/(\pi Ld))/R_{N.} \tag{4.4}$$

Here, R_N and $ASR/(\pi Ld)$ are total cell resistance including anode resistance, and true cell resistance, respectively.

(b) Double terminal (DT) model

Figure 4.21 (a) shows the schematic image of the double terminal (DT) current collecting method. Current collection from the anode side is made from both terminals of the anode tube. The anode tube was divided into 2^{N-1} and each of them was assigned to an equivalent circuit (highlighted in Figure 4.21 (b)). R_1, R_{a1}, R_2, R_{a2}, $R_{a2,new}$ … , R_N shown in Figure 4.21 were given by following equations:

$$
\begin{aligned}
R_1 &= ASR/\Delta A_{cell} \\
R_{a1} &= \Delta x/(A_{collector\sigma}) \\
R_2 &= {R_1}^2/(2R_1 + R_{a1}) \\
R_{a2} &= R_1 R_{a1}/(2R_1 + R_{a1}) \\
R_{a2,new} &= 2R_{a2} + R_{a1} \\
&\vdots \\
R_{a,N} &= R_{N-1} R_{a,N-1,new} / (2R_{N-1} + R_{a,N-1,new}) \\
R_N &= {R_{N-1}}^2 / (2R_{N-1} + R_{a,N-1,new}).
\end{aligned} \tag{4.5}
$$

Total cell resistance including anode resistance can be written as:

$$R_T = R_N + \Sigma R_{a,n}/2. \tag{4.6}$$

In this calculation, the formula of Δ-Y conversion (Figure 4.21(c)) was used to simplify the equivalent circuit. Performance loss was determined using the following relation:

$$\text{Performance loss} = (R_T - ASR/(\pi L d) / \mathrm{R}_T, \tag{4.7}$$

where $ASR/(\pi L d)$ is true cell resistance.

Results of Calculation

Figure 4.22 shows the calculated performance loss in Eqs. 4.4 and 4.7 as a function of the number of divisions N for the ST model and 2^{N-1} for the DT model, respectively, using the operating condition of $0.5\,\mathrm{W\,cm^{-2}}$ at 550°C. It was shown that the performance loss saturates when the number of division is over 100 and, thus, the values of $N = 1000$ for the ST model and $2^{N-1} = 1024$ ($N = 11$) for the DT model were selected for further calculation.

Figure 4.23 shows performance loss as a function of anode tube length (L) using the ST and DT models for (a) 1.6 mm and (b) 0.8 mm diam. tubes at 550°C, (c) 1.6 mm and (d) 0.8 mm diam. tubes at 500°C, respectively. A dashed line was drawn at 3% performance loss, which is considered to be the limit for practical use. As can be seen, the performance loss increased as the tube length increased due to increased

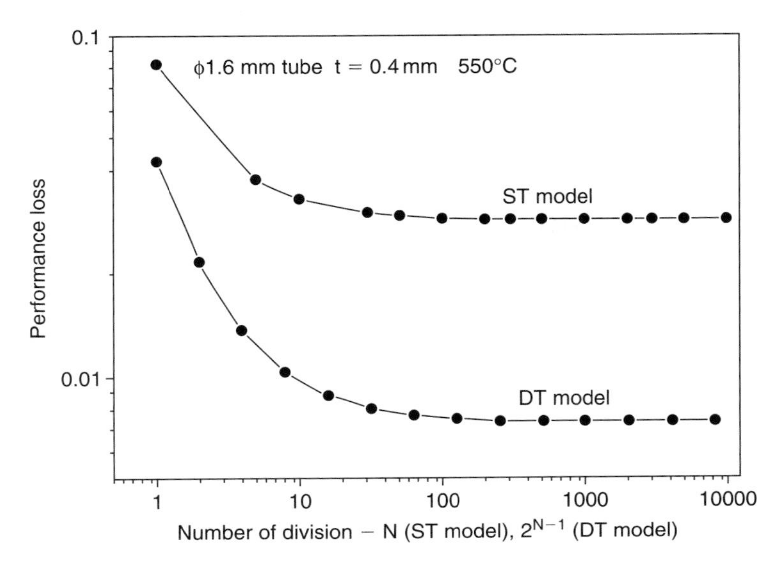

FIGURE 4.22 Relationship between calculated performance loss and the number of divisions, N for the ST model, and 2^{N-1} for the DT model.

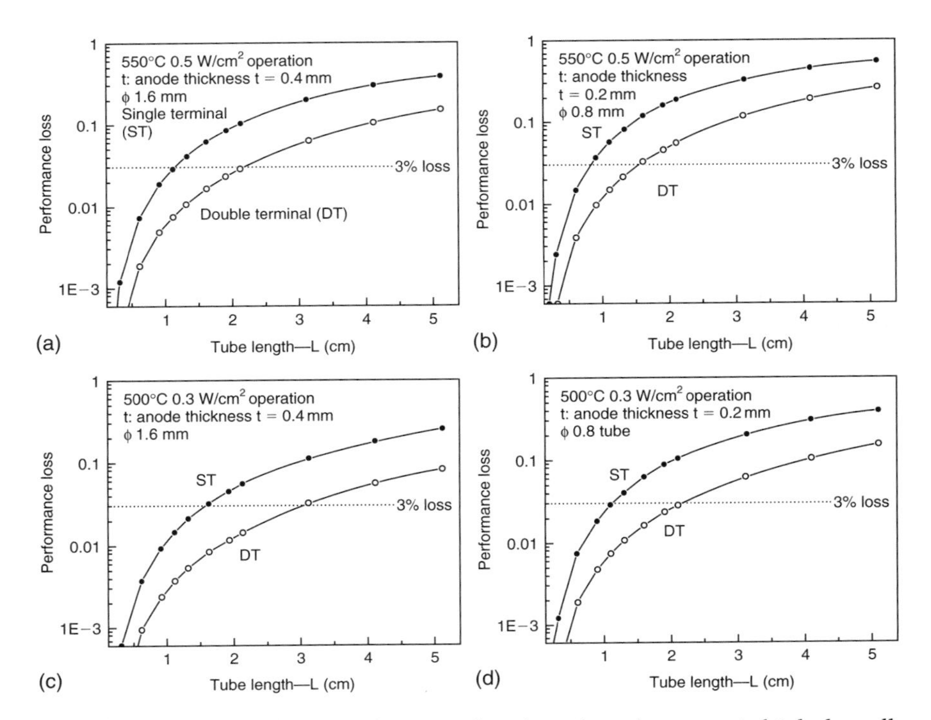

FIGURE 4.23 The performance loss as a function of anode supported tubular cell length (a) 1.6 mm diam. tubular cell, (b) 0.8 mm diam. (550°C), and (c) 1.6 mm diam., (d) 0.8 mm diam. (500°C).

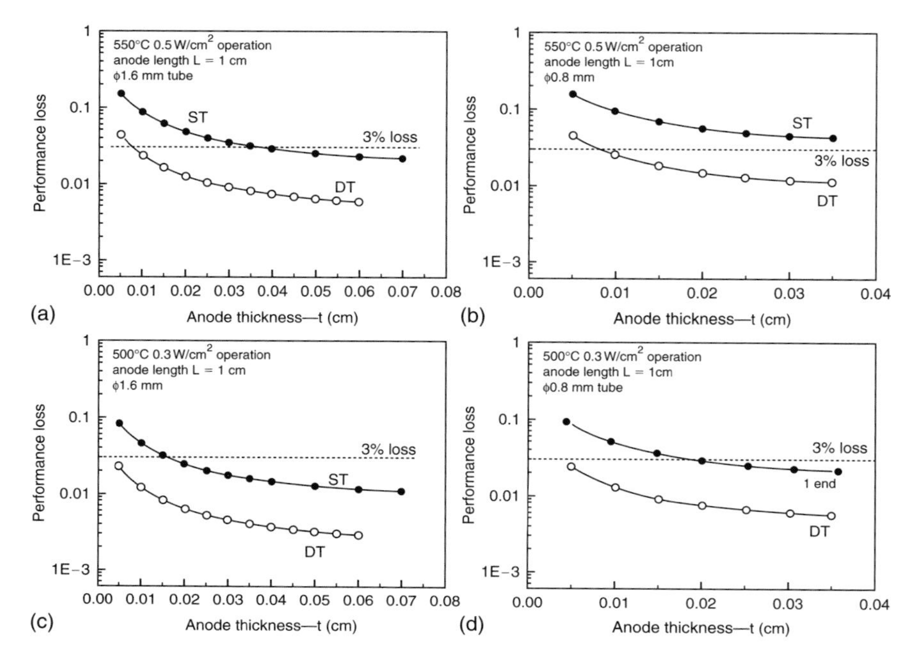

FIGURE 4.24 The performance loss as a function of anode tube thickness (a) 1.6 mm diam. tubular cell, (b) 0.8 mm diam. (550°C), and (c) 1.6 mm diam., (d) 0.8 mm (500°C).

anode resistance as a current collector. In addition, the performance loss increased at higher operating temperature due to two reasons, according to Eqs. 4.4 and 4.7, decreased cell resistance (increased power density) and increased anode tube (metallic) resistance.

Comparison of the two models (ST and DT) clearly shows the advantage of the DT current collecting method. At each anode tube length, the performance loss obtained from the DT model was about 2–4 times lower than that obtained from the ST model. The DT model becomes more effective when the operating temperature and the tube length increase.

Figure 4.24 shows performance loss as a function of anode tube thickness (t) for 1 cm length cells (a) 1.6 mm and (b) 0.8 mm diam. tubes at 550°C, (c) 1.6 mm and (d) 0.8 mm diam. tubes at 500°C, respectively. The thickness of the anode tube does also strongly affect the performance loss, especially for 0.8 mm diam. tubular SOFCs. As shown in Figure 4.24, the DT current collecting method is also effective for reducing the loss caused by changing the thickness of the anode tube. It appears that the performance loss was estimated at over 3% at 550°C operating temperature for actual experimental conditions (1.6 mm diam. cell with $t = 0.4$ mm), while the loss can be negligible under 500°C operating temperature using the DT current collecting method.

From these calculations, it was shown that the following can be effective in decreasing the loss: (i) decreasing operating temperature, (ii) increasing anode thickness, (iii) increasing conductivity of the anode (less porous structure), and (iv) decreasing electrode length. Since (i) and (iii) sacrifice the performance of the cell, (ii) and/or (iv) are considered to be more effective. Thus, it was shown that the selection of the anode tube length and current collecting method is crucial to minimize the performance loss. This simulation can be useful and beneficial for designing the cell stacks and modules.

4.4 BUNDLE AND STACK DESIGN FOR MICRO TUBULAR SOFCs

4.4.1 Fabrication and Characterization of Micro Tubular SOFC Bundle

Micro Tubular SOFC "Cube" Concept

Figure 4.25 shows the concept of micro tubular SOFC bundles and stacks designed for mass production. The micro tubular SOFC bundle consists of micro tubular SOFCs and supports which were prepared from cathode materials. This cathode support, cathode matrix, contains several grooves to hold micro tubular SOFCs and allows simple accumulation of micro tubular SOFCs. Then, the cubes can be easily integrated to be a stack after applying gas seal layers and current collecting layers as shown in Figure 4.25.

The cathode matrix has three roles: (i) collecting current, (ii) supporting micro tubular SOFCs, and (iii) providing gas (air) flow paths. Therefore, optimization of the cathode matrix microstructure to obtain sufficient electrical conductivity and gas permeability is most important in realizing sufficiently high volumetric power density (~2 W cm^{-3}).

Fabrication of the Cathode Matrix

Fabrication procedure of the cathode matrix is shown in Figure 4.26. The cathode matrix was made from LSCF powder (Daiichi Kigenso Kagaku Kogyo Co., Ltd.), poly methyl methacrylate beads

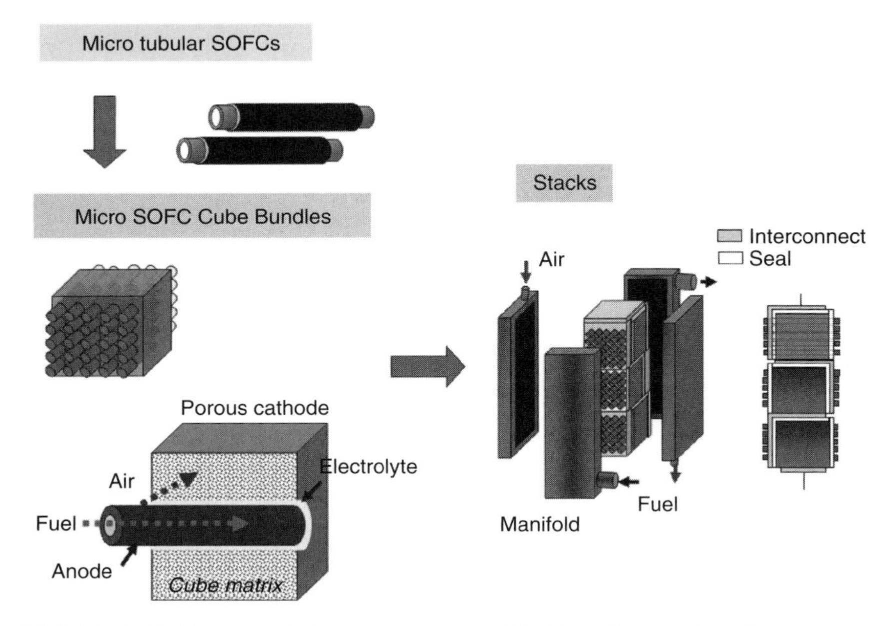

FIGURE 4.25 Concept of the micro tubular SOFC, bundles, and stacks.

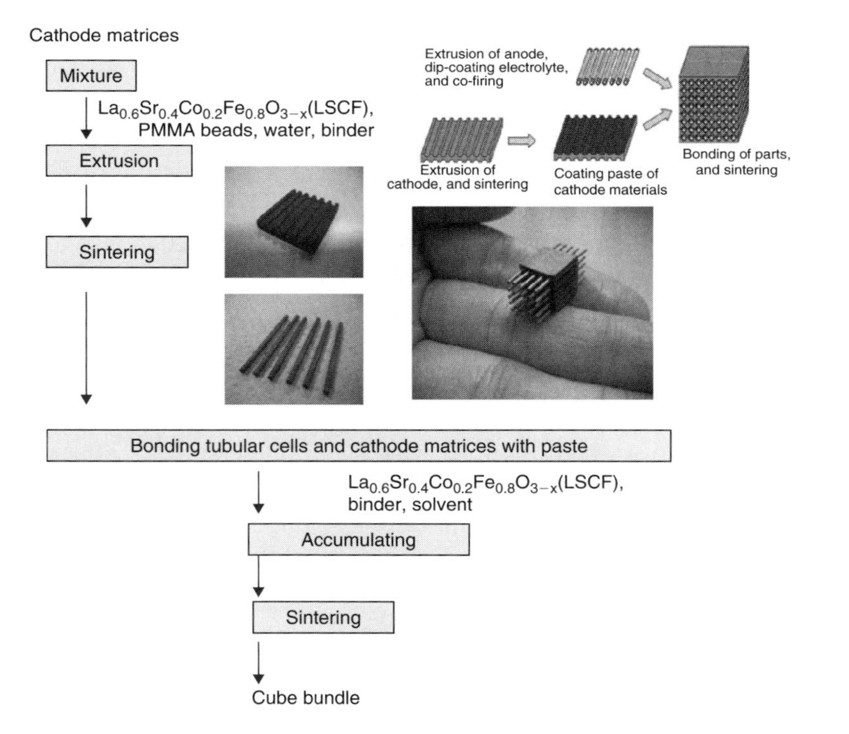

FIGURE 4.26 Fabrication process of the cathode matrices and bundles.

(PMMA) and cellulose (Yuken Kogyo Co., Ltd.). Three different particle sizes of the LSCF powders (0.05, 2, and 20 μm, respectively) were used to investigate the effect on the microstructure as well as the electrical property. These powders were mixed and extruded from a metal mold using a screw cylinder-type extruder (Miyazaki Tekko Co., Ltd.). The microstructure of the cathode matrices was controlled by changing the amount and diameter of the pore former, the grain size of the starting LSCF powder, and the sintering temperatures [25].

For connecting tubular SOFCs and cathode matrices, a bonding paste was used, prepared by mixing an LSCF powder, a binder (cellulose), a dispersant (polymer of an amine system), and a solvent (diethylene glycol monobutyl ether). The paste was screen-printed on the surface of the cathode matrix, followed by the placement of the tubular cells and sintered at 1000°C for 1 h in air.

Gas Permeability and Electrical Conductivity of the Cathode Matrix

The gas penetration test of the cathode matrices was conducted using the experimental setup shown in Figure 4.27. The gas pressure was applied to one side of the sample, and the amount of penetrated gas was measured using a soap film flow meter. The cathode matrices of foursquare plates of 2.5 mm thick without the grooves were used for this measurement. The electrical conductivity of the cathode matrices was investigated by using a DC power supply (ADVANTEST R6234) and a digital multimeter (KEITHLEY 2700) in a DC 4 point probe measurement. The Ag wire was used for four terminals, which were fixed by Ag paste.

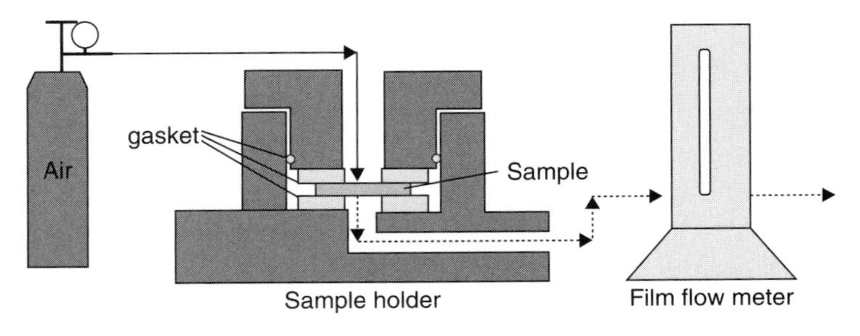

FIGURE 4.27 The experimental apparatus for measuring gas penetration.

TABLE 4.1 Properties of the cathode matrices prepared from different grain size LSCF powders.

	LSCF0.05	LSCF2	LSCF20
Grain size of LSCF (mm)	**0.05**	**2**	**20**
Porosity (%)	60.7	68.6	78.2
Gas permeability (mL cm^{-2} sec^{-1} Pa^{-1})	8.5×10^{-6}	8.4×10^{-5}	6.2×10^{-4}
Conductivity (S/cm)	139	108	49
Amount of PMMA 70 vol.%	Sintering temperature 1400°C		

Table 4.1 summarizes the properties of the cathode matrices prepared from different LSCF powders with grain sizes 0.05, 2, and 20 μm, and each sample was named as LSCF0.05, LSCF2, and LSCF20, respectively. As can be seen, the microstrucure of the cathode matrix can be effectively controlled by changing the grain size of the starting cathode material. The porosity varied from 60–78%, corresponding to the variation of gas permeability from 8.5×10^{-6}–6.2×10^{-4} mL cm cm^{-2} sec^{-1} Pa^{-1}. The electrical conductivity at 500°C was lowered to 49 S cm^{-1} for the sample with the highest porosity (bulk conductivity ~ over 300 S cm^{-1} [36]).

Figure 4.28 shows the relationship between maximum gas flow obtained in the cathode matrices under given pressure differences (0.01–0.03 MPa) and the gas permeability for samples in Table 4.1. As can be seen, under a given pressure difference, only LSCF20 can be used to obtain sufficient gas (air) flow to achieve 2 W cm^{-3} (dashed line in Figure 4.28), assuming that air utilization is 30%. These results suggest that an air compressor may be needed to feed air when LSCF2 and LSCF0.05 are used for 2 W cm^{-3} operation.

Fabrication and Performance of the Cube-type SOFC Bundle

A cube-type SOFC bundle of $1 \times 1 \times 1$ cm was fabricated by arranging nine anode tubes in 3×3 configuration (2.0 mm diameter SOFC) and cathode matrices of LSCF20. Figure 4.29 (a) shows the appearance of the cube-type SOFC bundle using the preparation method as shown in Figure 4.26, which allows simple fabrication of highly accumulated bundles. By applying a current collector, gas seal and gas manifold, the bundle performance was investigated. Figure 4.29 (b) shows the performance of the micro tubular SOFC

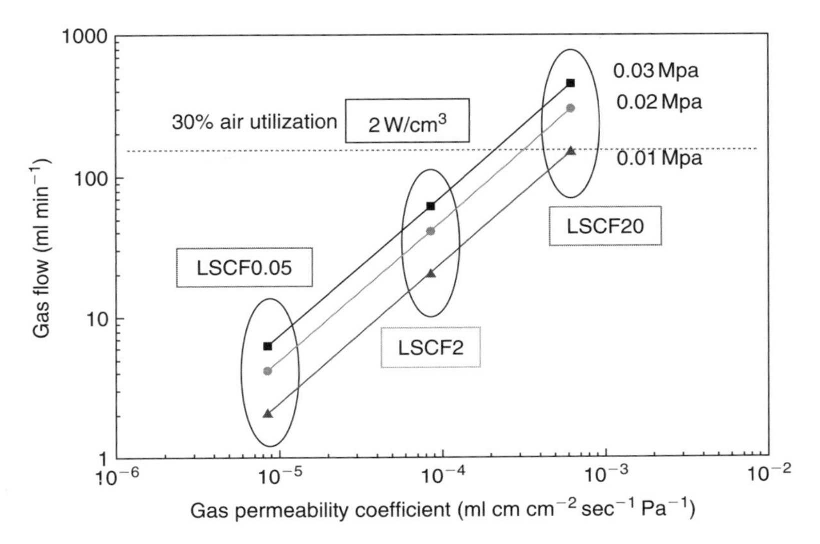

FIGURE 4.28 Possible gas flow rate in the cathode matrices as a function of gas permeability under the given gas pressure difference of 0.01, 0.02, and 0.03 MPa, estimated for LSCF0.05, LSCF2, and LSCF20, respectively. Dashed line shows the gas flow rate needed to achieve 2W/cm^3 performance at 30% air utilization.

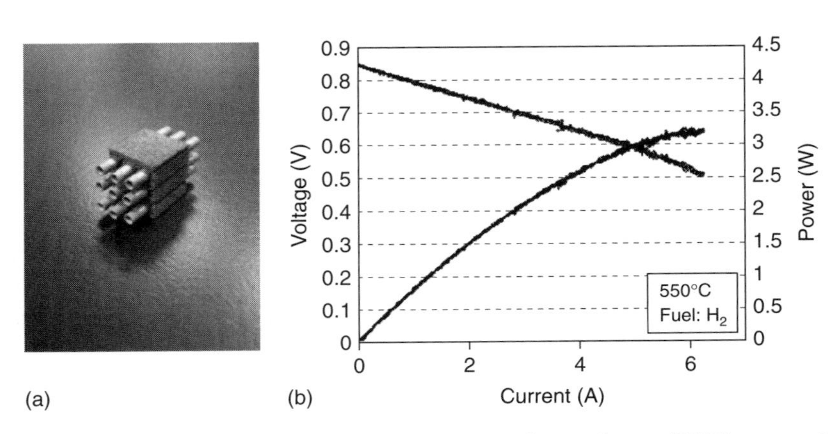

FIGURE 4.29 "Cube" bundle performance obtained at 550°C operating temperature.

bundle obtained at a 550°C operating temperature, obtained 2.1 W @ 0.7 V at 550°C [37]. Lower OCV of 0.85 V could be related to the increase of cell temperature in the cube because of high current flow over 5 A in the 1 cm^3 of the cube bundle. However, over 2 W @ 0.7 V of the bundle performance at 550°C operating temperature

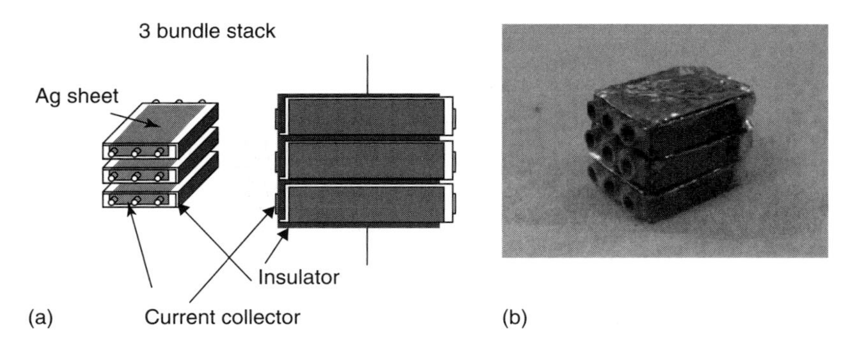

FIGURE 4.30 (a) A schematic image of the tubular SOFC stack (3 bundles in series) and (b) an image of actual tubular SOFC stack with the volume of $\sim 1\,cm^3$.

was achieved. Currently, simulation was conducted for the analysis of temperature distribution in the cube during operation, and optimization of the bundle design is under consideration.

4.4.2 Fabrication and Performance of the Cube-Type SOFC Stack

Since the single cube bundle can generate more than 2 W @ 0.7 V at 550°C, the current collection loss could have a significant impact on the performance of the SOFC bundle, since the electrical conductivity of the cathode matrix is as low as $59\,S\,cm^{-1}$ (Table 4.1) [25]. Thus, alternative bundle/stack design may be beneficial for highly efficient small SOFC systems and a new stack design was proposed to obtain higher OCV (and lower current flow) by utilizing present bundle fabrication technology. A schematic image and a photo of the stack made of three SOFC bundles are shown in Figure 4.30. Each bundle consists of three tubular SOFCs and the cathode matrix. Since each bundle has current collectors for the anode (attached on the top of the bundle) and for the cathode (whole bottom area of the bundle) as shown in Figure 4.30 (a), it allows simple assembly of the bundles in a series electrical connection to make a stack. The volume of the stack turned out to be about $1\,cm^3$.

The stack performance test was conducted using the experimental setup shown in Figure 4.31. Thermocouples were placed at the inlet and outlet of each gas, and at the bottom of the stack. The discharge characterization was investigated by using a Parstat 2273 (Princeton Applied Research) in DC 4 point probe measurement.

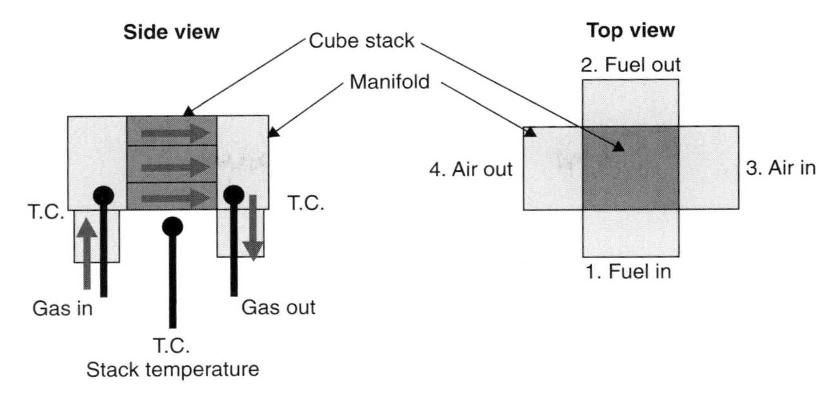

FIGURE 4.31 Experimental setup for stack performance test.

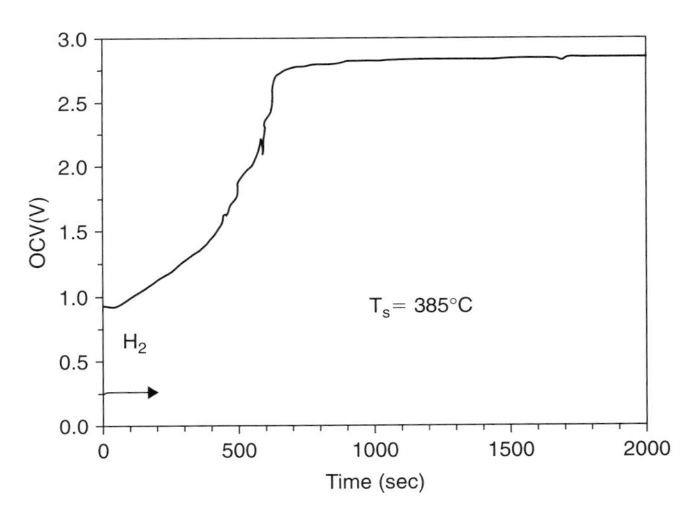

FIGURE 4.32 Initial startup behavior of the tubular SOFC stack (3 bundles in series) at 385°C.

The Ag wire was used for collecting current from the anode and cathode sides, which were both fixed by Ag paste. Hydrogen (humidified by bubbling water at room temperature) was flowed at the rate of 100 mL min^{-1} and the air was flowed at the rate of 1000 mL min^{-1} at the cathode side.

Figure 4.32 shows initial startup behavior of the stack at 385°C stack temperature. As can be seen, even at under 400°C, the stack can be started without reducing anode at higher temperature, and

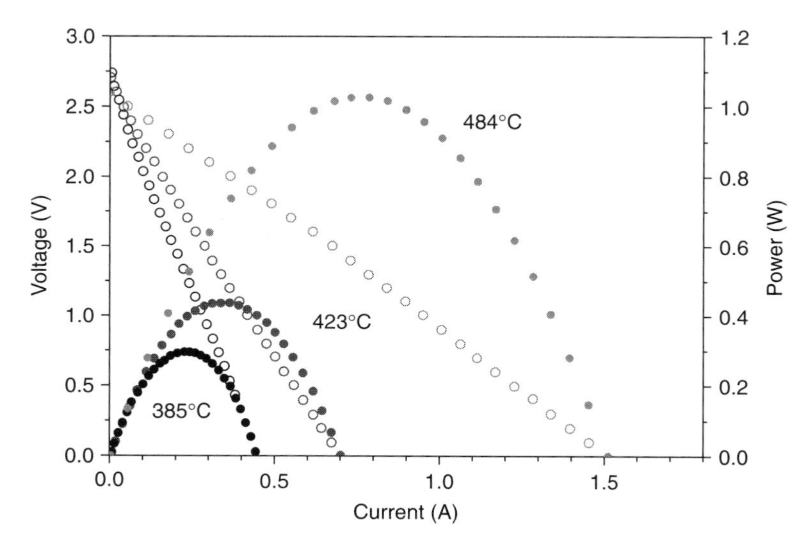

FIGURE 4.33 Performance of the SOFC stack operated from 385 to 484°C.

showed over 2.8 V after about 10 min. Maximum OCV obtained in this stack (three bundles in series connection) was 2.85 V.

Figure 4.33 shows the performace of the stack, with a volume of about 1 cm^3. The maximum output powers of 0.3, 0.44, 1.0 W were obtained for 385, 423, and 484°C stack temperatures, respectively. These results indicated that a portable SOFC system operable under 500°C can be realized using the developed tubular SOFC stacks. Total electrode area of the tubular SOFCs is 5.65 cm^2, and thus, the power density of 0.18 W cm^{-2} was obtained at 484°C operating temperature. In fact, this was lower than that of the single cell, so further improvement of the stack performance can be expected by optimizing gas flow rates, interfacial resistance between the tubular SOFC and the cathode matrix, as well as improving the cathode matrix and sealing technology.

Currently, integration technology of the tubular SOFC bundles is being examined to obtain higher output voltage. Figure 4.34 shows the design of a 72 V output stack module with the volume (exclude manifold) of 30 cm^3 stack volume, which is expected to produce 30 W @ 48 V maximum output under 500°C operating temperature. This bundle design allows stacks to perform any output power and voltage, and therefore, use of the SOFC bundles for stack fabrication could be useful, especially for portable SOFC systems.

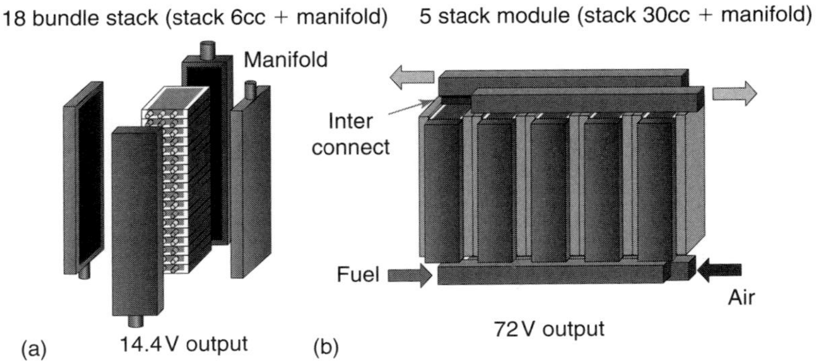

18 bundle stack (stack 6cc + manifold) 5 stack module (stack 30cc + manifold)
Manifold
Inter connect
Fuel
Air
(a) 14.4 V output
(b) 72 V output

FIGURE 4.34 Schematic images of stacks and modules using the tubular SOFC bundles.

Finally, we would like to comment on potential problems in the development of the micro tubular SOFC systems. For practical realization of the micro tubular SOFC system, further investigations of the tubular cells and the bundles are necessary to gain the knowledge necessary for understanding the following issues:

(a) Stack design (manifold design)
(b) Thermal distribution in the stack and thermal management (simulation)
(c) Gas pressure loss in the stack/system and optimization of gas flow
(d) Fuel utilization
(e) Current collecting loss from the cathode matrix
(f) Sealing technology

These design challenges are the focus of ongoing detailed investigations.

4.5 CONCLUDING REMARKS

In this chapter, the recent development of fabrication/integration technology for micro tubular SOFCs was shown; the tubular cells with diameters of 0.8–1.8 mm were successfully prepared by a newly developed preparation method. This technique gives the freedom to design the anode microstructure (supported tube), which was shown to be important for the improvement of cell performance.

New bundle design for accumulating micro tubular SOFCs was also proposed and fabricated using a cathode material. The proposed structures (cathode matrices) acted as supports of the tubular cells, current collectors, and air gas paths, which needed well-controlled microstructure. The bundle performance was shown to be over $2\,W\,cm^{-3}$ @ 0.7 V at 550°C operating temperature.

In addition, a new stack design using the tubular SOFC bundle was proposed and fabricated. The performance was shown to be ~2.8 V (OCV) and 1 W @ 1.6 V under 500°C operating temperature with the stack volume of about $1\,cm^3$.

A number of potential problems regarding the fabrication of micro tubular SOFC systems were addressed. Further accumulation of the tubular SOFC bundles needs to be considered to obtain stacks with higher output voltage and power at various operating temperatures. Development of sealing technology and design techniques including thermal management, optimization of gas manifold, and so on is also currently underway.

The newly developed micro SOFC bundle/stack could be a promising candidate for advanced SOFC applications.

ACKNOWLEDGMENTS

The authors thank Ms. M. Kobayashi for her aid in preparing samples and Drs. T. Otake and K. Nagai from Toho Gas Co., Ltd, Nagoya, Japan, for conducting cube bundle performance tests. This work has been supported by NEDO, as part of the Advanced Ceramic Reactor Project.

References

[1] O. Yamamoto, Electrochim. Acta 45 (2000) 2423.
[2] N.Q. Minh, J. Am. Cer. Soc. 76 (1993) 563–798.
[3] S.C. Singhal, Solid State Ionics 152–153 (2002) 405.
[4] S. de Souza, S.J. Visco, L.C. DeJohnge, J. Electrochem. Soc. 144 (1997) L35.
[5] A.V. Virkar, J. Chen, C.W. Tanner, J.W. Kim, Solid State Ionics 131 (2000) 189.
[6] B.C.H. Steele, Mat. Sci. Eng. B 13 (1992) 79.
[7] T. Ishihara, H. Matsuda, Y. Takita, Solid State Ionics 79 (1995) 147.
[8] H.G. Bohn, T. Schober, J. Am. Cer. Soc. 83 (2000) 768.
[9] Z. Shao, S.M. Haile, Nature 431 (2004) 170–173.
[10] B.C.H. Steele, Solid State Ionics 129 (2000) 95.

[11] K. Kuroda, I. Hashimoto, K. Adachi, J. Akikusa, Y. Tamou, N. Komado, T. Ishihara, Y. Takita, Solid State Ionics 132 (2000) 199.
[12] S. Yoon, J. Han, S.W. Nam, T.H. Lim, I.H. Oh, S.A. Hong, Y.S. Yoo, H.C. Lim, J. Power Sources 106 (2002) 160.
[13] B.C.H. Steele, A. Heinzel, Nature 414 (2001) 345–352.
[14] S. Simner, J.F. Bonnett, N.F. Canfield, K.D. Meinhardt, V.L. Sprenkle, J.W. Stevenson, Electrochem. Solid-State Lett. 5 (2002) A173.
[15] H. Huang, M. Nakamura, C. Su, R. Fasching, Y. Saito, F.B. Prinz, J. Electrochem. Soc. 154 (1) (2007) B20–B24.
[16] K. Eguchi, T. Setoguchi, T. Inoue, H. Arai, Solid State Ionics 52 (1992) 165.
[17] T. Hibino, A. Hashimoto, K. Asano, M. Yano, M. Suzuki, M. Sano, Electrochem. Solid-State Lett. 5 (2002) A242.
[18] J. Yan, H. Matsumoto, M. Enoki, T. Ishihara, Electrochem. Solid-State Lett. 8 (8) (2005) A389–A391.
[19] K. Kendall, M. Palin, J. Power Sources 71 (1998) 268–270.
[20] K. Yashiro, N. Yamada, T. Kawada, J. Hong, A. Kaimai, Y. Nigara, J. Mizusaki, Electrochemistry 70 (12) (2002) 958–960.
[21] N.M. Sammes, Y. Du, Int. J. Appl. Ceram. Technol. 4 (2007) 89–102.
[22] N.M. Sammes, Y. Du, R. Bove, J. Power Sources 145 (2005) 428–434.
[23] P. Sarkar, L. Yamarte, H. Rho, L. Johanson, Int. J. Appl. Ceram. Technol. 4 (2007) 103–108.
[24] P. Bance, N.P. Brandon, B. Girvan, P. Holbeche, S. O'Dea, B.C.H. Steele, J. Power Sources 131 (2004) 86–90.
[25] Y. Funahashi, T. Shimamori, T. Suzuki, Y. Fujishiro, M. Awano, ECS Trans. 7 (1) (2007) 643–649.
[26] T. Yamaguchi, T. Suzuki, S. Shimizu, Y. Fujishiro, M. Awano, J. Membr. Sci. 300 (2007) 45–50.
[27] T. Suzuki, T. Yamaguchi, Y. Fujishiro, M. Awano, J. Electrochem. Soc. 153 (2006) A925–A928.
[28] B.C.H. Steele, Solid State Ionics 129 (2004) 95–110.
[29] H. Yokokawa, T. Horita, N. Sakai, K. Yamaji, M.E. Brito, Y.-P. Xiong, H. Kishimoto, Solid State Ionics 174 (2004) 205–221.
[30] T. Suzuki, Y. Funahashi, T. Yamaguchi, Y. Fujishiro, M. Awano, Electrochem. Solid-State Lett. 10 (2007) A177.
[31] E.D. Wachsman, Solid State Ionics 152–153 (2002) 657.
[32] Y. Xiong, K. Yamaji, N. Sakai, H. Negishi, T. Horita, H. Yokokawa, J. Electrochem. Soc. 148 (2001) E489.
[33] A. Tomita, S. Teranishi, M. Nagao, T. Hibino, M. Sano, J. Electrochem. Soc. 153 (6) (2006) A956–A960.
[34] R.T. Leah, N.P. Brandon, P. Aguiar, J. Power Sources 145 (2005) 336–352.
[35] T. Suzuki, T. Yamaguchi, Y. Fujishiro, M. Awano, J. Power Sources 160 (2006) 73–77.
[36] L.W. Tai, M.M. Nasrallah, H.U. Anderson, D.M. Sparlin, S.R. Sehlin, Solid State Ionics 76 (1995) 273.
[37] Y. Funahashi, T. Suzuki, Y. Fujishiro, K. Nagai, T. Otake, T. Shimamori, M. Awano, Proc. 2007 Fuel Cell Seminar & Exposition (Eds. D. Rastler and M. Hicks), San Antonio, TX, 2007, 223–226.

CHAPTER

5

Enzymatic Biofuel Cells

Michael C. Beilke, Tamara L. Klotzbach, Becky L. Treu, Daria Sokic-Lazic, Janice Wildrick, Elisabeth R. Amend, Lindsay M. Gebhart, Robert L. Arechederra, Marguerite N. Germain, Michael J. Moehlenbrock, Sudhanshu and Shelley D. Minteer

Department of Chemistry
St. Louis University

This chapter details the development of enzymatic biofuel cells over the last four decades. The chapter compares enzymatic biofuel cells to traditional fuel cells and discusses the types of enzymes employed at the anode and cathode of biofuel cells, along with strategies for immobilization of those enzymes at electrode surfaces. A detailed comparison of mediated electron transfer and direct electron transfer, along with discussion of the advantages and disadvantages of both types of electron transfer mechanisms are discussed. Finally, this chapter describes the importance of metabolic pathways in enzymatic biofuel cell development and fuel cell design and engineering issues for enzymatic biofuel cells.

5.1 INTRODUCTION AND BACKGROUND

Global energy demands continue to increase every year. Although fossil fuels meet present energy demands, pollution and global warming concerns are causing an increase in a search for alternative energy sources [1]. As energy demands continue to increase, research is being directed toward alternative energy sources that will be more sustainable, better for the environment than fossil fuels, and able to meet current and future power demands [1–3].

Currently, electrochemical devices, which convert chemical or light energy directly into electrical energy, are being heavily researched [4]. There are three types of these devices: batteries, fuel cells, and solar cells [4]. Batteries are electrochemical devices that contain fuel inside of the compartment and allow a reaction to take place at the anode and the cathode [3]. A fuel cell is analogous to a battery except the device contains the fuel outside of the reaction compartment. Batteries are recharged electrically, whereas fuel cells can be recharged by adding additional fuel [3]. A solar cell uses light which is converted into electrical energy [4].

There are two different types of batteries: primary and secondary. The primary battery cannot be recharged due to irreversible chemical reactions which produce electricity [5]. Alkaline batteries, like silver oxide/zinc, carbon/zinc, manganese oxide/zinc, and mercury oxide/zinc, are all examples of primary batteries [5]. The secondary battery is reusable and rechargeable by application of an external power source after it is discharged [5]. Recharge occurs by reversible chemical reactions. Nickel-cadmium, lithium-ion, lead-acid, and nickel-metal-hydride batteries are all examples of secondary batteries [5]. Since batteries contain their reactants inside their reaction compartments, they will eventually expire when they have reacted away all of the reactants. Secondary batteries are plagued by hysteresis, which limits their ability to be recharged to their original state [5]. Hysteresis is a cumulative process that degrades battery performance over time. The deficiency of primary and secondary batteries can be overcome by using fuel

TABLE 5.1 Comparison of the six types of traditional fuel cells.

Fuel Cell Type	Mobile Ion	Operating Temperature
Alkaline (AFC)	OH^-	50–200°C
Proton exchange membrane (PEMFC)	H^+	30–100°C
Direct methanol (DMFC)	H^+	20–90°C
Phosphoric acid (PAFC)	H^+	~220°C
Molten carbonate (MCFC)	CO_3^{2-}	~650°C
Solid oxide (SOFC)	O^{2-}	500–1000°C

cells, which continue to produce power as long as fuel is being added to them.

William Grove demonstrated the first hydrogen fuel cell in 1839 [5]. Fuel cells were first developed for use in space vessels, as their energy demand could not be delivered by traditional batteries. Electrical power was needed in order to run equipment collecting the scientific data in space and transmitting the data back to earth. The demand of the spacecraft and equipment was at least 200–300 hours of constant power delivery, but batteries lasted, on average, up to 30 hours [5]. Fuel cells attracted the attention of researchers worldwide due to their potentially high efficiency. Fuel cells find their use in very diverse areas, but they can be categorized by their power output capacity, electrolytes utilized, and lifetime. The following chart, shown as Table 5.1, compares the six types of traditional fuel cells [5].

Both batteries and fuel cells contain two electrodes that are in contact with an electrolyte solution [3]. In a conventional fuel cell, fuel is oxidized at the anode to produce electrons, which flow through an external circuit where a load is applied. The electrons are then transferred to the cathode where they are used, along with free protons, to reduce oxygen to water.

Traditional fuel cells employ expensive, nonrenewable metal catalysts which oxidize many fuels including hydrogen gas, methane, and alcohols [5]. These metal catalysts are nonselective [5, 6].

Generally, they suffer from poor efficiency due to the passivation or poisoning of the catalysts from crossover of fuel through the polymer electrolyte membrane (PEM), fuel impurities, and difficulty in miniaturizing and stacking the technology.

There are a few different types of traditional fuel cells. The alkaline fuel cell contains concentrated KOH or NaOH as the electrolyte and is used for low-temperature applications. This type of fuel cell must stay hydrated with water. If the water evaporates faster than it is produced at the cathode, the PEM dehydrates and is unable to ion exchange. Phosphoric acid fuel cells use concentrated phosphoric acid as their electrolyte, which causes the water vapor pressure to decrease, reducing the problems associated with water maintenance inside of the cell. Phosphoric acid fuel cells operate efficiently only at high temperatures. Molten carbonate fuel cells contain an electrolyte that is a combination of alkali carbonates such as Na, K, and Li. These are retained in a ceramic matrix of $LiAlO_2$. Molten carbonate fuel cells operate at very high temperatures. Another type of fuel cell is the solid oxide fuel cell. This type of fuel cell contains an electrolyte that is a solid, nonporous metal oxide. Solid oxide fuel cells can operate at even higher temperatures than the molten carbonate fuel cell [7]. These different classes of fuel cells have different applications due to their various properties.

Like traditional PEM fuel cells, biofuel cells consist of an anode and cathode separated by a membrane [8]. The difference is that biofuel cells eliminate the dependence on precious metal catalysts by replacing them with biological catalysts. Biofuel cells can be categorized as microbial fuel cells and/or enzymatic biofuel cells. Microbial fuel cells utilize living microorganisms to oxidize the fuel, whereas enzymatic fuel cells employ isolated enzymes.

The first microbial fuel cell was demonstrated in 1912 by Potter who used yeast fed with glucose to produce electrical energy [8, 9]. Microbial fuel cells have certain advantages over enzymatic fuel cells; by utilizing the whole microorganism, this eliminates the need to isolate enzymes [10]. This allows the microorganism to work efficiently since the enzymes are able to live in their natural environment [10]. Also, the use of microorganisms allows biofuels to be more thoroughly oxidized [1]. This is attributable to the microorganism which contains all the enzymes necessary to partially or fully oxidize biofuels. This property increases the energy efficiency and specific energy density of the fuel cell as well as ensuring there are no toxic byproducts for disposal. Research has shown that microbial fuel cells have lifetimes up to five years [11]. Unfortunately, microbial fuel cells suffer from low volumetric catalytic activity of the whole organism and low power densities due to slow mass transport of the fuel across the cell wall [12, 13].

Fifty-two years after the first microbial fuel cell was demonstrated, the enzyme-based biofuel cell was established [14]. A schematic of a biofuel cell is shown in Figure 5.1. The development of the enzymatic biofuel cell eliminated the fuel transport problems that caused low power densities in microbial fuel cells. Using isolated enzymes also allows for greater specificity of biofuels. Despite the benefits of the enzymatic biofuel cell, there are drawbacks. These include short lifetimes and incomplete oxidation of the biofuel [11]. Advances in research are leading to solutions of these issues.

Isolated enzymes are attractive catalysts for fuel cells due to their high catalytic activity and selectivity. One significant problem associated with the biofuel cells, however, is that although the electron supply is readily produced by the enzymes, it cannot be exploited unless the electrons can be transferred to the electrode. To eliminate this issue, the use of mediators has been employed. Mediators, typically organometallic complexes or organic dyes, are

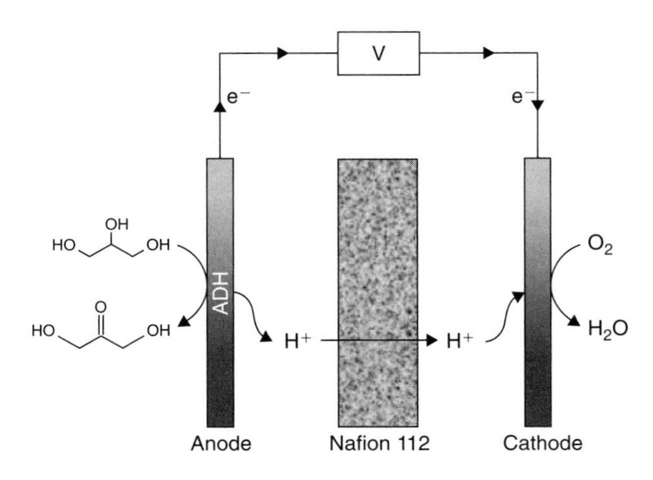

FIGURE 5.1 Schematic of a PEM biofuel cell [151].

responsible for transferring the electrons from oxidized fuel to the surface of the electrode [11]. Many traditional mediators are dyes. Dyes are large complex aromatic molecules that have the ability to undergo oxidation and reduction in much the same way metal bipyridines do. Some popular dye mediators that are currently being used are methylene green, neutral red, and methylene blue.

Mediators are utilized along with cofactors. Cofactors are molecules that transfer charge between the enzyme and the mediator or current collector, thereby allowing the charge to be used to create a flow of current. A mediator is a molecule that can easily and reversibly switch between oxidation states, so that cofactor can be regenerated. The most common cofactors found in biofuel cells are nicotinamide adenine dinucleotide (NAD^+), nicotinamide adenine dinucleotide phosphate ($NADP^+$), pyrroloquinoline quinine (PQQ), and flavin adenine dinucleotide (FAD). These molecules are reduced when the substrate molecule is oxidized by the enzyme.

The small molecule mediators must either be dissolved into the fuel solution or co-immobilized into the enzyme immobilization membrane. If co-cast into the membrane, they tend to leach out, so extra mediator must be added to the fuel solution. However, it is undesirable to have substances other than oxidized fuel and water in solution due to the expense and the environmental impact of treating the waste.

Enzymes are extremely complex biomolecules. Enzymes are proteins and their three-dimensional structure is vital to their ability to catalyze a reaction. The three-dimensional structure typically contains an active site for the substrate (fuel) to dock [152]. If the active site changes geometry (even slightly), the fuel will neither dock nor react as shown in Figure 5.2. The complex structure of the enzyme is fragile, so that if a sample of pure enzyme

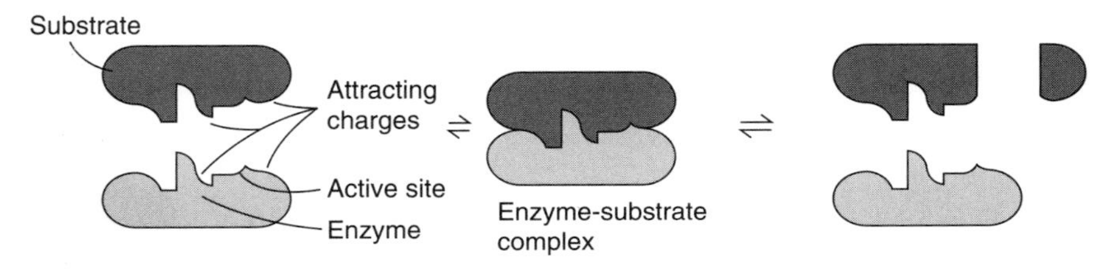

FIGURE 5.2 Lock and key model [152].

is allowed to reach room temperature or higher, it will begin to denature. Denaturing/unfolding destroys the three-dimensional structure of the enzyme, so it no longer has an intact active site for catalysis. Denaturing can occur with high temperatures, extreme pHs, or by interactions with some solvents.

When enzymes were first applied to an anode in their free form, they only lasted a few hours before denaturing [15]. Previous biofuel cells used enzymes contained in solution rather than enzymes immobilized in polymers on the electrode surface [15]. This technique only allowed the enzymes to be stable for a couple of days. Recent research has focused on finding novel immobilization techniques which keep the enzymes intact and prolong their lifetime. There are three commonly used techniques for immobilizing enzymes: "wired," sandwich, and entrapment.

Along with enzyme immobilization techniques, cell design and catalyst support are widely studied topics. Good fuel cell design and engineering is essential to achieve optimal performance with any PEM-based fuel cell. Many biofuel cells are also PEM-based and contain gas diffusion cathodes allowing them to benefit from the performance gains achieved from cell design and engineering optimization. High surface area catalyst supports, such as carbon black and carbon nanotubes, have traditionally been used in hydrogen and direct methanol fuel cells (DMFCs). They have also led to large performance increases in biofuel cells [16]. Both design and catalyst support optimizations lead to enhanced power outputs that approach other fuel cell technologies such as direct methanol and direct alcohol fuel cells that contain precious metal catalysts. As energy demands increase, researchers have been trying to develop technologies that can efficiently meet these demands, while making minimal impact on the environment and keeping the cost of these technologies low.

5.2 SIMILARITIES AND DIFFERENCES TO TRADITIONAL FUEL CELL CATALYSTS

One of the fundamental aspects of fuel cell technology is the use of various catalysts. Catalysts serve to effectively speed up a chemical process yet will not undergo any permanent chemical change while doing so. They are not consumed in a reaction; however, the surface of some solid catalysts may physically change by becoming roughened or brittle. This indicates that the catalyst participates in the reaction process but is being regenerated [17, 18].

A catalyst's participation in a reaction is solely to increase the rate at which it progresses, not how much product is generated. They accomplish this by working in one of two different reaction schemes: homogenous or heterogeneous. Homogenous catalysis exists within a homogenous mixture, or a mixture that contains the catalyst, reactants, and products all throughout the solution. All aspects of the reaction exist in the same physical state or phase with one another [17, 18].

In contrast, heterogeneous catalysis is unique in that the catalyst is present in a different phase of matter from the reactants and products. This type of catalysis is used in many important industrial processes and usually involves the activity of gases or liquids on the surface of a solid catalyst. The heterogeneous catalyst's purpose is to attach, or adsorb, the reactants from a gaseous or liquid phase on its surface. The advantage of heterogeneous catalysis is the catalysts can be used repeatedly. Once the reactants are adsorbed, they move along the surface via diffusion and dock on an active site. The product is generated upon the catalyst, undergoes desorption, and is released back into solution. In heterogeneous catalysis, catalytic activity is associated with many transition metal elements and their compounds. The precise mechanism of heterogeneous catalysis is not totally understood, but it has been proposed that the availability of electrons in the *d* orbitals in the surface atoms may play a role [17].

In conventional fuel cells, heterogeneous catalysis is most common using various precious metals as the catalyst. These metals act as the solid catalyst required for heterogeneous catalysis, thus remaining in a separate physical state from the rest of the solution. These types of catalysts are used for fuel cells, employing the following fuels: methanol, ethanol, hydrogen, and other alcohols [19].

5.2.1 Metals as Catalysts

Metals are chosen to function as catalysts in fuel cells due to their multifaceted molecular orientations. These facets serve as active sites for rapid and efficient oxidation and reduction reactions. Each metal has its own unique structure and chemical properties. This gives metals some aspect of selectivity for certain fuels. For example, the facets of gold crystals in particular have been found to be specific for oxidation of intermediate alcohol oxidation products such as carbon monoxide to carbon dioxide and water. On the other hand, gold is not considered useful in the electro-oxidation of lower alcohols such as methanol or ethanol. It can only play a part in methanol oxidation if it is blended homogenously with a group VIII noble metal, or heterogeneously blended for ethanol oxidation [19]. Another reason metals are used in catalysis is due to their easy adaptation for specialization by the infusion of various precious metals with other more affordable metals as alloys. As alloyed metals are put into the cell, they tend to work in conjunction with each other to form more specialized catalysts. The faceted structure is changed when creating an alloy, thereby giving the catalyst an advantage over a standard metal alone, as well as generating a catalyst that has a specifically refined active site surface for reactions. Precious metals in particular are the most common types of metals used in fuel cells due to their lower oxidation rates relative to the more inexpensive metals such as copper [19, 20].

5.2.2 Metals Used in Hydrogen and Direct Methanol Fuel Cells

Most hydrogen and DMFCs employ the precious metal platinum as a catalyst. In a DMFC, carbon monoxide is a problematic passivation impurity, so for the methanol cell, platinum is the catalyst of choice due to its limited passivation by hydrogen or oxy-hydroxy adsorbates on the surface. Different metals have been sought for DMFCs, yet research shows that platinum, or platinum-based catalysts, are the only effective catalysts for the electro-oxidation of methanol [20].

Another type of fuel cell technology that scientists are developing for future use in automotive applications is the hydrogen fuel cell. Hydrogen fuel cells use PEMs, with a platinum catalyst at both the anode and the cathode. However, the use of pure platinum poses a problem for the efficacy of its operation. Although platinum is effective as a catalyst, its affinity for poisoning reduces efficiency and creates a more expensive product turn-around for the industry. Additionally, a platinum oxide layer forms on the electrode surface, dramatically slowing the reaction rate, causing eventual electrode failure [21, 22]. The electrode's durability is compromised and the lifetime of the fuel cell is weakened.

5.2.3 Alloys

The transfiguration of precious metals with other metals as alloys can customize the electrode interface for synergistic results and improved reactions. These catalysts have the potential to lessen the effects of poisoning through oxidation and material loss. Also, the introduction of other transition metals as a blend reduces the cost of electrodes. For example, recent advances in the hydrogen fuel cell include a platinum-nickel (Pt-Ni) or platinum-cobalt (Pt-Co) alloy as an improved catalyst [22]. This technique works by creating a platinum "skin" atop either a Pt-Ni or Pt-Co alloy. The alloy coated with the platinum "skin" has been shown to improve cathode performance. This alloy has decreased catalyst poisoning by activating the encased Ni molecules which inhibit the formation of an oxide layer [22]. Researchers identified the Pt-Ni alloy configuration $Pt_3Ni(111)$ as displaying the highest oxygen reduction reaction activity detected on a cathode catalyst, which is ten times better than a single crystal surface of pure Pt(111), and 90 times better than the Pt-C catalysts which are currently used [22]. Platinum alone or nickel alone is not durable enough to handle oxidation, yet when combined, they create an advantage over other potential catalysts. By using an alloy, it is possible to develop smaller electrodes, thereby reducing cost and increasing efficiency. Other alloy combinations with platinum such as titanium, vanadium, chromium, manganese, molybdenum, and rhenium have shown DMFC enhancement, with ruthenium exhibiting optimal performance enhancement [20].

Although precious metals are more resistant to oxidation than other transition elements such as copper or tin, they still exhibit weaknesses that affect the fuel cell's performance. In a DMFC, a platinum-tin alloy solves passivation issues when blended. The responsibility of tin in this relationship is to catalyze the removal of the surface poison generated during methanol electro-oxidation. This type of alloy has proven useful in ethanol fuel cells as well [23].

In certain ethanol fuel cells, precious metals are blended with carbon dioxide absorbing ceramics to reduce passivation. This blend is also quite heat tolerant, which gives hope for future automotive developments [22–24]. Developments for general fuel cells include the use of gold nanoparticles on the surface of pure platinum [21]. Gold is also an inert element, not participating in the chemical reaction, thereby boosting a catalyst's durability. Using a small dusting of gold nanoparticles upon a platinum catalyst's surface reduces the risk of platinum oxidation by creating a protective coating on it. This increases the cell's lifetime and overall activity [21]. Metal blends have shown in lab tests to be superior to precious metals alone, therefore alloys are pertinent to fuel cell durability and lifetime [21–23].

5.2.4 Disadvantages of Metal Catalysts

Although precious metals have unprecedented activity when used as catalysts in fuel cells, their use in modern times has become somewhat antiquated in comparison to newer, more versatile techniques. These new techniques shed light on the inadequacies of metal catalysts, and improve upon the problems associated with them.

One of the predominant disadvantages of metal catalysts is the high price of large scale production. Presently, the price of platinum is upwards of about $36,000 per kilogram. Similarly, the price of gold, copper, and other catalytic metals fluctuate with market demand. When metals are employed as catalysts and undergo oxidation loss, they need to be replaced. Since availability is limited, the continual replacement of precious metal catalysts poses the problem of high replacement costs and rapid resource scarcity [21].

Another focal disadvantage that occurs with the use of metal catalysts is the easy passivation that occurs in the presence of impurities. The surface of the solid catalyst is susceptible to adsorption by the impurity rather than the needed reactant, thereby eventually rendering the catalyst inactive. This vulnerability leads to catalyst poisoning and ultimately a replacement is necessary [17, 20].

Metals cannot adapt to fuel cell problems that arise without physical interception via transfiguration. Their use is very specific for particular fuels as well, creating a need for multiple types of metals that need to be examined for their maximum power potential. Modern research involves the use of more selective biological materials to replace the use of the more expensive, problematic precious metals. Proteins such as enzymes have active, biological selectivity that a material such as a transition metal cannot compete with.

5.3 ENZYMATIC BIOELECTROCATALYSIS

In general, catalysts stabilize the transition state relative to the ground state, and this decrease in energy is responsible for the rate acceleration of the reaction. Platinum-based catalysts are utilized in current fuel cell technology to facilitate oxidation of fuels (e.g., hydrogen gas, methane, and methanol) in the electrochemical cell. Biofuel cell technology employs nontraditional fuels (ethanol, carbohydrates, and fatty acids) along with nonconventional biocatalysts (enzymes).

Living organisms are able to undergo metabolism due to the enzymes contained within the cells. Enzymes are composed of numerous complex proteins that are produced by living cells and catalyze specific biochemical reactions while not being consumed themselves [25]. The majority of enzymes have optimal reaction kinetics around 37°C. Since enzymes are extremely efficient catalysts, do not experience surface passivation effects in the presence of impure fuels, and do not experience the diffusional limitations associated with transport of fuel across cell membranes in whole cells, recent research has focused on the possibility of building fuel cells with enzymes as catalysts (biofuel cells) [26].

5.3.1 Enzymes as Biocatalysts

Enzymes display a number of remarkable properties when compared with other types of catalysts such as high catalytic power, specificity, and the extent to which their catalytic activity can be regulated. Enzymes decrease the activation energy necessary for reactions to take place and can be extremely selective when used as biocatalysts. The interaction with the substrate can be highly specific to that substrate alone, thus avoiding interferences with other substances [27]. Enzymes can have molecular weights of several thousand to several million, yet catalyze transformations of molecules as small as carbon dioxide or nitrogen. The catalytic effects of enzymes can increase the rate of a reaction by as much as 10^{14} the normal rate [28]. Most enzymes are highly specific both in the nature of the substrate(s) they utilize and the reaction they catalyze.

5.3.2 Enzyme Isolation

Enzymes along with other biocatalysts (organelles, proteins, DNA, etc.) are derived from natural sources (animal and plant tissues and microorganisms). Natural sources are renewable and more cost effective when compared to traditional heavy metal catalysts used in fuel cells. Generally, natural sources produce low quantities of enzymes due to the fact that their systems are not optimized for overproduction/elevated levels of enzymes. Often the gene for the desired enzyme can be transferred and overexpressed in wild type or recombinant microorganisms by the manipulation of the transcriptional, translational, and post-translational processes that influence the yield of active enzyme. This can be achieved with and without genetic engineering methods [29].

Enzymes of interest must be isolated to the desired purity for their application as a biocatalyst. Purification of extracellular enzymes is much simpler when compared to intracellular (membrane bound) enzymes, as extracellular enzymes can be isolated directly from the growth media without disrupting the cell. Membrane bound enzymes require cell lysis and subsequent sample preparation techniques before the enzyme of interest can be further purified [29].

All purification schemes include some form of chromatography following sample preparation and extraction. Most purification schemes require multiple steps in order to achieve the purity required of the sample. For any purification scheme, in order to maximize enzyme yield, it is desirable to develop a scheme that has minimal steps and the simplest design due to product loss with each step. As enzymes are comprised of proteins, the physical and chemical properties of proteins (and the amino acids they are comprised of) can be exploited in developing purification schemes. Protein properties and the corresponding type of chromatography used during purification include charge-ion exchange chromatography, size-gel filtration, hydrophobicity-hydrophobic interaction/reverse phase chromatography, and biorecognition (ligand specificity)-affinity chromatography [29]. Availability of natural sources from which enzymes can be derived coupled with the development of purification schemes has allowed for a wide variety of available biocatalysts for employment in enzymatic biofuel cells.

5.3.3 Enzyme Classification

Enzymes are principally classified and named according to the reaction they catalyze. The system used by the Enzyme Commission (EC) to classify known enzymes consists of four numbers (E.C. 1.1.1.1). The first number indicates one of the six possible reaction types that the enzyme can catalyze: (1) oxidoreductases (oxidation/reduction), (2) transferases (group transfer reactions), (3) hydrolases (hydrolysis), (4) lyases (non hydrolytic bond breaking reactions), (5) isomerases (isomerization reactions), and (6) ligases (bond formation reactions). The second number defines the chemical structures that are changed in the process, such as a -CH-OH- bond (1.1), a -C = O- bond (1.2), or at a -C = C- bond (1.3) for oxidoreductase enzymes. In transferases, C1 groups (2.1), aldehyde or keto groups (2.2), acyl groups (2.3), or glycosyl groups (2.4) can be reacted upon. Hydrolysis reactions catalyzed by hydrolases take place on ester bonds (3.1), glycoside bonds (3.2), ether bonds (3.3), peptide bonds (3.4), or amide bonds (3.5). Lyases undergo chemical reactions at -C-C- (4.1) bonds and -C-O- bonds (4.2). Isomerases catalyze the transformation of racemizations (5.1), cis-trans isomerizations (5.2), and intramolecular oxidoreductases (5.3) are catalyzed by isomerases. Ligases will form -C-O- bonds (6.1), -C-S- bonds (6.2), -C-N- bonds (6.3), and -C-C- bonds (6.4). The third number defines the properties of the enzyme involved in the catalytic reaction or further characteristics of the catalyzed reaction and the fourth number is a running number assigned to each enzyme characterized by the previous three digits [30]. The EC classification system aids in the determination of the correct enzyme for specific applications.

The most commonly utilized enzymes in biofuel cell technology are oxidoreductase enzymes. Oxidoreductase enzymes, namely dehydrogenases and oxidases, are of interest to biofuel cell research due to the wide variety of possible substrates/enzymatic systems that can be incorporated. The range of specificity varies between enzymes making them extremely versatile for many applications. In addition, the catalytic activity of enzymes may be activated or inhibited by small ions or other molecules, temperature, and pH [28].

5.3.4 Enzyme Function

During catalysis, an enzyme substrate complex forms by binding the substrate to a small cavity in the enzyme known as the active site. Only a dozen or so amino acid residues may make up the active site. Of these amino acid residues, only two or three may be involved directly in substrate binding and/or catalysis, but in some enzymes cofactors are also present [28]. Most oxidoreductase enzymes contain a prosthetic group which often includes one or more metal atoms and requires an additional organic molecule or transition metal complex known as a coenzyme or cofactor at the active site to assist in electron transfer. Coenzymes are essential for the catalytic action of enzymes dependent upon them by binding either covalently or ionically to the enzyme's active site [28].

Enzyme catalysis is characterized by two features: specificity and rate enhancement. Since catalysis takes place in the active site, there are many hypotheses regarding the function of the remainder of the enzyme [28]. One suggestion is that the most effective binding of the substrate to the enzyme results from close packing of the atoms within the protein, and the remainder of the enzyme outside the active site is required to maintain the integrity of the active site for effective catalysis [31]. The protein may also serve the function of channeling substrate into the active site. Storm and Koshland suggested that the active site optimally aligns with the orbitals of substrates and catalytic groups on the enzyme for conversion to the transition state structure [32]. Jenks proposed that the fundamental feature distinguishing enzymes from simple chemical catalysts is the ability of enzymes to utilize binding interactions away from the site of catalysis [33]. These binding interactions facilitate reactions by positioning substrates with respect to one another and with respect to the catalytic group's active site.

5.3.5 Enzyme Kinetics

The stability of the enzyme substrate (ES) complex is related to the affinity of the substrate for the enzyme which is measured by its K_s (dissociation constant for the ES complex). E represents the enzyme of interest, S represents the substrate and P represents the products [27].

$$E + S \underset{k_{-1}}{\overset{k_1}{\longleftrightarrow}} ES \xrightarrow{k_2} E + P \tag{5.1}$$

$$K_s = K_m = \frac{k_{-1} + k_2}{k_1} \tag{5.2}$$

The overall rate of formation of products during enzymatic catalysis is given by the Michaelis-Menton Equation [27].

$$[E_o] = [E] + [ES] \tag{5.3}$$

$$v = \frac{k_2[E_o][S]}{[S] + K_m} \tag{5.4}$$

When $k_2 >> k_{-1}$, we refer to the term k_2 as the k_{cat} (catalytic rate constant) and the dissociation constant (K_s) is called the K_m (Michaelis-Menton Constant).

$$V_{max} = k_2[E_0] \tag{5.5}$$

$$v = \frac{V_{max}[S]}{K_m + [S]} \tag{5.6}$$

The maximum rate, or V_{max}, would be achieved when all of the enzyme molecules have substrate bound.

The k_{cat} or turnover number represents the maximum number of substrate molecules (in μ moles) converted to product molecules per active site per unit of time (minutes). Typical values for

TABLE 5.2 Brief summary of oxidoreductase enzymes utilized in enzymatic biofuel cells.

Substrate	Enzyme	Coenzyme Required	Product
Anode			
Methanol	Alcohol dehydrogenase	NAD^+	Formaldehyde
Formaldehyde	Formaldehyde dehydrogenase	NAD^+	Formate
Formate	Formate dehydrogenase	NAD^+	CO_2
Ethanol	Alcohol dehydrogenase	NAD^+	Acetaldehyde
	Alcohol dehydrogenase	PQQ (bound)	Acetaldehyde
Acetaldehyde	Aldehyde dehydrogenase	NAD^+	Acetate
	Aldehyde dehydrogenase	PQQ (bound)	Acetate
Glucose	Glucose oxidase	FAD (bound)	Gluconolactone + peroxide
	Glucose dehydrogenase	NAD^+/PQQ	Glucose-6-phosphate
	Glucose dehydrogenase	PQQ (bound)	Glucose-6-phosphate
Fructose	Fructose dehydrogenase	NAD^+	5-dehydro-D-fructose
Lactate	Lactate dehydrogenase	NAD^+	Pyruvate
Lineolic Acid	Lipoxygenase	Bound	(9Z,11E)-(13S)-13-hydroperoxyoctadeca-9,11-dienoate
Cathode			
Oxygen	Bilirubin oxidase	Copper (bound)	H_2O
	Laccase	Copper (bound)	H_2O
Peroxide	Horseradish peroxidase	Heme (bound)	H_2O
	Microperoxidase-11	Heme (bound)	H_2O
	Cyclooxygenase	Heme (bound)	H_2O

k_{cat} are on the order of 10^3 s^{-1} or about 1000 molecules of substrate converted to product per second per enzyme. Oxidoreductase enzymes (namely dehydrogenases and oxidases), can experience turnover numbers on the order of 10^3 s^{-1} [28].

5.3.6 Biofuel Cell Catalysts

A biofuel cell harnesses the energy from enzymatic activity on the substrate by manipulating the coenzyme reaction that occurs simultaneously. Interest in utilizing glucose found *in vivo* as fuel spurred the growth of the field of enzyme-based biofuel cells. The enzymes most commonly employed at the anode of a glucose biofuel cell are FAD-dependent oxidases [34–38]. Along with FAD-dependent enzymatic systems, $NAD(P)^+$-dependent enzymes have been widely studied for use in glucose as well as methanol, ethanol and lactate biofuel cells [13, 39–42]. Currently, there is growing interest in oxidoreductase enzymatic systems which do not require additional coenzymes to carry out catalysis for use at the anode and cathode of a biofuel cell. These enzymes contain multiple metal centers capable of transferring electrons directly from the redox reaction to the electrode surface [43–45]. Table 5.2 is a brief summary of commonly used oxidoreductase systems currently being employed in enzymatic biofuel cell technology [4].

Chemical energy can be harnessed into electrical energy through the use of enzymes as catalysts. The super-selectivity of enzymes allow for enhanced specificity when compared to traditional heavy metal catalysts. The kinetics of the enzymatic system chosen plays a large role in the catalysis of substrates and performance of a biofuel cell. A wide variety of enzymes and coenzymes expand the potential for nontraditional fuels to be employed in biofuel cells.

5.4 MEDIATED ELECTRON TRANSFER VS. DIRECT ELECTRON TRANSFER

5.4.1 Mediated Electron Transfer

There are two ways of coupling an electrode process to an enzyme reaction. They include using low molecular weight redox mediators (MET) and using direct electron transfer (DET) as

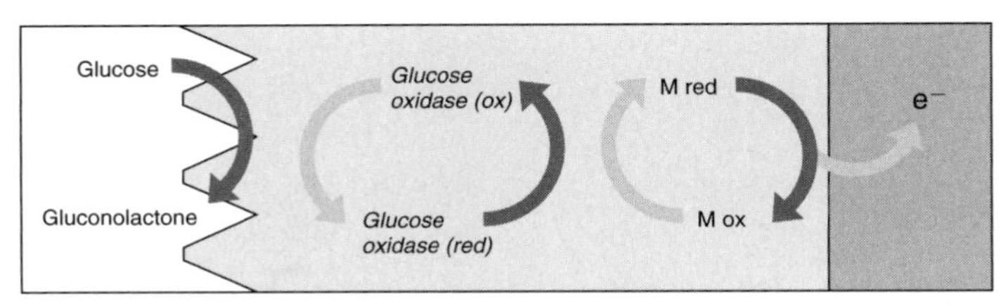

MET of glucose oxidase

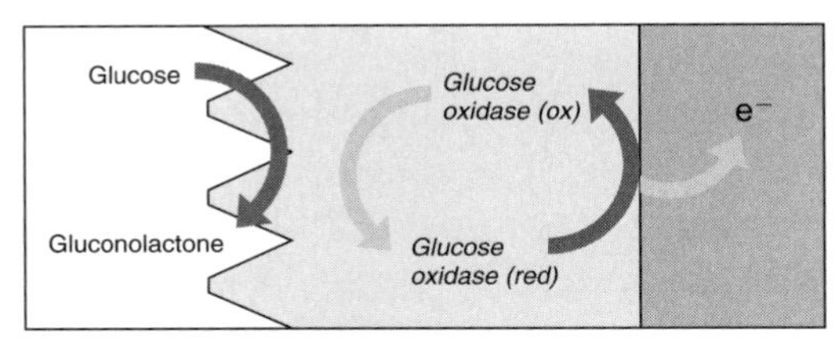

DET of glucose oxidase

FIGURE 5.3 MET vs. DET at a glucose oxidase bioelectrode.

shown in Figure 5.3. During biocatalysis at an electrode surface, most oxidoreductase enzymes require additional redox compounds or electron transfer mediators, which is referred to as mediated electron transfer (MET). These electron transfer mediators are employed to facilitate transfer of electrons from the enzyme to the electrode surface. Mediators are low molecular weight artificial electron transferring compounds that can readily participate in the redox reaction with the biological component and help with rapid electron transfer [46]. During the catalytic reaction, the mediator first reacts with the reduced enzyme or coenzyme and then diffuses to the electrode surface to undergo rapid electron transfer. Various organic and inorganic compounds along with some redox proteins have been used as mediators. For biofuel cell applications, it is advantageous to incorporate the mediator within the enzyme immobilization membrane which contains the enzyme.

An ideal mediator should be able to react rapidly with the reduced enzyme or coenzyme. It should exhibit reversible heterogeneous kinetics. The overpotential for the regeneration of the oxidized mediator should be low and pH independent. The mediator should have stable oxidized and reduced forms, and the reduced form should not react with oxygen. Some commonly used mediators for biofuel cell applications are organic dyes which include: methylene blue, phenazines, methyl violet, alizarin yellow, prussian blue, thionine, azure A, azure C, and toluidine blue [47–52]. Organic dyes have been shown to be problematic due to poor stability and pH dependence of their redox potential. Inorganic redox ions such as ferricyanide have also been used but have proven to be problematic due to poor solubility, stability, and difficulty in tuning their redox potentials [46].

5.4.2 Direct Electron Transfer

DET has been one of the most frequently studied aspects of biofuel cells in the last few years. During the 1980s, the fundamentals of bioelectrocatalysts capable of DET were beginning to be explored. During this time, a number of enzymes were found to be capable of interacting directly with an electrode while catalyzing the oxidation of a substrate. DET generates catalytic current and catalytic reduction of the reaction overvoltage without the use of additional mediators [44]. Electroanalytical applications of these unique enzymes began to appear in the late 1980s [53–57]. During DET, the redox enzyme (oxidoreductase) acts as an electrocatalyst facilitating the electron transfer between the electrode and substrate molecule with no mediator involved in the process. The ability of a number of oxidoreductase enzymes to catalyze an electron transfer from the electrode surface to the substrate molecule (or vice versa) has been demonstrated [25, 53, 58–63].

5.4.3 MET at Bioanodes

Glucose Oxidase

The different locations of enzyme active centers affect electron transfer within the enzyme structure as shown in Figure 5.4. FAD-dependent enzymes have strongly bound redox centers surrounded by a glycoprotein shell which inhibits DET. Direct electron transfer of electrons to and from the active center is hindered by kinetics, requiring the use of a redox mediator capable of having contact with the enzyme in order to transport charge [4]. The most widely studied type of biofuel cells employ the FAD-bound

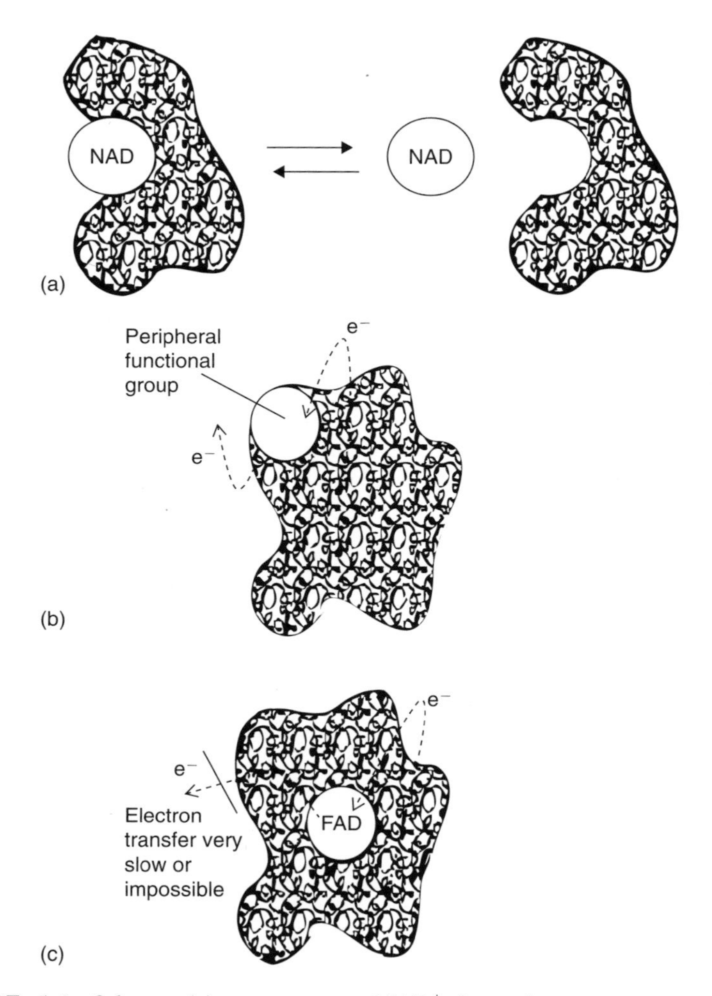

FIGURE 5.4 Scheme (a) represents a NAD^+-dependent enzyme capable of transferring electrons to and from the enzymes active site. Scheme (b) represents a DET capable enzyme containing a conducting cofactor. Scheme (c) represents a FAD-bound enzyme incapable of DET [4].

enzyme, glucose oxidase (GOD) as a biocatalyst at the anode. Many different MET GOD-based biofuel cells have been reported in literature. Willner et al. and Katz et al. incorporated the coenzyme PQQ into a monolayer along with GOD at the anode of a glucose biofuel cell with a PQQ-FAD tether acting as the mediator between the enzyme and the electrode surface [34–36, 64, 65]. Pizzariello et al. employed a GOD and ferrocene in a solid binding matrix at the anode of a MET glucose biofuel cell with ferrocene acting as an inorganic mediator [37].

Many authors describe the use of osmium-based mediator complexes incorporated within redox polymers at the anode of MET glucose biofuel cells. Tsujimura et al. employed the complex [Os(4,4′-dimethyl-2,2′-bypyridine)$_2$Cl]$^{+/2+}$ incorporated within a poly(1-vinylimidazole) layer containing GOD to assist in the electron transfer at a MET glucose biofuel cell [38]. Chen et al. described a MET GOD-based biofuel cell which employs GOD crosslinked with poly(*N*-vinyl imidazole[Os(4,4′-dimethyl-2,2′-bypyridine)$_2$Cl]$^{+/2+}$-co-acrylimide) where the osmium complex is acting as a membrane bound mediator [66]. Mano et al. also employed an osmium-based mediator, [Os(*N*,*N*′-dialkylated-2,2′-bis-imidazole)$_3$]$^{2+/3+}$ incorporated within a redox polymer, poly(vinyl pyridine), acting as the mediator between GOD and the electrode surface in a MET glucose biofuel cell [66–69]. Kim et al. incorporated GOD in a redox polymer poly(1-vinylimidazole), containing the complex [Os(4,4′-diamino-2,2′-bypyridine)$_2$Cl]$^{+/2+}$, which is employed as the mediator at the anode in a MET glucose biofuel cell [70]. Mano and Heller report a MET glucose biofuel cell employing GOD within a poly(vinyl pyridine) redox polymer containing the osmium-based mediator complex [Os(4,4′-dimethoxy-2,2′-bypyridine)$_2$Cl]$^{+/2+}$ in order to transfer electrons from the active site of GOD to the electrode surface [71]. In addition, Heller also employed GOD incorporated within a redox polymer poly(vinyl pyridine) with [Os(*N*,*N*′-dimethyl-2,2′-biimidazole)$_3$]$^{2+/3+}$ acting as the redox mediator in a MET glucose biofuel cell [72].

NAD$^+$-Dependent Enzymes

Select enzymes have active centers containing NAD$^+$ or NADP$^+$ which diffuse out of the enzyme and travel back to the electrode acting as the mediator. Additional electrocatalysts are necessary for NADH oxidation due to the high overpotential observed at electrode surfaces. A large number of dehydrogenase enzymes utilize NAD$^+$ as a cofactor. Therefore, its electrochemical oxidation at low potentials is of interest. NADH is generated by the reduction of NAD$^+$; however, NADH oxidation on platinum and carbon electrodes has poor reaction kinetics and occurs at a large overpotential [73].

The term overpotential is defined as the deviation of the electrode/cell potential from its equilibrium value, $\eta = E - E_{eq}$; it can be either positive or negative [74]. The various types of overpotential include: activation, concentration, reaction, transfer, and resistance. Activation overpotential is voltage lost due to the limited speed of charge transport. Concentration overpotential is voltage differences which are caused by diffusion, such as the current flow using up necessary substrates faster than more substrate can diffuse to the electrode. Reaction overpotential is due to a related chemical reaction which causes a change to the chemical environment. Transfer overpotential is a function of kinetic electron transfer whereby electrolytes are transported to the electrode surface, and resistance overpotential is not chemically related, but is simply the voltage drop across resistive cell components [7]. The overpotential of NADH is attributed to all of these factors.

Palmore et al. described a MET methanol biofuel cell employing NAD^+-dependent alcohol (ADH), aldehyde (AldDH), and formate (FDH) dehydrogenases in which benzoviologen acts in conjunction with diaphorase in solution as the mediator reducing NAD^+ to NADH [40]. Takeuchi and Amao describe a photo-operated MET glucose biofuel cell which employs zinc chlorine-e_6 as mediator for NAD^+-dependent glucose dehydrogenase (GDH) [75].

Cyclic voltammetric studies of poly(methylene green)-coated glassy carbon electrodes have shown that poly(methylene green) can act as an electrocatalyst for NADH [76]. Akers et al. have employed this polymer-based electrocatalyst to regenerate NAD^+ and to shuttle electrons from the reduction of NADH to the electrode at a MET alcohol biofuel cell employing ADH, AldDH, FDH, and formaldehyde dehydrogenase [13]. Ansari et al. have also described the use of poly(methylene green) as an electrocatalyst layer for use in a MET sucrose biofuel cell employing NAD^+-GDH and fructose dehydrogenase [77].

5.4.4 MET at Biocathodes

The two primary molecules of interest for reduction at a biocathode include oxygen (air breathing system) and peroxide (a byproduct of glucose oxidation via GOD). Palmore et al., Akers et al., Ansari et al., and Takeuchi and Amao utilize enzymatic bioanodes coupled to platinum cathodes to form a complete biofuel cell [13, 40, 75, 77]. Platinum will catalyze the reduction of oxygen to water, but requires the anodic compartment to be separated from the cathodic compartment by a PEM. This is due to the fact platinum is not only selective towards oxygen, but it can be passivated by crossover of fuel from the anodic compartment. The PEM adds to the resistance of the biofuel cell along with an increase in cost, so it is of interest to fabricate an enzymatic cathode to couple with an enzymatic anode. Biofuel cells that operate with both enzymatic anodes and cathodes can function without the PEM due to the super selectivity of the enzymes used as catalysts.

Peroxide Reduction

In addition to MET being employed at bioanodes, literature shows the use of mediators in enzymatic systems employed at biocathodes. Pizzariello et al. described a MET glucose biofuel cell in which horseradish peroxidase (HPR) was used as the catalyst in order to reduce peroxide to water with ferrocene acting as the mediator at the electrode surface [37]. Katz et al. utilized cyclooxygenase (COx) as a catalyst with cytochrome c (cyt c) acting as the mediator in a maleimide monolayer containing GOD crosslinked with glutaraldehyde for stability in a MET glucose biofuel cell in order to reduce peroxide to water [36].

Oxygen Reduction

Bilirubin Oxidase Osmium-based mediators are reported for use at cathodes which employ bilirubin oxidase (BOD) and laccase, both of which reduce oxygen to water. The majority of literature for MET shows use of the osmium-based mediator $[Os(4,4'\text{-dichloro-}2,2'\text{bipyridine})_2Cl_2]^{+/2+}$ incorporated within a redox polymer which facilitates the transfer of electrons in solution to the cathode surface. BOD was also employed as a biocatalyst as described by Mano et al., Kim et al., and Mano and Heller et al. utilizing the mediator complex $[Os(4,4'\text{-dichloro-}2,2'\text{bipyridine})_2Cl_2]^{+/2+}$ along with a redox copolymer of polyacrylamide and poly(N-vinyl-imidazole) to reduce oxygen to water at the cathode [67–69]. Tsujimura et al. developed a cathode utilizing BOD complexed with $[Os(2,2'\text{-bipyridine})_2Cl]$ and attached with bromoethylamine to a poly(4-vinylpyridine) redox polymer to mediate the transfer of electrons from solution to the cathode electrode surface [38].

While most BOD-based biocathodes discussed in literature utilize osmium-based mediators, there are several disadvantages associated with these complexes. Osmium is highly toxic in its free form, and it is extremely pH dependent, making it a poor choice for possible *in vivo* applications. In 2003, Topcagic et al. reported the use of $Ru(bpy)_3^{2+/3+}$ as a mediator for a BOD-based biocathode. BOD is immobilized within a modified Nafion® membrane containing the coenzyme bilirubin along with the mediator $Ru_2(bpy)_3^{2+/3+}$ [6]. Ruthenium-based complexes are less toxic and are not pH dependent, which allows for a wider range of possible applications as mediators.

Laccase Literature also shows the use of osmium-based mediators for MET cathodes employing laccase. Laccase reduces oxygen to water, but the enzyme has optimal activity at a decreased pH (typically a pH around 5) when compared to BOD. At the cathode of a MET glucose biofuel cell, Chen et al. employed laccase along with the mediator [Os(4,4′-dimethyl-2,2′,6′,2″-terpyridine)]$^{2+/3+}$ with one fifth of the pyridine rings complexed within the redox polymer poly-(*N*-vinyl imidazole) [66]. Heller describes a laccase-based biocathode which incorporates the osmium-based complex Os[4,4′-dimethyl-2,2′-bipyridine)$_2$(4-aminomethyl-4′-methyl-2,2′-bipyr-idine] reacted to form amides with *N*-(5-carboxypentyl)pyridinium functions of poly(4-vinyl pyridine) in order to reduce oxygen to water [72].

While mediators add diversity to the range of oxidoreductase enzymes that can be employed in biofuel cells, they can add complexity to the enzymatic system. Current research has been focused on eliminating the need for mediators and having direct electron transfer between the enzyme and electrode surface.

5.4.5 DET at Bioanodes

Most bioanodes in literature employ FAD or NAD^+-dependent enzymes which require mediators to facilitate the transfer of electrons between the enzyme and electrode surface. There has been recent interest in studying enzymes which contain the bound cofactor PQQ. PQQ-containing enzymes belong to the group of quinoproteins recently discovered [53, 78–80]. Many PQQ-dependent enzymes are heme-containing or quinohemo-containing enzymes. A heme is a porphyrin-based molecule which exists in a number of reduced and oxidized states. The heme serves as a universal mediator of internal electron transfer from the specific catalytic site to the electrode. As an individual moiety, heme exhibits various catalytic properties which create ideal opportunities for heme-incorporated enzymes in DET enzymatic schemes.

Literature has shown that DET for PQQ-containing enzymes alcohol dehydrogenase and D-fructose dehydrogenase is feasible on different carbon electrodes and also on bare and modified gold, silver, and platinum electrodes [43, 53, 55, 81, 82]. Literature suggests that catalytic substrate formation occurs on the PQQ active site. Electrons are transferred through the protein molecule to the heme redox site, then further transferred to the electrode as shown in Figure 5.5 [82]. PQQ-dependent enzymes contain multiple heme c moieties which allow the enzyme to undergo direct electron transfer when immobilized on certain electrode materials depending on their surface properties. The electron transfer rate depends on both the electrode material and the nature of the enzyme allowing for a reagentless bioanode.

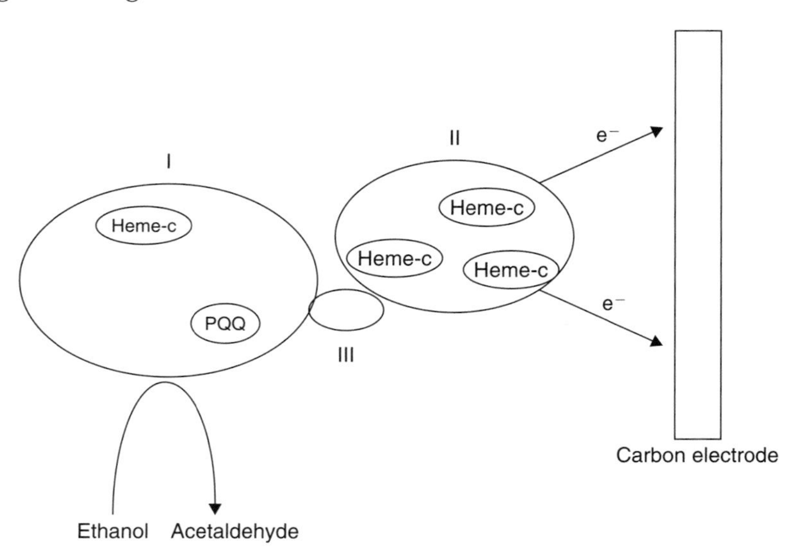

FIGURE 5.5 Schematic of the structure of PQQ-ADH.

PQQ-Dependent Glucose Dehydrogenase

There has been an increased interest in DET glucose biofuel cells. This is accomplished by utilizing PQQ-dependent glucose dehydrogenase (PQQ-GDH) in lieu of NAD^+-GDH. In 2005, Yuhashi reported on a PQQ-GDH catalyzed glucose biofuel cell [83]. State-of-the-art technology is reported by Ansari et al. of a glucose biofuel cell using a DET PQQ-GDH enzymatic system in 2006 [77]. Bioelectrodes which incorporate enzymes capable of DET will yield higher current outputs due to lower resistivity from lack of additional mediators along with closer and faster electron transfer mechanisms between the electrode surface and enzyme.

PQQ-Dependent Alcohol Dehydrogenase

Treu et al. reported the use of PQQ-dependent alcohol dehydrogenase (PQQ-ADH)-based bioanodes for use in a DET ethanol biofuel cell. PQQ-ADH was immobilized in a modified Nafion® membrane and cast onto AvCarb® paper without the incorporation of additional mediators [84]. The PQQ-ADH ethanol-based biofuel cell outperformed current state-of-the-art MET ethanol biofuel cell technology, NAD^+-ADH-based fuel cell developed by Akers et al. [13, 84]. For the PQQ-ADH bioanode coupled with a traditional platinum cathode, the PQQ-ADH-based enzymatic system increased the overall lifetime by greater than 711%, increased open circuit potential by 67%, and increased power density by 251%. The most important phenomenon to examine is the increase in lifetime of PQQ-ADH-based bioanodes. The PQQ-based bioanodes are more stable than NAD^+-based bioanodes due to the fact that PQQ is a stable bound cofactor, and NADH does not leach out of the system.

5.4.6 DET at Biocathodes

PQQ-dependent enzymes utilized at bioanodes are heme-containing enzymes where iron complexes facilitate in DET of substrates at the electrode surface. Select enzymes commonly employed as catalysts at biocathodes are also capable of undergoing DET due to their multiple metal centers. HPR, laccase, and BOD, the most commonly employed cathodic enzymes, all contain metal centers and are capable of DET. HPR is a heme-containing enzyme similar to PQQ-dependent enzymes employed at bioanodes, but it has not been utilized in a DET system at a biofuel cell cathode [85]. Willner and Katz describe a DET for the reduction of peroxide which employs microperoxidase-11 (MP-11). MP-11, a heme-containing enzyme, was coupled to cystamine on a gold electrode in order to form a self-assembled monolayer at the cathode of a glucose biofuel cell [64].

Laccase and BOD contain multiple copper centers which enable the transfer of electrons directly from the enzyme to the electrode surface [45, 86, 87]. Gupta et al. showed evidence of oxygen electroreduction catalyzed by laccase on monolayer modified gold electrodes [86]. Duma et al. have shown the presence of DET for a BOD-based biocathode. Literature described by Duma et al. showed a DET enzymatic system comprised of BOD immobilized directly at the electrode surface for use as a biocathode in an ethanol biofuel cell [88]. BOD was immobilized within a modified Nafion® membrane and cast directly onto carbon felt paper without the incorporation of additional mediators.

5.5 FAD-DEPENDENT ENZYMES

(FAD)-dependent enzymes join the class of oxidoreductase enzymes, which act to transfer electrons from a reductant to an electron acceptor called an oxidant. FAD is a cofactor that is required for some redox reactions to take place. FAD serves as the initial electron acceptor in the reaction and is reduced to $FADH_2$. The $FADH_2$ later is oxidized back to FAD by an electron acceptor (oxygen). The electron acceptor is then reduced to hydrogen peroxide, which serves as one of the final products. In addition, the substrate is oxidized in the reaction. Figure 5.6 shows a representation of a bioanode containing an FAD-dependent enzyme.

Unlike NAD^+-dependent enzymes, FAD-dependent enzymes incorporated into a fuel cell do not require an electrocatalyst to produce electricity as the necessary cofactor is bound to the enzyme. The coenzyme that these enzymes require for activity is FAD itself, therefore a production step in preparing an anode for implementation into a biofuel cell is unnecessary [89]. This is advantageous when compared to the NAD^+-dependent biofuel

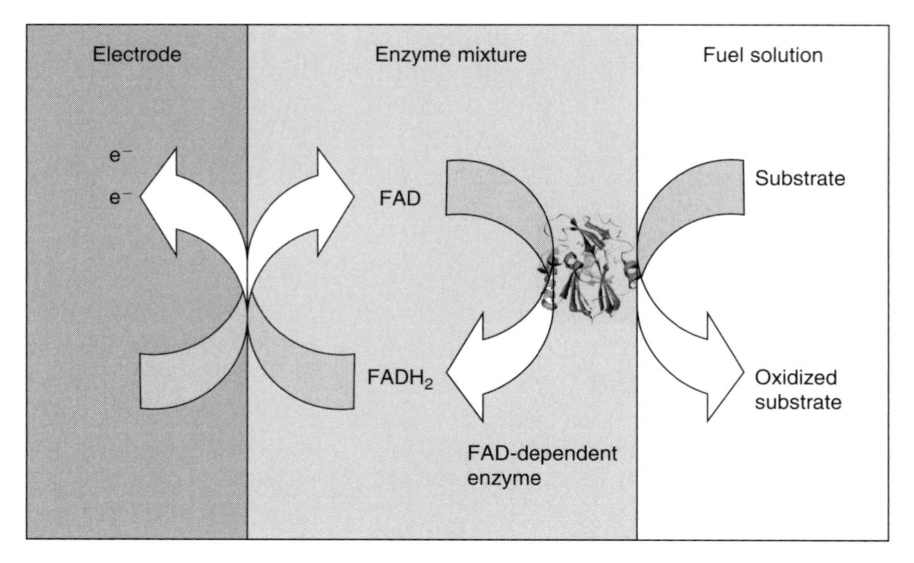

FIGURE 5.6 Representation of a bioanode containing a FAD-dependent enzyme.

cell as time loss is minimal for this system. There has been a great deal of research devoted to this additional step found in NAD^+-dependent electrode preparation. For instance, Willner et al. tether dehydrogenase enzymes to a gold electrode using FAD or PQQ [41, 90]. Since there is no need for an electron mediator on the surface of the electrode, the enzyme itself can transfer electrons directly to the electrode [91]. Also, a higher surface area increases the efficiency of the electrode. The enzyme is able to undergo DET with high surface area material, which causes the increased performance in the anode. High surface area supports include single walled carbon nanotubes (SWCNT) [92] and also carbon fiber [93]. While FAD-containing enzymes do not require mediated electrodes, electrodes containing FAD are improved with the addition of a mediator. A mediator provides the electrode with lower overpotential that must be overcome for the electrode to function. Mediated FAD-containing electrodes utilize mediators such as ferrocene [94] or quinone derivatives, [95] ruthenium complexes, [96] or phenoxazine compounds [97].

Many types of FAD-dependent enzymes exist. The most prevalent enzyme that is used in research is GOD [65]. However, others like alcohol oxidase can be used [98]. The main reaction that occurs is the oxidation of the necessary substrate by the enzyme, while FAD is reduced to $FADH_2$ with the subsequent release of electrons. The released electrons are what lend to the power output of the biofuel cell. Hydrogen peroxide is also formed as a byproduct. The formation of hydrogen peroxide is both an advantage and a disadvantage for use in a biofuel cell. The disadvantage occurs if, as it builds up, it becomes detrimental to the performance of the enzyme. A potential benefit arises when the FAD-dependent electrode is paired with a cathode enzyme which reduces hydrogen peroxide [99]. The hydrogen peroxide-using cathode then uses the excess hydrogen peroxide as fuel, which simultaneously lowers the hydrogen peroxide concentration. Research in this area has typically focused on peroxidases and catalase for peroxide consuming cathodes [100].

5.6 DEEP OXIDATION OF BIOFUEL CELLS

In the last few decades, fuel cell research reported in literature concentrated on developing fuel cells that use readily available biomass such as glucose, alcohols (methanol, ethanol), lactate, and, more recently, glycerol as fuel. One of the key issues in developing effective enzymatic biofuel cells is the successful immobilization of multi-enzyme systems that can completely oxidize fuel to carbon dioxide. By doing so, the overall performance of the fuel cell is increased. Most current enzymatic cells employ a single oxidoreductase and are able to only partially oxidize the fuel. In contrast, living systems are able to completely oxidize biofuels to carbon dioxide and water. Living organisms are able to undergo this metabolism due to the enzymes contained within the cells. Two major metabolic pathways that living systems employ for complete oxidation are glycolysis and the Kreb's cycle [101, 102]. Both of these pathways include electron-producing enzymes, referred to as oxidoreductase enzymes, and enzymes such as kinases and mutases which catalyze chemical reactions. In order to mimic this naturally occurring process, cascades of enzymes taken from these pathways can be immobilized at an electrode surface to completely oxidize biofuels [103, 104].

5.6.1 Metabolic Pathways

Glycolysis

There are several starting points in the glycolysis pathway. The most common one involves the oxidation of glucose or glycogen to produce glucose-6-phosphate. However, other modified sugars can enter into the pathway at various points along the reaction scheme. Since glucose is readily available from the environment and is inexpensive, it offers an added benefit for use in a biofuel cell.

The glycolysis cycle can be seen as consisting of two separate phases. The first phase is the chemical priming phase requiring energy in the form of ATP. The second phase is the energy-yielding phase. In the priming phase, two equivalents of ATP are used to convert glucose to fructose 1,6-bisphosphate. In the energy-yielding phase, fructose 1,6-bisphosphate is degraded to pyruvate with the production of four equivalents of ATP and two equivalents of NADH. Several enzymes are involved in the glycolysis cycle as can be seen in Figure 5.1; however, only the glyceraldehyde-3-phosphate dehydrogenase (GAPDH) enzyme is responsible for energy production. Other enzymes are used for conversion of molecules into substrates useable in further reactions. GAPDH immobilized on an electrode is responsible for the reduction of NAD^+ to NADH which is then oxidized by an electrocatalyst layer which regenerates NAD^+ and ultimately produces power therefore serving as the energy producing step.

Kreb's Cycle/Citric Acid Cycle

The glycolysis pathway can be used in conjunction with the Kreb's cycle containing many dehydrogenases, allowing for further oxidation and, ultimately, more power production. In living cells, the Kreb's cycle is the main energy-generating mechanism. It oxidizes acetyl groups from many sources. However, under aerobic conditions, pyruvate in most cells is further metabolized via this cycle. The Kreb's cycle involves a series of enzymatic steps through which the end product of glycolysis (pyruvate) after conversion to acetate, is oxidized to carbon dioxide and water, as can be seen in Figure 5.7. The oxaloacetate that is consumed in the first step of the Kreb's cycle is regenerated in the last step of the cycle. Thus, the whole Kreb's cycle acts as a multi-step catalyst that can oxidize an unlimited number of acetyl groups.

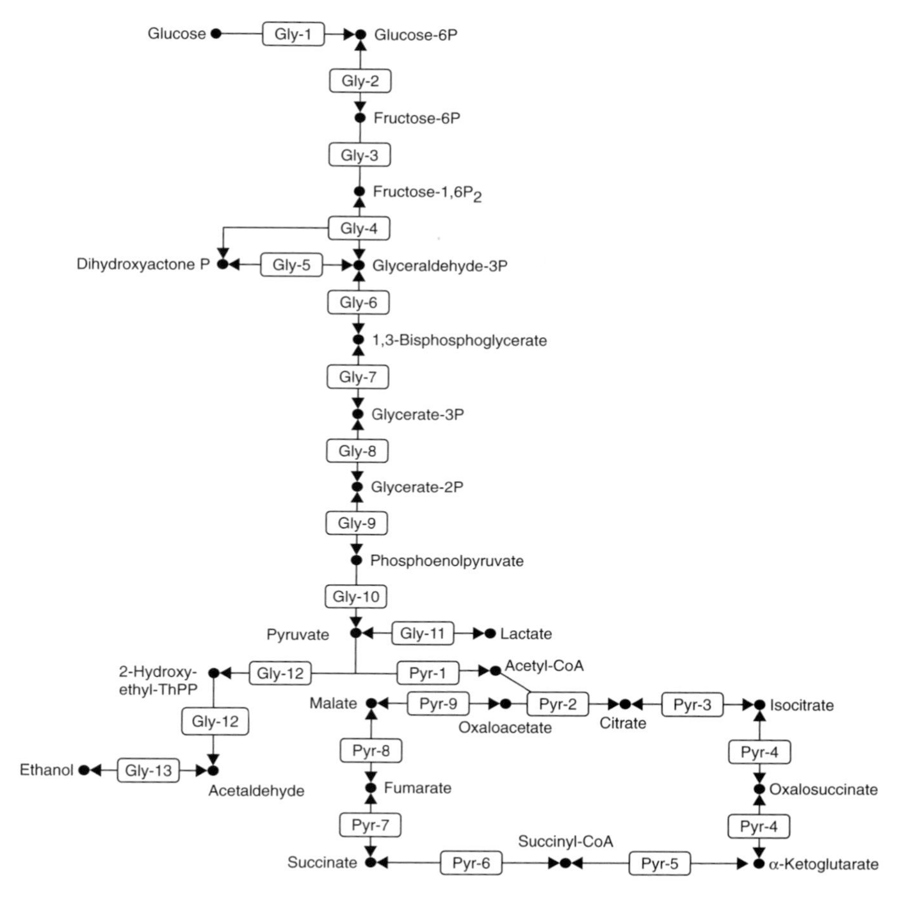

FIGURE 5.7 Schematic of enzyme cascade used for the oxidation of common biofuels. The labels are as follows: Gly-1, Hexokinase; Gly-2, Phosphoglucose Isomerase; Gly-3, Phosphofructose Kinase; Gly-4, Aldolase; Gly-5, Triose Phosphate Isomerase; Gly-6, Glyceraldehyde-3 Phosphate Dehydrogenase; Gly-7, Phosphoglycerate Kinase; Gly-8, Phosphoglycerate Mutase; Gly-9, Enolase; Gly-10, Pyruvate Kinase; Gly-11, Lactic Dehydrogenase; Gly-12, Pyruvate Decarboxylase; Gly-13, Alcohol Dehydrogenase; Pyr-1, Pyruvate Dehydrogenase; Pyr-2, Citrase; Pyr-3, Aconitase; Pyr-4, Isocitrate Dehydrogenase; Pyr-5, α-Ketoglutarate Dehydrogenase; Pyr-6, Succinyl CoA Synthatase; Pyr-7, Succinate Dehydrogenase; Pyr-8, Fumarase; Pyr-9, Malate Dehydrogenase.

There are a variety of ways the Kreb's cycle can be utilized in a biofuel cell. One is the combination of the Kreb's cycle enzymes with the enzymes of the glycolysis cycle. A second involves entering the Kreb's cycle via ADH, AldDH, s-acetyl coenzyme A synthetase, and oxaloacetate to form acetate. Ultimately this would yield complete oxidation of another type of inexpensive fuel substrate, ethanol. The main challenge with utilizing these metabolic pathways on an electrode is understanding how these metabolic pathways are controlled. Greater understanding is gained through metabolic flux analysis utilizing NMR to identify the enzymes that catalyze the rate determining steps [105–107].

5.6.2 Glucose Oxidation

Research interest in enzymatic biofuel cells utilizing glucose as fuel was boosted by the attempts to develop an implantable power supply for an artificial heart. Implantable applications do not require the complete oxidation of glucose, because the fuel (glucose) is plentiful and oxidation byproducts can be absorbed and further oxidized within the body. However, in any type of *ex situ* application, there are no other means by which glucose can be oxidized outside of the fuel cell, so unless biocatalysts employed are able to completely oxidize the fuel then the remaining energy density of the fuel is wasted and the overall energy density of the fuel cell would be minimal. The first enzyme-based glucose/O_2 biofuel cell operating at a neutral pH was described by Yahiro et al. in 1964 [14].

Since glucose is a readily available fuel, many methods have been developed to oxidize it. Willner et al. have developed a glucose biofuel cell that uses enzymes for both anode and cathode reactions [34]. The anodic reaction is glucose oxidation to gluconic acid by GOD reconstituted onto a PQQ-FAD monolayer, and the cathodic reaction is the reduction of hydrogen peroxide by MP-11. They estimated that the GOD monolayer reconstituted in this manner operates at a turnover rate of electron transfer exceeding that of native GOD with oxygen as the electron acceptor. PQQ-FAD-GOD monolayer electrode drives the bioelectrocatalyzed oxidation of the glucose fuel substrate, whereas the MP-11 acts as the cathode for the biocatalyzed reduction of hydrogen peroxide. This biofuel cell produced a current density of 114 $\mu A\ cm^{-2}$ and maximum power density of 160 $\mu W\ cm^{-2}$ [34]. The power output decreased by *ca.* 50% after *ca.* 3 hours of operation, which could originate from the depletion of the fuel substrate, among other causes [34]. This particular biofuel cell design was further developed by Katz et al. replacing the MP-11-based cathode reaction with a tethered, ordered and crosslinking stabilized cyt c/cytochrome oxidase complex [36]. The open circuit potential obtained was 110 mV, and the maximum power density derived was 5 μW cm^{-2} [36]. In addition to previously mentioned modification, Katz et al. developed another variation on the GOD/MP-11 system which involved a two-phase separation of analyte and catholyte. Both biocatalytic electrodes, Au/PQQ-FAD-GOD and Au/MP-11, were integrated into one system, creating a biofuel cell using glucose and cumene peroxide as the fuel substrate and the oxidizer, respectively. This fuel cell system yielded a maximum power output of *ca.* 4.1 mW cm^{-2} and a short-circuit current density of *ca.* 830 μA cm^{-2} [35].

In 2001, Katz et al., demonstrated the oxidation of glucose using GOD and the oxidation of lactate using lactate dehydrogenase (LDH) by studying the magnetic field effects on bioelectrocatalytic reactions of these enzyme systems [108]. The first biofuel cell included the biocatalytic anode based on the LDH/NAD^+-integrated electrode oxidizing lactate as a fuel. The maximum power output values in the absence and presence of a magnetic field (B = 0.92 T) were 4.1 μW cm^{-2} and 12.4 μW cm^{-2}, respectively [108]. The second biofuel system studied included GOD/FAD-PQQ-electrode as an anode for the oxidation of glucose. The presence of the magnetic field had only a minor effect on power density output which originates from the fact that the oxygen-reducing cytochrome c/cytochrome oxidase electrode exhibits lower biocatalytic activity as compared to the GOD/FAD-electrode. The power density outputs in the absence and the presence of the magnetic field were about 20.5 μW cm^{-2} and 25.4 μW cm^{-2} respectively [108].

In addition to GOD-based biofuel cells, GDH can also be used as a catalyst for glucose biofuel cells with the possibility of being a MET system or DET system based on the coenzyme of the dehydrogenase. In 1985, Persson et al. reported a MET enzymatic system with an open circuit potential of 0.8 V, current density of 0.2 mA cm^{-2}, and power output was not reported [109]. Takeuchi et al. modified a nanocrystalline titanium oxide electrode with zinc chlorine-e_6 acting as an electron acceptor from NADH [75]. The cell was illuminated and achieved an open circuit potential of 444 mV and maximum power density of 16 nW cm^{-2}. The cell, as described, requires far more light energy input to activate it than it returns through the consumption of fuel. In 2006, Ansari et al. developed a biofuel cell for a MET GDH system that has an open circuit potential of 0.83 V, current density of 3.94 mA cm^{-2}, and a power density of 1.27 mW cm^{-2} [77].

A recent state-of-the-art pathway reported by Beilke et al. demonstrated a biomimic of glycolysis on a bioanode for glucose oxidation [105]. This glucose biofuel cell utilized glyceraldehyde-3-phosphate dehydrogenase, aldolase, phosphofructokinase, phosphoglucose isomerase, and hexokinase. Each enzyme was mixed with its respective modified polymer. The first enzyme immobilized onto the bioanode surface was glyceraldehyde-3-phosphate dehydrogenase mixed with tetrabutylammonium bromide (TBAB) modified Nafion® polymer suspension. After the bioanode dried, each of the other enzymes mixed with their respective optimum modified polymers was immobilized consecutively. The maximum potential recorded for multi-enzyme system was 0.965 V while the average maximum current was 0.347 ± 0.006 mA cm^{-2}. The listed potential and current densities findings yielded power densities of 0.0713 ± 0.0092 mW cm^{-2}. This system was compared to a single enzyme system containing glyceraldehyde-3-phosphate dehydrogenase which obtained a power output of 0.0740 ± 0.0192 mW cm^{-2}. The decrease in performance of the multi-enzyme system compared to single enzyme system indicates that there is a loss in the cell performance for the oxidation of glucose. A very possible reason for this is the increased resistance of substrate diffusing through multiple polymer layers before it is able to react with the enzymes on the carbon surface.

5.6.3 Oxidation of Alcohols

Alcohols, specifically methanol and ethanol, are also very desirable biofuels which can be completely oxidized. Alcohol-based enzymes chosen most frequently for the bioanode involve NAD^+-ADH, which oxidizes alcohols to aldehydes. Combining this enzyme with NAD^+-AldDH allows for further oxidation of aldehyde. The first biofuel-based multistep oxidation of alcohols was demonstrated by Palmore et al. who employed alcohol dehydrogenase, aldehyde dehydrogenase, and formate dehydrogenase

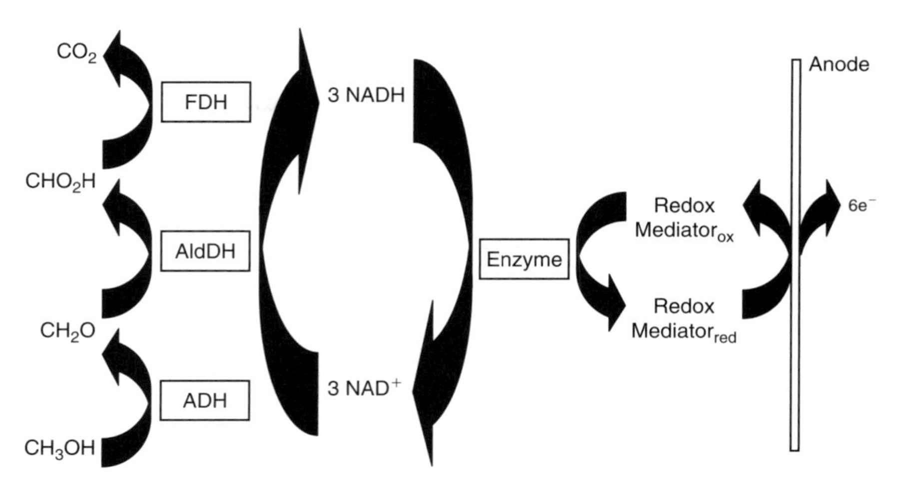

FIGURE 5.8 The oxidation of CH_3OH to CO_2 is catalyzed by NAD^+-dependent alcohol- (ADH), aldehyde- (AldDH), and formate- (FDH) dehydrogenases. Regeneration of NAD^+ is accomplished electro-enzymatically with an enzyme coupled to the anode via a redox mediator.

to completely oxidize methanol to carbon dioxide and water (Figure 5.8) [40]. Since these dehydrogenases are dependent upon NAD^+ reduction, a fourth enzyme diaphorase was introduced to regenerate the NAD^+ by reducing benzyl viologen. This dehydrogenase catalyzed methanol/O_2 biofuel cell produced an open circuit potential of 0.8 V and power density of 0.68 mW cm^{-2}.

Akers et al. studied the two-step oxidation of ethanol to acetate using ADH and AldDH in a novel membrane assembly (MEA) configuration [13, 110]. Since dehydrogenase enzymes are NAD^+ -dependent, and NADH oxidation on platinum and carbon electrodes has poor reaction kinetics and occurs at large overpotentials, a polymer-based electrocatalyst, poly(methylene green), was used to regenerate NAD^+ and to shuttle electrons from NADH to the electrode [102]. Bioanodes undergoing one-step oxidation were compared to bioanodes undergoing two-step oxidation. The ethanol-based biofuel cell undergoing only one-step oxidation with ADH immobilized in a TBAB modified Nafion® membrane has shown open circuit potentials ranging from 0.60 to 0.62 V and an average maximum power density of 1.16 ± 0.05 mW cm^{-2}. In contrast, the open circuit potential and the maximum power density for the ethanol-based biofuel cell employing a mixture of both

ADH and AldDH immobilized in a TBAB/Nafion® membrane were 0.82 V and 2.04 mW cm^{-2} respectively. For comparison reasons, a methanol/O_2 biofuel cell was fabricated, and it produced an average power density of 1.55 ± 0.05 mW cm^{-2}. Although the results were successful, the system permitted only a 33% oxidation of the fuel which represents a low fuel utilization and energy density.

5.6.4 Glycerol Oxidation

Most enzyme-based biofuel cells rely on traditional biofuels such as the ones mentioned previously in this section. In 2007, Arechederra et al. reported glycerol as a new fuel for biofuel cells that allows for fuel concentrations up to 98.9% without swelling the Nafion® membrane, which traditionally has been problematic in most alcohol-based fuel cells [16]. Glycerol has a higher energy density (6.331 kWhr/L pure liquid) compared to ethanol (5.442 kWhr/L pure liquid), methanol (4.047 kWhr/L pure liquid), or glucose (4.125 Whr/L saturated solution) making it energetically a very attractive fuel. In addition, glycerol is a renewable energy source that is produced in large quantities as a byproduct of biodiesel. Employing PQQ-ADH on a bioanode immobilized in a trimethyloctylammonium bromide (TMOA) modified Nafion® membrane has led to open circuit potentials ranging from 0.55 to 0.73 V at 20°C and average maximum power densities of 1.13±0.15 mW cm^{-2} [111]. Compared to the same bioanode in the ethanol/O_2 biofuel cell, the bioanode using glycerol as fuel showed 62.5% increase in power as glycerol can undergo five oxidations before it diffuses out of the membrane; ethanol can only undergo two oxidations.

Most current biofuel cells in literature only employ single-step oxidation. With multi-step oxidation, biofuel cell performance can be enhanced along with flexibility in fuel substrates. The two major metabolic pathways which living systems utilize for complete oxidation are glycolysis and the Kreb's cycle. The glycolysis cycle can be used in conjunction with the Kreb's cycle and allow for further oxidation and ultimately more power production. Although many advances have been accomplished in enzymatic-based biofuel cells, there is still room for improvement on strategies for immobilizing cascades of enzymes at the electrode surface, to decrease transport limitations, and allow for complete oxidation of fuels.

5.7 BIOCATHODES

5.7.1 Importance of the Cathode

Although the majority of research on bioelectrodes has covered the anode, the cathode's importance should not be overlooked. Optimizing a biocatalyst's oxidative ability is crucial, but without an equally efficient reductive biocatalyst, the fuel cell's capacity is diminished.

5.7.2 Selectivity

Traditionally, fuel cells have been designed with a PEM which separates the anode and cathode into two distinct environments. This PEM allows ion transfer between the two compartments and prevents charge build-up on either side of the cell. As metals (platinum, palladium, ruthenium, etc.) have been the cathode catalyst of choice, the separation of the anode and cathode chamber environments has been necessary as metals are nonselective catalysts. As the cathode's role in the fuel cell is to reduce a substrate using the electrons obtained from the cell circuit, the cathodic enzyme's oxidation of fuel would severely inhibit the potential necessary for useful power density. The PEM, however, is a significant source of resistance overpotential, and maximization of current and power densities combined with minimal size depends upon a single-chamber anode/cathode compartment. Due to the inherent selectivity of enzymes, they are an ideal choice as their preferred substrates, environmental conditions, and reduction potentials can be discriminately chosen to complement a specific biocathode.

Specifically, a PQQ-ADH bioanode and BOD biocathode have been utilized in membraneless format. In this study, initial testing was performed in a membrane separated fuel cell and then repeated in a membraneless cell. The maximum power density reported for the membrane cell was 0.95 mW cm^{-2}, and open circuit potential was 0.82 V. Maximum power density reported for the membraneless cell was 2.44 mW cm^{-2} with a maximum open circuit potential of 1.09 V, representing a 181% increase in power density and 67% increase in open circuit potential [112].

5.7.3 Enzyme Characterization

Of the vast number of enzymes available, the class of enzymes predominantly studied is the oxidoreductase EC 1.1.3 category (oxidoreductases which utilize oxygen as an electron acceptor). Specifically of interest are those enzymes capable of direct electron transfer with the electrode and a direct 4-electron reduction mechanism which reduces oxygen completely to H_2O: laccase, BOD, and HRP, for example. Despite the fact that many enzymes are redox active, very few are capable of DET.

5.7.4 Mediators

Despite the use of DET-capable enzymes, mediators are often employed to increase current density. In the quest for an ideal cathode, overpotential limitations must be addressed. Because the active sites of the most commonly used oxidoreductase enzymes have active sites buried deep within the enzyme, electron tunneling between it and the electrode is a significant source of overpotential [113]. To increase kinetics, a mediator functions as an electron shuttle whereby it accepts one or more electrons from the surface of the electrode and delivers them to the active site of the enzyme. An ideal mediator should have fast electron transfer kinetics and a potential similar to that of the active site center to reduce the effects of overpotential [114].

Due to laccase's low substrate specificity and slow DET rate, it is ideal for use with mediators [114]. ABTS (2,2′-azinobis-(3-ethylbenzothiazoline-6-sulfonate) diammonium salt) has been utilized successfully as a mediator in conjunction with laccase by Palmore and Kim, [114] Ikeda, [115] and Farneth [116]. In the study conducted by Palmore and Kim, [114], the objective to find a mediator with a formal potential similar to that of laccase's T1 active site, led to testing of ABTS, syringaldazine, ferrocene, $K_3Fe(CN)_6/K_4Fe(CN)_6$ and I_2/I_3^-. Syringaldazine and ferrocene were insoluble/unstable in acidic solution, and cyanide and the halides irreversibly bound to laccase (preventing the binding of oxygen). Electrochemical tests were performed on the fuel cell which utilized a PEM, a platinum gauze anode, and platinum, glassy carbon or carbon in a laccase/ABTS solution (laccase was not immobilized). When electrochemically tested against either a platinum or glassy carbon cathode, open circuit voltage of the laccase/ABTS solution showed significantly better results at 0.53 V compared with 0.34 V for platinum and 0.28 V for glassy carbon (all versus SCE). Similarly, under a 1 kΩ load, the maximum power density of the biofuel cell was 42 μW cm^{-2} at 0.61 V, 15 μW cm^{-2} for platinum, and 2.9 μW cm^{-2} for glassy carbon [114].

An alternative form of mediator is that developed by Heller, et al. [117]. While the Palmore design utilized a free-floating laccase/ABTS solution in the cathode chamber of a PEM cell, Heller's cathode design features a laccase enzyme wired to an osmium complex/redox polymer within a miniature, membraneless glucose/O_2 biofuel cell. This "wired" redox polymer has flexible "tethers" between the osmium complex and the polymer backbone which serve to sweep electrons from the hydrated redox polymer (Figure 5.9). By wiring the enzymes' reaction centers to the polymer and adding the eight-atom long tethers, the apparent electron diffusion coefficient was increased 100fold with an overpotential of −0.07 V as compared to 0.37 V overpotential of platinum [117].

5.7.5 High Current Applications

The supply of oxygen to the cathode presents a particular challenge in terms of concentration overpotential as there is a very low percentage of dissolved molecular oxygen in solution. Increased oxygen supplies must be confined to the cathode region to prevent anodic enzyme oxidation. In addressing this issue, the strategy used by Tingrey was to supply oxygen separate from the electrolyte by modifying a porous carbon electrode, whereby oxygen was continually diffused from the inner to the outer surface of the cathode. With oxygen diffusion through the cathode, maximum power density reached 18 μW cm^{-2}. By comparison, the maximum power density obtained when oxygen was bubbled into the cathode compartment was 9 μW cm^{-2} [118].

As previously detailed, a variety of immobilization methods have been developed in an attempt to increase current densities,

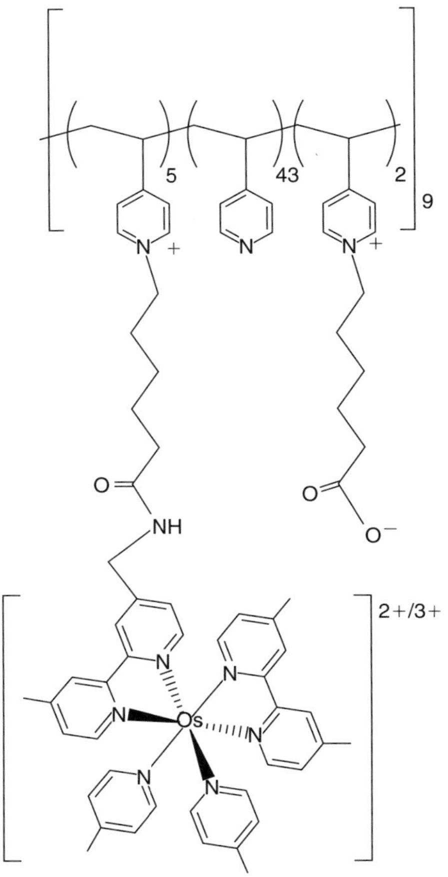

FIGURE 5.9 Osmium complex with flexible polymer tethers.

stabilize enzyme activity, and increase lifetimes, with encapsulation and wired methods being most predominant in cathode optimization.

Encapsulation of various oxidoreductases has been successfully implemented by a number of researchers. Farneth and D'Amore demonstrated efficacy of such a cathode by impregnating carbon paper with laccase and ABTS then dipping into a tetramethoxysilate-phosphate buffer mixture creating a silica coating within which enzymes are immobilized. Although enzymes remain encapsulated, ABTS mediator leakage from the pores is problematic [119].

In an effort to circumnavigate mediator leakage issues, Minteer and Dunn have, independently, presented effective immobilization techniques which exploit copper-center oxidoreductases' ability to do DET. Minteer's procedure involves the modification of Nafion® whereby co-casting with tetrabutylammonium bromide results in a pore size sufficient for enzyme immobilization [88]. Dunn's approach involves a sol-gel/carbon nanotube composite which immobilizes enzymes and increases the effective surface area of the electrode [120]. Both systems provide enzyme conformation preservation, substrate diffusion, and retention of a micro environment in which an optimal catalytic environment can be sustained.

Viable wired approaches have been put forth by Heller (as described above) and Armstrong. Armstrong's bound laccase was achieved through the modification of an aryl amine to form anthracene-2-diazonium covalently bound to an electrode. The modification of laccase occurs when it attaches to the anthracene through its active site. As electrons flow from the electrode through the covalently attached anthracene and directly into the laccase active site, this technique displays a sort of "plug-in-socket" effect [121].

5.8 ENZYME IMMOBILIZATION

Enzyme immobilization is the process by which the enzyme catalyst is trapped at the bioanode or biocathode surface. There are three common techniques for immobilizing enzymes with the most common being the wired and sandwich techniques. The third technique is microencapsulation. Figure 5.10 shows a pictorial representation of each type.

Enzyme immobilization has many advantages in the development of biofuel cells. One of the most significant advantages of immobilization is the increase in stability and lifetime as compared to the free enzyme. The lifetime of the free enzyme in solution is typically only a few hours to a couple of days, as changes in pH or temperature cause it to denature, [5] whereas immobilized enzymes can have lifetimes upwards of a year or more protected

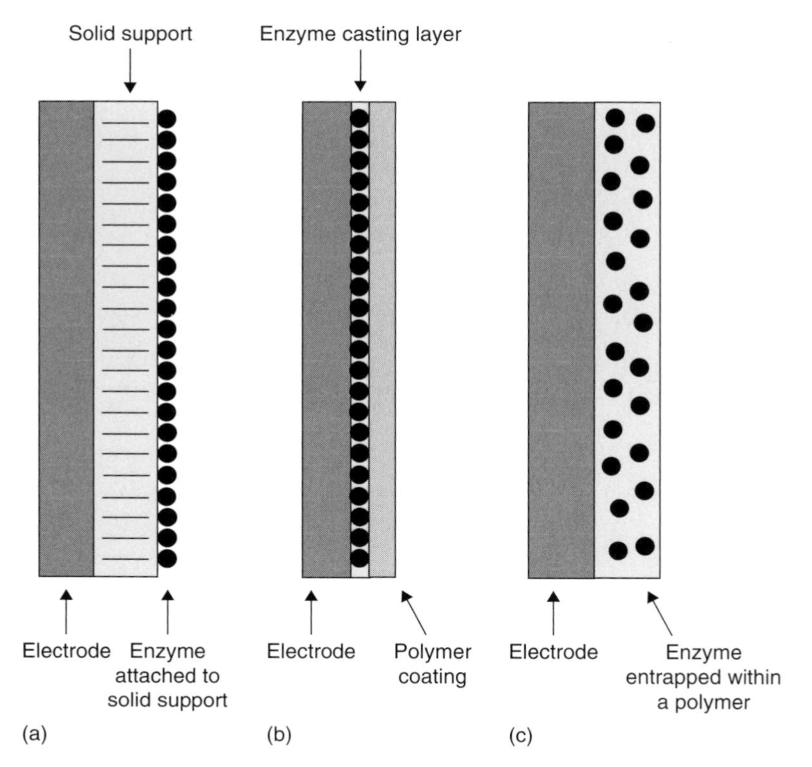

FIGURE 5.10 Techniques for enzyme immobilization. (a) Wired; (b) Sandwich; (c) Entrapment.

from changes in the surrounding environment [13, 88]. Secondly, by capturing the enzyme at the surface of the electrode, the efficiency of the biofuel cell is improved as the electrons can more easily transport to the electrode surface.

The wired technique involves creating an ionic or covalent interaction between the enzyme and the solid support. Cai et al. were able to bind negatively charged enzymes to single-walled carbon nanotubes layered with positively charged polyelectrolyte poly(dimethyldiallylammonium chloride) [122]. They studied the secondary structure using AFM, Raman, FTIR, UV-Vis, and electrochemical techniques after the enzymes had been immobilized and found that the electrodes maintained stability and reproducibility for more than 30 days for GOD and more than 25 days for ADH [122, 123]. Ryu et al. utilized multi-walled carbon nanotubes in their immobilization process by first carboxylating the nanotubes in order to improve the efficiency of the immobilization and using a carbodiimide coupling reagent to bind GOD to the nanotubes [124]. Katz and Willner have also reported reconstituting apo-GOD on a layer of PQQ-FAD which was tethered to the anode, while the cathode was covalently functionalized with MP-11 for the reduction of peroxide to water [65, 125].

The sandwich technique involves the enzyme being deposited on the electrode surface, followed by a layer of polymer. Sahney et al. immobilized urease by first coating an electrode with tetramethylorthosilicate (TMOS) derived sol-gel matrix. It was then dipped into enzyme stock solution, followed by application of a top layer of TMOS [126]. Electrodes showed activity for at least 30 days, but with significantly decreased activity. Nafion® polymers can also be used to coat the enzyme modified electrode [127, 128]. However, due to the acidic side chains, it does not provide the ideal environment for enzymes.

The third technique is microencapsulation, which involves entrapping the enzyme within the pores of a membrane at the electrode surface. Minteer et al. have sought to develop a membrane using Nafion®, which is an ion exchange polymer. In its unmodified form, Nafion® is an acidic, hydrophilic, micellar polymer with pores of about 40 Å. Because of the sulfonic acid sites on the polymer backbone, it is too acidic to provide an ideal environment for the immobilization of enzymes within its pores. Therefore, a tetraalkylammonium bromide salt is exchanged with the proton of the sulfonic acid site, which allows for the polymer to be more conducive to enzyme immobilization [129].

One advantage of the modification is that by altering the size of the ammonium salt used, the pore size can be engineered to fit the enzyme [129]. A second advantage is that the acidic proton is exchanged for a near neutral ammonium cation, and the polymer has a higher affinity for the hydrophobic cation than the proton. Therefore, the polymer environment is buffered against external changes in pH [129].

Aside from the advantages of using Nafion® as the enzyme immobilization membrane, there are several disadvantages. Nafion® is expensive, and because of its perfluorinated polymer backbone, it is neither biocompatible nor biodegradable [130].

Chitosan has therefore been investigated as a possible polymer backbone for hydrophobic modification [130]. Chitosan is the product of the deacetylation of chitin, which is a widely available and very inexpensive biomass obtained from crustaceans. It is a linear polysaccharide, which makes it both biodegradable and biocompatible [131, 132]. The chitosan backbone is easily hydrophobically modified by reductive amination using aldehydes with long alkyl chain lengths [131]. Chitosan also benefits from high mechanical strength, so it will not break down under stress, as well as being nontoxic which allows it to have a wide variety of applications in the medical field [130, 132]. Previous research shows that chitosan can be used to modify electrodes for sensing applications to preconcentrate both cations and anions [133–135]. Chitosan is also able to form polymer micelles, which makes applications in enzyme immobilization more ideal [136–138].

While the microencapsulation technique provides the most stable environment for enzymes while enhancing lifetimes, it must be able to accommodate active fuel transport through the membrane [9]. Kim et al. have developed a way to encapsulate single enzyme molecules within "armored" nanoparticles [139]. The activity of the enzymes encapsulated within the nanoparticles showed only a slight percent decrease in activity over a period of at least twelve days [139]. This is ideal, but future research will need to be investigated to find optimal methods for mass and electron transport in these systems. Dunn et al. have also been successful in immobilizing GOD and BOD within a carbon nanotube/sol-gel polymer matrix [140]. They were able to construct a biofuel cell that produced 0.120 mW cm^{-2} [140].

Research by Heller et al. also shows a technique for enzyme encapsulation [26]. The enzymes are immobilized by mixing the enzyme with a redox hydrogel and casting it onto an electrode surface [71, 141]. The properties of the anode have been altered by changing the polymeric structure as well as the ligands on the osmium complex, which is incorporated in the hydrogel.

Despite the advantages to immobilizing the enzymes onto the electrode surface, there are still several challenges for all three of the techniques. One major challenge is to modify the electrode surface with as minimal resistivity as possible, so as to allow the most efficient performance of the biofuel cell [5]. Overall, the challenge remains to develop the immobilization technique which allows for the generation of the maximum power output for each enzyme/fuel system, while at the same time minimizing the stress on the enzyme.

5.9 ENZYMATIC FUEL CELL DESIGN

Cell design in the optimization of biofuel cells is often thought of as a secondary concern. Although often taken for granted, optimal cell design may greatly advance the performance and reproducibility of the fuel cell while also allowing it to be streamlined and directly applicable to small commercially available products. Most research efforts have focused on the analysis of fuel cell systems that have their design based on a traditional H-shaped electrochemical test cell (H-cell). However, recent developments have opened doors to innovative cell designs based on adaptations to this traditional design or utilizing microfluidics.

The traditional H-cell was initially developed to separate the two half reactions of an electrochemical cell while maintaining ion transport between cathodic and anodic solutions, thus completing a circuit. Three primary manifestations of this H-cell design have been utilized in biofuel cell design. The first, shown in Figure 5.11a, consists of the traditional H-cell-containing cathodic and anodic solutions separated by a porous frit [142]. This porous frit allows minimal mixing of the two solutions while still allowing ion transport. The shortcoming of this design is that, over time, the cathodic and anodic solution will diffuse across the frit and mix, causing unwanted side reactions. Another design consists of two entirely separate containers for the anode and cathode which are connected by an electrolyte bridge. The bridge is a tube containing electrolyte solution that is maintained by plugging the ends with a semipermeable material to prevent mixing. The drawback of this design is that the electrolyte solution in the bridge eventually diffuses out through the semipermeable material into both anodic and cathodic solutions [142]. The third and most common electrochemical test cell used in the development of biofuel cells uses a Nafion® polymer electrolyte membrane. This membrane has the ability to serve as a salt bridge while maintaining adequate

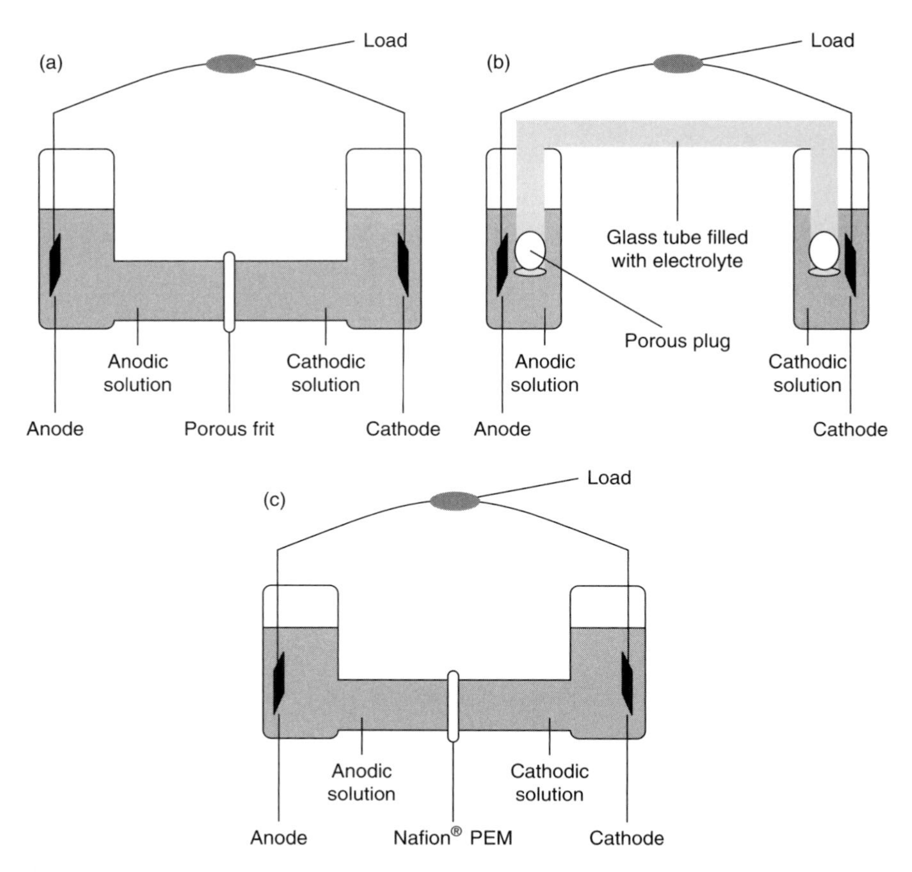

FIGURE 5.11 Demonstration of the three most common and simplistic test cells for the development of biofuel cells.

separation of the anodic and cathodic solutions. Due to the Nafion® polymer's cation exchange abilities and small porosity, it provides an ideal boundary that improves the performance and lifetime of biofuel cells. However, in the presence of some solvents (i.e., ethanol and methanol), Nafion® swells considerably, allowing crossover of the two solutions which results in the poisoning of the cathode catalyst.

Although these test cells provide an adequate platform for the initial development of enzymatic and microbial systems, their performance and commercial feasibility are not optimal. One major limitation on these designs is the lack of oxygen present at the cathode for reduction to form water, the major byproduct of most biofuel cells. Under typical laboratory settings, most cathodic solutions contain approximately 7 μg mL^{-1} oxygen. This translates to a maximum power density of only 1 to 2 mW cm^{-2} [142]. Another limitation lies in the simple geometry of the cell. Most H-cells separate the anode and the cathode by at least 10 cm [142]. This greatly limits the current produced, as the ions must travel through a greater distance. In order to alleviate this, most commercially available fuel cells incorporate electrodes that are directly pressed onto the Nafion® membrane. However, this design is not practical for research laboratories because the electrodes and fuel solutions are not capable of being changed frequently.

Recent efforts have eliminated the need for the physical separation of cathodic and anodic solutions through the replacement of the traditional platinum catalyst with an enzymatic catalyst system, thereby imparting selectivity of the cathode. Enzymes commonly used include BOD, laccase, and cytochrome c oxidase which are capable of assisting in the reduction of oxygen [6, 88, 118, 143–145]. The efforts in this area have been discussed in a prior section. This development improves reaction kinetics at the cathode, making the cell solely mass transport limited. It is also capable of matching and improving the cell performance whose cathodic reactions are based on platinum catalysts while eliminating an expensive and easily poisoned material. This design also allows the cell to function best at physiological pH, at which a platinum electrode will tend to corrode.

The ability to use mixed fuel solutions and eliminate physical barriers between solutions has resulted in the development of more efficient cell designs. A modification to the H test cell design has been described which reduces the distance between cathode and anode. The modification also allows for simple interchangeability of fuel and anode. In this setup, the cathode is in direct contact with air, thus eliminating limitations that arise from lack of oxygen. This design has been termed the I-cell due to its shape, shown in Figure 5.12. This design utilizes a gas diffusion electrode which is hot pressed to a Nafion® 112 membrane [142]. This gas diffusion electrode is clamped between two glass tubes, one of which contains the fuel solution. Although this is a marked improvement on traditional test cell design, it remains impractical to link multiple cells in series to power conventional devices.

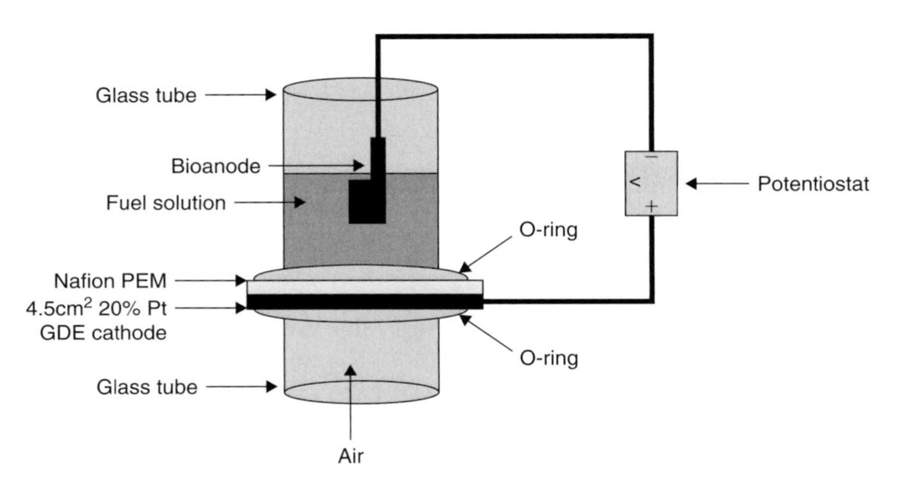

FIGURE 5.12 Representation of the I-cell design.

Although the previously mentioned design does improve the geometry of the cell, this improvement pales in comparison to the efforts made in developing smaller, more efficient traditional fuel cells. Power density, as described by Eq. 5.7, is not only a function of the thermodynamics of the system (denoted by the equilibrium potential of the cell, V°_{cell}) but also the kinetics of the system (demonstrated by the resistance of the cell, R_{cell}) [11].

$$P^{-} = (V^{o}{}_{cell} - I^{-} \cdot R_{cell}) \cdot I^{-} = V^{o}{}_{cell} \cdot I^{-} - I^{-2} \cdot R_{cell} \quad (5.7)$$

R_{cell} incorporates various attributes including charge transfer resistance of the electrode material, electrode connections, membrane conductance, and electrolyte conductance. All of these terms are heavily affected by the geometry of the cell [11]. One method for reducing the effect of cell resistance is a stack cell design. Such designs are commonplace in traditional fuel cell research efforts; however, only one research group has applied this idea to biofuel cell design. Teodorescu et al. have described a small-scale design shown in Figure 5.13 [146]. This design utilizes not only the common stacking technique but also incorporates the use of microelectrodes to overcome energy losses due to activation, ohmic resistance, and slow mass-transport. The cell stack's extremely small distance between fuel cell components is optimally applicable

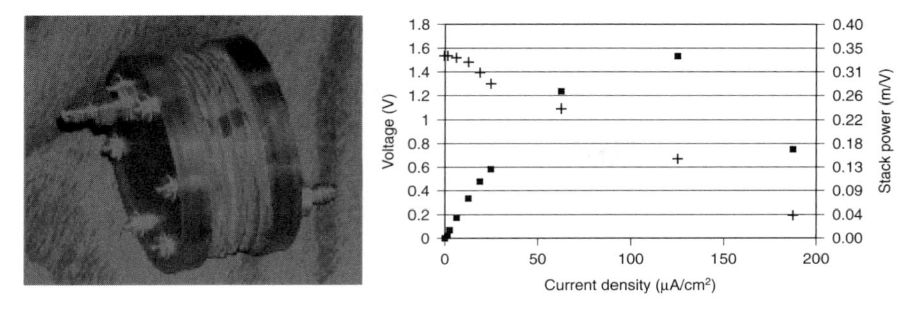

FIGURE 5.13 Picture of the only known biofuel stack cell design and its resultant open circuit potential and power curve [146].

to small-scale electronics. Also the efficiency of the cell is improved from the reduction in resistance and transport limitations to both the electrode surface and immobilized enzymes. This preliminary design was demonstrated to generate an open circuit potential of 1.57 V and a power density of nearly 0.35 mW cm^{-2} with six cells stacked together, which still falls significantly short of the performance seen in recent traditional cells [146]. Due to the confined assembly of the design and the material it was fabricated with, diffusion of oxygen to the cathode surface becomes a limiting factor, thereby reducing the performance of the cell.

Other works have focused on reducing the footprint of their biofuel cell and optimizing its performance and stability. Often using compartment-less designs, groups have described fuel cells as small as 1.4 cm^2. One group in particular has demonstrated an ability to place their component and have a functional biofuel cell within a grape [147]. However, this reduction in size comes at the expense of cell performance. Although run off a single grape, the power produced is only 2.4 μW at 0.52 V [147]. Future work will need to focus on improved cathode performance and higher surface area electrode materials in order for the cell to perform well enough to be utilized as a viable renewable energy source for commercial items [146].

The field of microfluidics lends a possible avenue to solve problems with geometric design and transport limitations. Fluid handling and the dimensionality of these devices could be an ideal substrate for efficient, inexpensive biofuel cell production. Devices have been fabricated in near limitless design patterns with many different substrates, most commonly polymers such as polydimethylsiloxane (PDMS), polymethylmethacrylate (PMMA), or polycarbonate. PDMS offers a great deal of potential as a substrate for fuel cells because of its inexpensive production through common soft-lithography techniques. Most importantly, it has the ability to readily allow diffusion of gas through its polymer structure. This combined with tunable small dimensions may improve on previously mentioned cathode limitations due to a lack of oxygen at the surface of the electrode. Microchannels containing fuel, which could be hydrodynamic or static, could provide efficient fuel transport and a more complete fuel use.

Another advantage of microchip-based fuel cell design is the dominating presence of laminar flow. Little work has been done with microfluidic biofuel cell applications; however, recent work has shown an ability to utilize laminar flow boundary layers for separating cathodic and anodic solutions, eliminating the need for membrane materials which increase the resistance of the fuel cell. Palmore et al. have utilized a laminar boundary layer to deliver separate flow streams of catholyte (containing laccase) and anolyte (containing ABTS) over gold microelectrodes. This device pictured in Figure 5.14 demonstrates a maximum power density of 26 μW cm^{-2} and an open circuit potential of 0.4 V [148]. Similar efforts in the utilization of laminar flow for a microfluidic biofuel cell have also been described for an immobilized enzyme system which demonstrates an added stability and longevity of the enzyme systems [149]. Further advancement on this platform for the production

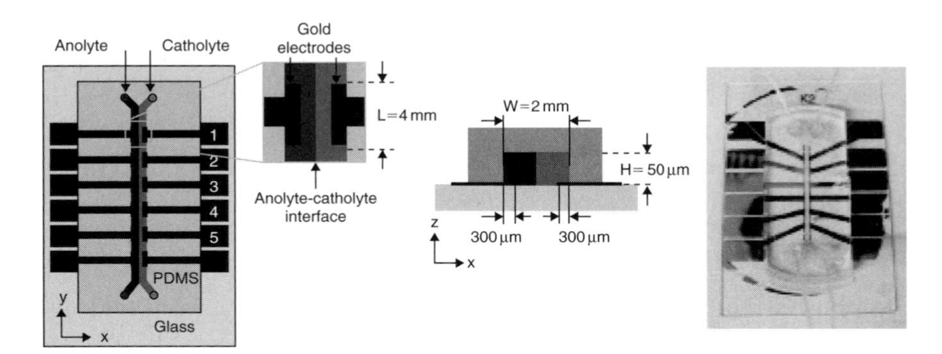

FIGURE 5.14 A design for a microfluidic biofuel cell and a picture of the actual device [148].

of biofuel cells may lead to unmatched power density production because of its extremely small size, ease of cell stacking, ability for fast oxygen diffusion, and intricate control of not only fuel delivery, but the environment of enzymes contained in the cell.

5.10 NANOMATERIALS IN ENZYMATIC BIOFUEL CELLS

Until recently, enzymatic biofuel cells have utilized linear surface area supports such as simple carbon electrodes [38, 129]. By incorporating high surface area, conductive nanomaterials, such as carbon black or nanotubes, power densities can be substantially increased by at least one order of magnitude [16, 150]. The reason for the improvement in the power density from the nanostructures is the increased surface area for enzyme attachment. Such support for enzyme immobilization has catapulted the usefulness of enzymatic biofuel cells into the power density range of traditional precious metal catalyzed low temperature alcohol fuel cells.

Another advantage to enzymatic biofuel cells is the need for additives in some cases. By incorporating additives, this allows for higher volumetric loadings. Controlled surface immobilization of enzymes, which enables high-rate DET, would eliminate the need for the mediator component, thus possibly leading to enhanced stability. Structures of controlled nanoporosity are necessary to obtain this surface immobilization at high volumetric enzyme loadings [26].

One specific example of nanostructures used in an enzymatic biofuel cell is a PQQ-ADH biofuel cell and the PQQ-GDH biofuel cell in which carbon black has been used to create a high surface area catalyst support matrix [151]. Carbon black has many varieties, but the one that is most commonly used in fuel cells is Vulcan XC-72 by Cabot Corporation. Vulcan XC-72 consists of 20 nm diameter spherical carbon nanoparticles. The carbon nanoparticles have a graphitic nature and resemble graphite onions (as shown in Figure 5.15), where there are multiple layers of graphite uniformly wrapped around a central core [153]. This allows the nanoparticles to be very conductive and maintain chemical inertness while electrochemical reactions are taking place. The reason

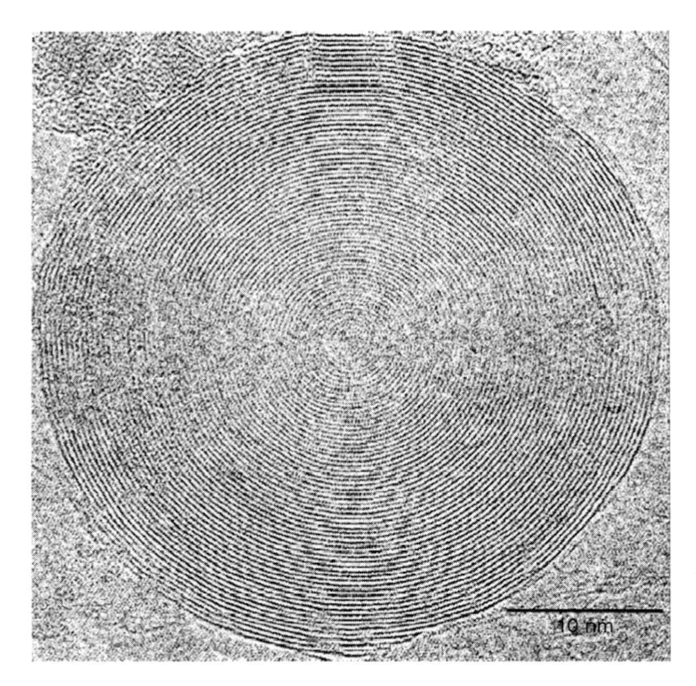

FIGURE 5.15 High resolution transmission electron micrograph of the graphitic onion that is the substructure of a carbon black nanoparticle [153].

Vulcan XC-72 is used so widely in fuel cells is because it is inexpensive to produce, has optimal bulk conductivity, has high surface area, and is easily dispersed. Vulcan XC-72 is called a high temperature furnace black or highly uniform acetylene black. It is produced by the inefficient burning of a hydrocarbon fuel in a heated low oxygen environment. The particle size is determined by the amount of oxygen present, temperature of the furnace, and duration in the furnace. Carbon black can be very hydrophobic depending on the synthesis conditions, meaning it does not like to disperse in water. This is from the graphitic qualities that particles maintain. This makes producing a usable fuel cell ink that can then be painted or sprayed onto an electrode challenging. To aid in the dispersion of the carbon black, frequently the carbon will be treated with a strong oxidizing acid such as nitric acid, making it much easier to disperse in water. When the graphitic surface of the carbon nanoparticles is subjected to the acid it reacts with the

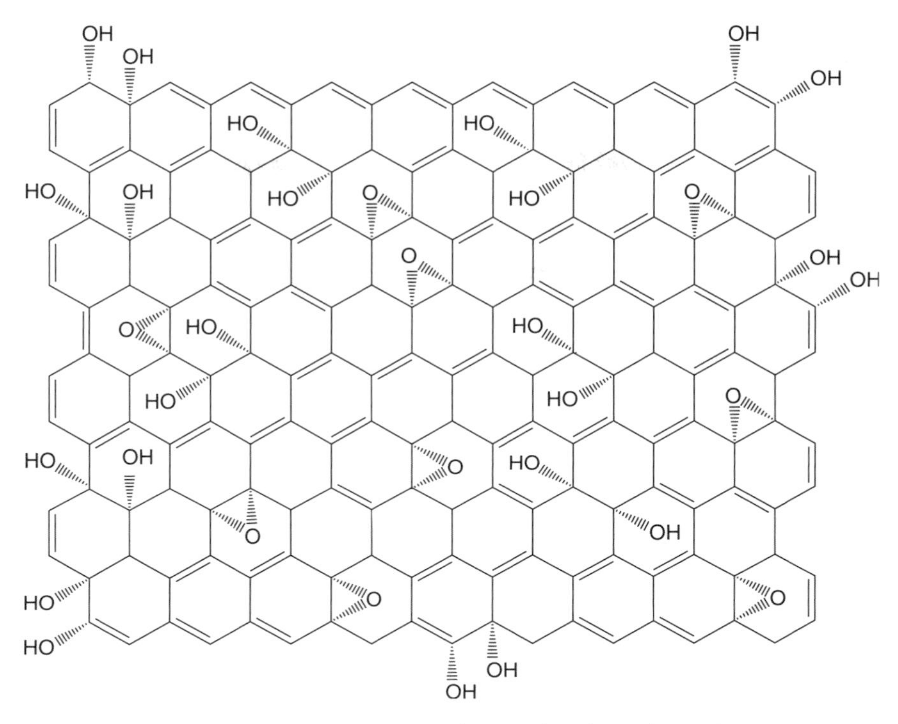

FIGURE 5.16 Representative structure of an oxidized graphene sheet.

fused aromatic rings of the graphene sheet to attach hydroxyls and carboxylic acid functionalities (as shown in Figure 5.16). This does not change the conductivity or surface area of the carbon; it only makes it more hydrophilic.

When the Vulcan XC-72 was incorporated into the PQQ-ADH biofuel cell and the PQQ-GDH biofuel cell, it increased power densities five- to tenfold over what was initially being produced [151]. To produce these high surface area biofuel cell electrodes, the enzyme solution is mixed with carbon black and a quaternary ammonium bromide-modified Nafion® polymer to produce a fuel cell ink. The ink is then applied onto the electrode by either painting or spraying. Carbon nanotubes are also being explored to increase the surface area and conductivity of the biofuel cell electrodes [152, 153].

Carbon nanotubes are graphene sheets that are rolled up to form tubular structures. Depending on the conditions of the nanotube

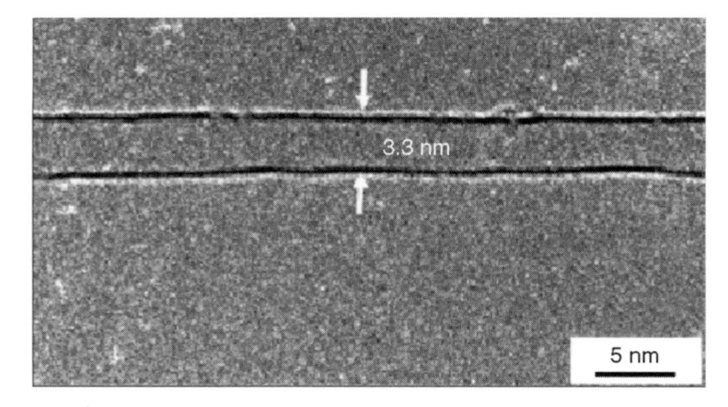

FIGURE 5.17 Transmission electron micrograph of a SWCNT [154].

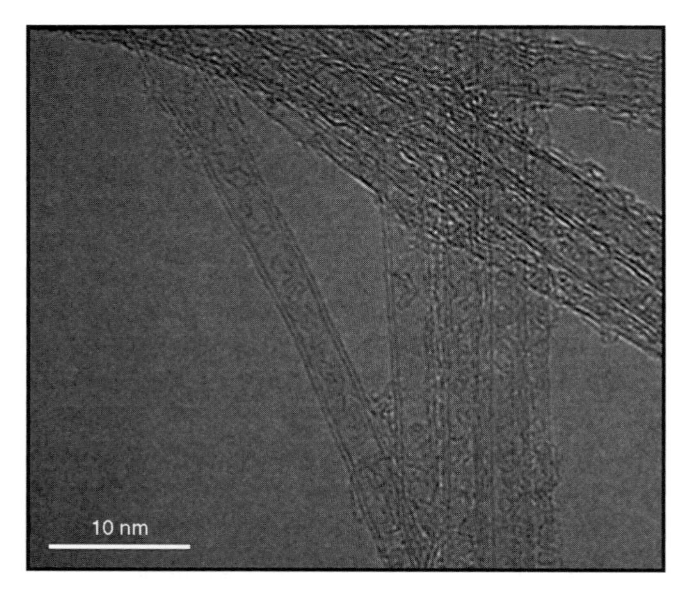

FIGURE 5.18 Transmission electron micrograph of a DWCNT [155].

synthesis, which include growth catalyst, temperature, and concentration of starting material, different types of carbon nanotubes can be created. The synthesis of such nanotubes is beyond the scope of this subject, but suffice it to say there are distinct types of carbon nanotubes. The first type is SWCNTs, which are shown in Figure 5.17. These nanotubes contain a single graphene sheet that is rolled up [154]. The next type of carbon nanotubes are the

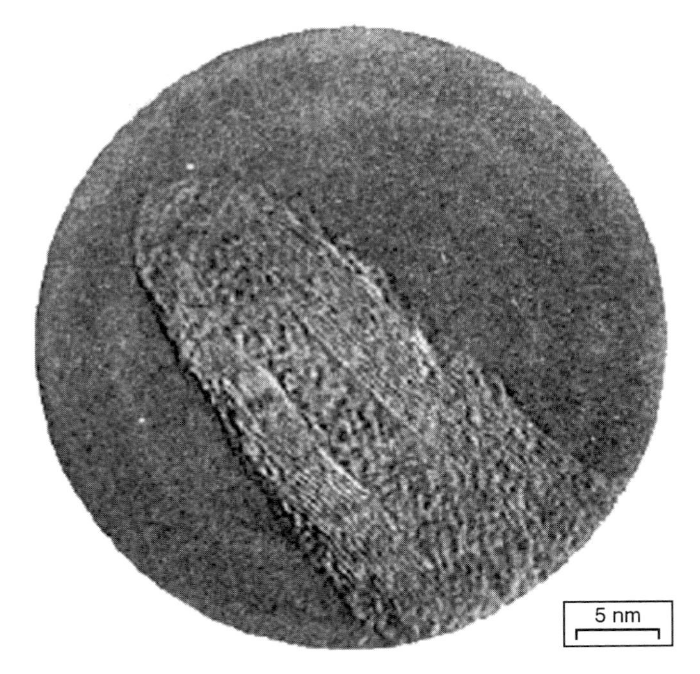

FIGURE 5.19 Transmission electron micrograph of MWCNT [156].

double-walled carbon nanotubes (DWCNT), which are shown in Figure 5.18 [155]. Likewise, multiwalled carbon nanotubes (MWCNT), which are shown in Figure 5.19, can have many layers of graphene sheets rolled into tubes [156]. The conductivity is directly related to the number of walls each carbon nanotube has [156]. Therefore, DWCNTs are more conductive than SWCNTs and MWCNTs are more conductive that DWCNTs.

When carbon black and nanotubes are used as a catalyst support matrix for a fuel cell, they are used to wire all the enzymes that are immobilized within the Nafion® to the current collector. In an ordinary low surface area electrode, only a small percentage of the enzymes that are immobilized in the polymer membrane are in contact with the electrode, leaving them to do all of the catalysis.

5.11 CONCLUSIONS

Although enzymatic biofuel cells have been a topic of research since the 1960s, it hasn't been until the last decade that researchers have started addressing the major issues associated with developing practical biofuel cells. Researchers have found strategies to increase enzyme and therefore biofuel cell operating lifetimes from hours/days to months/years, have increased power densities to a range that is reasonable for both implantable devices and portable electronics, and have started working with multi-enzyme systems to more completely oxidize the fuel, which results in an increase in efficiency. These technical developments have led to increased interest in enzymatic biofuel cells. Enzymatic biofuel cells can utilize a wide variety of high energy density fuels that can't normally be used with a traditional precious metal catalyzed fuel cell, including: sugars, ethanol, glycerol, soybean oil, etc. Modern enzymatic biocathodes can also operate in a number of modes: air breathing, dissolved oxygen, and peroxide consuming. Therefore, the versatility of biofuel cells will likely cause biofuel cells to find numerous applications in the future.

ACKNOWLEDGMENTS

The authors would like to thank the United Soybean Board, Air Force Office of Scientific Research, and Akermin for support.

References

[1] F. Davis, S.P.J. Higson, Biosens. Bioelectron. 22 (2007) 1224–1235.
[2] E. Kjeang, D. Sinton, D.A. Harrington, J. Power Sources 158 (2006) 1–12.
[3] M. Winter, R.J. Brodd, Chem. Rev. 104 (2004) 4245–4269.
[4] R.A. Bullen, T.C. Arnot, J.B. Lakeman, F.C. Walsh, Biosens. Bioelectron. 21 (2006) 2015–2045.
[5] S. Topcagic, B.L. Treu, S.D. Minteer, in: S.D. Minteer (Ed.), Alcoholic Fuels, CRC Press, Boca Raton, FL, 2006, pp. 215–231.
[6] S. Topcagic, S.D. Minteer, Electrochim. Acta. 51 (2006) 2168–2172.
[7] K. Kordesch, G. Simader, Fuel Cells and Their Applications, VCH Publishers, New York, 2001.
[8] K.R. Williams, An Introduction to Fuel Cells, Elsevier Publishing Company, New York, 1966.
[9] J. Kim, H. Jia, P. Wang, Biotechnol. Adv. 24 (2006) 296–308.
[10] E. Katz, A.N. Shipway, I. Willner, in: W. Vielstich, H.A. Gasteiger, A. Lamm (Eds.), Handbook of Fuel Cells-Fundamentals, Technology and Applications, vol. 1, John Wiley & Sons, Ltd, Jerusalem, 2003, pp. 1–27.
[11] S.D. Minteer, B.Y. Liaw, M.J. Cooney, Curr. Opinin. Biotechnol. 18 (2007) 228–234.

[12] C.M. Moore, S.D. Minteer, R.S. Martin, Lab on a Chip 5 (2004) 218–225.
[13] N.L. Akers, C.M. Moore, S.D. Minteer, Electrochim. Acta 50 (2005) 2521–2525.
[14] A.T. Yahiro, S.M. Lee, D.O. Kimble, Biochim. Biophys. Acta 88 (1964) 375–383.
[15] G.T.R. Palmore, G.M. Whitesides, ACS Sym. Ser. 566 (1994) 271–290.
[16] R.L. Arechederra, S.D. Minteer, Prep. Symp.–Am. Chem. Soc., Div. Fuel Chem. 52 (2007) 175–176.
[17] R.H. Petrucci, W.S. Harwood, F.G. Herring, General Chemistry Principles and Modern Applications, 8th ed., Prentice Hall, Upper Saddle River, NJ, 2002.
[18] G. Burton, Salters Advanced Chemistry, 2nd ed., Harcourt Heinemann, 2000.
[19] N.R.K.V. Reddy, E.J. Taylor, Generation of electricity with fuel cells using alcohol fuel (PCT International Patent Application #WO9202965A1), vol. 5, Physical Sciences, Inc., United States, 1992, 6 vol. 132, 193.
[20] A.B. Anderson, E. Grantscharova, S. Seong, J. Electrochem. Soc. 143 (1996) 2075–2082.
[21] S. Perkins, Science News 171 (2007) 21–22.
[22] V.R. Stamenkovic, B. Fowler, B.S. Mun, G. Wang, P.N. Ross, C.A. Lucas, N. M. Markovic, Science 315 (2007) 493–497.
[23] A.O. Neto, M.J. Giz, J. Perez, E.A. Ticianelli, E.R. Gonzalez, J. Electrochem. Soc. 149 (2002) A272–A279.
[24] Y. Iwasaki, Y. Suzuki, T. Kitajima, M. Sakurai, H. Kameyama, J. Chem. Eng. Jpn. 40 (2007) 178–185.
[25] H.A.O. Hill, I.J. Higgins, Philos. Trans. R. Soc. London 302 (1981) 267–273.
[26] S. Calabrese-Barton, J. Gallaway, P. Atanassov, Chem. Rev. 104 (2004) 4867–4886.
[27] N.C. Price, L. Stevens, Fundamentals of Enzymology, 3rd ed., Oxford University Press, New York, 1989.
[28] R.B. Silverman (Ed.), The Organic Chemistry of Enzyme Catalyzed Reactions, Academic Press, San Diego, CA, 2000.
[29] AmershamPharmaciaBiotech, Protein Purification Handbook, AB ed., New Jersey, 1999.
[30] K. Buchholz, V. Kasche, U.T. Bornscheuer (Eds.), Biocatalysts and Enzyme Technology, Wiley-VCH, Weinheim, 2005.
[31] F.M. Richards, Annu. Rev. Biophys. Bioengin. 6 (1977) 151–176.
[32] D.R. Storm, D.E.J. Koshland, Proc. Natl. Acad. Sci. USA 66 (1970) 43–52.
[33] W.P. Jencks, Adv. Enzym. Relat. Areas Mol. Biol. 43 (1975) 219–410.
[34] I. Willner, E. Katz, F. Patolsky, A.F. Buckmann, J. Chem. Soc., Perkin Trans. 2: Phys. Org. Chem. (1998) 1817–1822.
[35] E. Katz, B. Filanovsky, I. Willner, New J. Chem. 23 (1999) 481–487.
[36] E. Katz, I. Willner, A.B. Kotlyar, J. Electroanal. Chem. 479 (1999) 64–68.
[37] A. Pizzariello, M. Stred'ansky, S. Miertus, Bioelectrochemistry 56 (2002) 99–105.
[38] S. Tsujimura, K. Kano, T. Ikeda, Electrochemistry 70 (2002) 940–942.
[39] M. Togo, A. Takamura, T. Asai, H. Kaji, M. Nishizawa, Electrochim. Acta 52 (2007) 4669–4674.
[40] G.T.R. Palmore, H. Bertschy, S.H. Bergens, G.M. Whitesides, J. Electroanal. Chem. 443 (1998) 155–161.
[41] E. Katz, O. Lioubashevski, I. Willner, J. Am. Chem. Soc. 127 (2005) 3979–3988.

[42] E. Simon, C.M. Halliwell, C. Toh, A.E.G. Cass, P.N. Bartlett, Bioelectrochemistry 55 (2002) 13–15.
[43] L. Gorton, A. Lindgren, T. Larsson, F.D. Munteanu, T. Ruzgas, I. Gazaryan, Anal. Chim. Acta 400 (1999) 91–108.
[44] A.L. Ghindilis, P. Atanasov, E. Wilkins, Electroanalysis 9 (1997) 661–674.
[45] S. Shleev, J. Tkac, A. Christenson, T. Ruzgas, A.I. Yaropolov, J.W. Whittaker, L. Gorton, Biosens. Bioelectron. 20 (2005) 2517–2554.
[46] A. Chaubey, B.D. Malhorta, Biosens. Bioelectron. 17 (2002) 441–456.
[47] B. Brunetti, P. Ugo, L.P. Moretto, C.R. Martin, J. Electroanal. Chem. 491 (2000) 166–174.
[48] T. Aoyagi, A. Nakamura, H. Ikeda, T. Ikeda, H. Mihara, A. Ueno, Anal. Chem. 69 (1997) 659–669.
[49] A.G. Dubinin, F. Li, Y. Li, J. Yu, Bioelectrochem. Bioenerg. 25 (1991) 131–135.
[50] A.A. Karyakin, O.V. Gitelacher, E.E. Karyakina, Anal. Chem. 67 (1995) 2419–2423.
[51] A.A. Karyakin, E.E. Karyakina, W. Schuhann, H.L. Schmidt, S.D. Varfolomeyev, Electroanalysis 6 (1994) 821–829.
[52] C.R. Molina, M. Boujtita, N.E. Murr, Anal. Chim. Acta 401 (1999) 155–162.
[53] T. Ikeda, Bull. Electrochem. 8 (1992) 145–159.
[54] L. Gorton, E. Csoregi, E. Dominguez, J. Emneus, G. Jonsson-Pettersson, G. Marko-Varga, B. Persson, Anal. Chim. Acta 250 (1991) 203–248.
[55] L. Gorton, G. Jonsson-Pettersson, E. Csoregi, K. Johansson, E. Dominguez, G. Marko-Varga, B. Persson, Analyst 117 (1992) 1235–1241.
[56] L. Gorton, Electroanalysis 7 (1995) 23–45.
[57] T. Ruzgas, E. Csoregi, J. Emneus, L. Gorton, G. Marko-Varga, Anal. Chim. Acta 330 (1996) 123–138.
[58] M.R. Tarasevich (Ed.), Bioelectrochemistry, vol. 10, Plenum Press, New York, 1985.
[59] H.A.O. Hill, N.J. Hunt (Eds.), Methods in Enzymology, vol. 227, Academic Press, San Diego, CA, 1993.
[60] S.D. Varfolomeev, Y.V. Savin, I.V. Berezin, J. Mol. Catal. 5 (1979) 147–156.
[61] V.J. Razumas, J.J. Jasaitis, J.J. Kulys, Bioelectrochem. Bioenerg. 12 (1984) 297–322.
[62] M.R. Tarasevich, V.A. Boganaovskaya, Russ. Chem. Rev. 56 (1987) 1139–1166.
[63] I.V. Berezin, Appl. Biochem. Microbiol. 18 (1982) 451–453.
[64] I. Willner, G. Arad, E. Katz, Bioelectrochem. Bioenerg. 44 (1998) 209–214.
[65] E. Katz, I. Willner, J. Am. Chem. Soc. 125 (2003) 6803–6813.
[66] T.S.C.B. Chen, G. Binyamin, Z. Gao, Y. Zhang, H.-H. Kim, A. Heller, J. Am. Chem. Soc. 123 (2001) 8630–8631.
[67] N. Mano, F. Mao, A. Heller, J. Am. Chem. Soc. 124 (2002) 12962–12963.
[68] N. Mano, F. Mao, A. Heller, J. Am. Chem. Soc. 125 (2003) 6588–6594.
[69] N. Mano, F. Mao, W. Shin, T. Chen, A. Heller, Chem. Commun. (Cambridge, United Kingdom) 4 (2003) 518–519.
[70] H.-H. Kim, N. Mano, Y. Zhang, A. Heller, J. Electrochem. Soc. 150 (2003) A209–A213.
[71] N. Mano, A. Heller, J. Electrochem. Soc. 150 (2003) A1136–A1138.
[72] A. Heller, Phys. Chem. Chem. Phys. 6 (2004) 209–216.

[73] W.J. Blaedel, R.A. Jenkins, Anal. Chem. 47 (1975) 1337–1343.
[74] C. Zoski, Handbook of Electrochemistry, Elsevier, 2007.
[75] Y. Takeuchi, Y. Amano, J. Jpn. Pet. Inst. 47 (2004) 355–358.
[76] D. Zhou, H. Fang, H. Chen, H. Ju, Y. Wang, Anal. Chim. Acta 329 (1996) 41–48.
[77] Y.A. Ansari, S.D. Minteer, Polym. Mater. Sci. Eng. 94 (2006) 375–376.
[78] V.L. Davidson, L.H. Jones, Anal. Chim. Acta 249 (1991) 235–240.
[79] V.L. Davidson (Ed.), Principles and Applications of Quinoproteins, Dekker, New York, 1993.
[80] J.A. Duine, J. Frank, J.A. Jongejan, Adv. Enzymol. Relat. Areas Mol. Biol. 59 (1987) 169–212.
[81] J. Paralleda, E. Dominguez, V.M. Fernandez, Anal. Chim. Acta 330 (1996) 71–77.
[82] H. Ikeda, D. Kobayashi, F. Matsushita, T. Sagara, D. Niki, J. Electroanal. Chem. 361 (1993) 221–228.
[83] N. Yuhashi, M. Tomiyama, J. Okuda, S. Igarashi, K. Ikebukuro, K. Sode, Biosens. Bioelectron. 20 (2005) 2145–2150.
[84] B.L. Treu, S.D. Minteer, Polym. Mater.: Sci. Eng. 92 (2005) 192–193.
[85] R.S. Freire, C.A. Pessoa, L.D. Mello, L.T. Kubota, J. Braz. Chem. Soc. 14 (2003) 230–243.
[86] G. Gupta, V. Rajendran, P. Atanassov, Electroanalysis 16 (2004) 1182–1185.
[87] S. Shleev, A. El Kasmi, T. Ruzgas, L. Gorton, Electrochem. Commun. 6 (2004) 934–939.
[88] R. Duma, S.D. Minteer, ECS Trans. 5 (2007) 117–127.
[89] J. Wang, J. Liu, L. Chen, F. Lu, Anal. Chem. 66 (1994) 3600–3603.
[90] A. Bardea, E. Katz, A.F. Bueckmann, I. Willner, J. Am. Chem. Soc. 119 (1997) 9114–9119.
[91] T. Ito, M. Kunimatsu, S. Kaneko, S. Ohya, K. Suzuki, Anal. Chem. 79 (2007) 1725–1730.
[92] B.R. Azamian, J.J. Davis, K.S. Coleman, C.B. Bagshaw, M.L.H. Green, J. Am. Chem. Soc. 124 (2002) 12664–12665.
[93] V. Vamvakaki, K. Tsagaraki, N. Chaniotakis, Anal. Chem. 78 (2006) 5538–5542.
[94] M.J. Green, P.I. Hilditch, Anal. Proc. 28 (1991) 374–376.
[95] J. Hu, A.P.F. Turner, Anal. Lett. 24 (1991) 15–24.
[96] N.A. Morris, M.F. Cardosi, B.J. Birch, A.P.F. Turner, Electroanalysis 4 (1992) 1–9.
[97] J. Kulys, H.E. Hansen, T. Buch-Rasmussen, J. Wang, M. Ozsoz, Anal. Chim. Acta 288 (1994) 193–196.
[98] M. Mayer, J. Ruzicka, Anal. Chem. 68 (1996) 3808–3814.
[99] E. Csoeregi, L. Gorton, G. Marko-Varga, A.J. Tuedoes, W.T. Kok, Anal. Chem. 66 (1994) 3604–3610.
[100] T. Tatsuma, K. Ariyama, N. Oyama, Anal. Chem. 67 (1995) 283–287.
[101] D. Voet, J.G. Voet, C.W. Pratt, Fundamentals of Biochemistry, upgrade ed., John Wiley & Sons, Inc., New York, 2002.
[102] F.A. Bettelheim, J. March, Introduction to General, Organic & Biochemistry, 5th ed., Harcourt Brace College Publishers, Philadelphia, 1998.

[103] M.D. Arning, B.L. Treu, S.D. Minteer, Polym. Mater.: Sci. Eng. 90 (2004) 566–569.
[104] M.C. Beilke, S.D. Minteer, Polym. Mater.: Sci. Eng. 94 (2006) 556–557.
[105] M.C. Beilke, Saint Louis University, 2007.
[106] W. Wiechert, Metab. Eng. 3 (2001) 195–206.
[107] S. Tran-Dinh, F. Beganton, T.T. Nguyen, F. Bouet, M. Herve, Eur. J. Biochem. 242 (1996) 220–227.
[108] E. Katz, A.F. Bueckmann, I. Willner, J. Am. Chem. Soc. 123 (2001) 10752–10753.
[109] B. Persson, L. Gorton, G. Johansson, A. Torstensson, Enzyme Microb. Technol. 7 (1985) 549–552.
[110] N.L. Akers, S.D. Minteer, Preprints of Symposia – American Chemical Society, Division of Fuel Chemistry 48 (2003) 895–896.
[111] R.L. Arechederra, B.L. Treu, S.D. Minteer, J. Power Sources 173 (2007) 156–161.
[112] S. Topcagic, B.L. Treu, S.D. Minteer, Proceedings – Electrochem. Soc. (2005) 2004-18, 230–242.
[113] D.L. Johnson, J.L. Thompson, S.M. Brinkmann, K.A. Schuller, L.L. Martin, Biochemistry 42 (2003) 10229–10237.
[114] G.T.R. Palmore, H.-H. Kim, J. Electroanal. Chem. 464 (1999) 110–117.
[115] S. Tsujimura, H. Tatsumi, J. Ogawa, S. Shimizu, K. Kano, T. Ikeda, J. Electroanal. Chem. 496 (2001) 69–75.
[116] W.E. Farneth, B.A. Diner, T.D. Gierke, M.B. D'Amore, J. Electroanal. Chem. 581 (2005) 190–196.
[117] V. Soukharev, N. Mano, A. Heller, J. Am. Chem. Soc. 126 (2004) 8368–8369.
[118] L. Brunel, J. Denele, K. Servat, K.B. Kokoh, C. Jolivalt, C. Innocent, M. Cretin, M. Rolland, S. Tingry, Electrochem. Commun. 9 (2007) 331–336.
[119] W.E. Farneth, M.B. D'Amore, J. Electroanal. Chem. 581 (2005) 197–205.
[120] J. Lim, N. Cirigliano, J. Wang, B. Dunn, Phys. Chem. Chem. Phys. 9 (2007) 1809–1814.
[121] C.F. Blanford, R.S. Heath, F.A. Armstrong, Chem. Commun. 17 (2007) 1710–1712.
[122] S. Liu, Y. Yin, C. Cai, Chin. J. Chem. 25 (2007) 439–447.
[123] S. Liu, C. Cai, J. Electroanal. Chem. 602 (2007) 103–114.
[124] S. Jung, Y. Chae, J. Yoon, B. Cho, K. Ryu, J. Microbiol. Biotechnol. 15 (2005) 234–238.
[125] I. Willner, E. Katz, Angew. Chem. Int. Ed. 39 (2000) 1181–1218.
[126] R. Sahney, S. Anand, B.K. Puri, A.K. Srivastava, Anal. Chim. Acta 578 (2006) 156–161.
[127] K. Zeng, H. Tachikawa, Z. Zhu, V.L. Davidson, Anal. Chem. 72 (2000) 2211–2215.
[128] N.V. Kulagina, L. Shankar, A.C. Michael, Anal. Chem. 71 (1999) 5093–5100.
[129] C.M. Moore, N.L. Akers, A.D. Hill, Z.C. Johnson, S.D. Minteer, Biomacromolecules 5 (2004) 1241–1247.
[130] T.L. Klotzbach, S.D. Minteer, Polym. Mater.: Sci. Eng. 94 (2006) 599–600.
[131] M. Yalpani, L. Hall, Macromol. 17 (1984) 272.

[132] J.H. Lee, J.P. Gustin, T. Chen, G.F. Payne, S. Raghavan, Langmuir 21 (2005) 26–33.
[133] X. Jinrui, L. Bin, Anal. 119 (1994) 1599.
[134] X. Wu, G. Lu, X. Zhang, X. Yao, Anal. Lett. 34 (2001) 1205.
[135] Y. Xin, L. Guanghan, W. Xiaogang, Z. Tong, Electroanalysis 13 (2001) 923.
[136] C. Esquenet, E. Buhler, Macromolecules 34 (2001) 5287–5294.
[137] C. Esquenet, P. Terech, F. Boue, Langmuir 20 (2004) 3583.
[138] J. Desbrieres, M. Rinaudo, L. Chtcheglova, Macromol. Symp. 113 (1997) 135.
[139] J. Kim, J.W. Grate, Nano Lett. 3 (2003) 1219–1222.
[140] J. Lim, P. Malati, F. Bonet, B. Dunn, J. Electrochem. Soc. 154 (2007) A140–A145.
[141] N. Mano, H. Kim, Y. Zhang, A. Heller, J. Am. Chem. Soc. 124 (2002) 6480–6486.
[142] T.L. Klotzbach, S.D. Minteer, Polymer Preprints 48 (2007) 1008–1009.
[143] Y. Yan, W. Zheng, L. Su, L. Mao, Adv. Mater. 18 (2006) 2639–2643.
[144] R. Duma, S.D. Minteer, Polym. Mater.: Sci. Eng. 94 (2006) 592–593.
[145] N. Mano, F. Mao, A. Heller, A. ChemBioChem 5 (2004) 1703–1705.
[146] S.G. Teodorescu, W.L. Gellett, M. Kesmez, J. Schumacher, 209th ECS Meeting Abstracts, Denver, CO, 2006.
[147] N. Mano, F. Mao, A. Heller, J. Am. Chem. Soc. 125 (2003) 6588–6594.
[148] K.G. Lim, G.T.R. Palmore, Biosens. Bioelectron. 22 (2007) 941–947.
[149] S. Yoon, M. Mitchell, R.S. Jayashree, S.D. Minteer, P.J.A. Kenis, 2006 ESC Meeting Abstracts, Denver, CO, 2006.
[150] D. Ivnitski, B. Branch, P. Atanassov, C. Apblett, Electrochem. Commun. 8 (2006) 1204–1210.
[151] R.L. Arechederra, S.D. Minteer, Polym. Mater.: Sci. Eng. 94 (2006) 558–559.
[152] S.S. Zumdahl, in: Chemistry, D.C. Heath and Company, Lexington, Massachusetts, and Toronto, 1986, p. 1023.
[153] T. Cabioc'h, E. Thune, M. Jaouen, F. Banhart, Philos. Mag. A: Phys. of Condens. Matter: Struct., Defects and Mech. Prop. 82 (2002) 1509–1520.
[154] B. Ha, J. Park, S.Y. Kim, C.J. Lee, J. Phys. Chem. B 110 (2006) 23742–23749.
[155] Y.D. Lee, H.J. Lee, J.H. Han, J.E. Yoo, Y.-H. Lee, J.K. Kim, S. Nahm, B.-K. Ju, J. Phys. Chem. B 110 (2006) 5310–5314.
[156] V. Svrcek, C. Pham-Huu, J. Amadou, D. Begin, M.-J. Ledoux, F. Le Normand, O. Ersen, S. Joulie, J. Appl. Phys. 99 (2006) 64306.

6

Glucose Biosensors—Recent Advances in the Field of Diabetes Management

Frank Davis and Séamus P. J. Higson
Cranfield Health Cranfield University at Silsoe

Diabetes mellitus is a world-wide health problem with estimates of 300 million expected sufferers throughout the world by 2045. Although a cure does not as yet exist for this condition, the lives of many sufferers have been greatly improved by the availability of inexpensive disposable electrochemical sensors for blood glucose levels. The ability to accurately determine glucose enables diabetics to control blood glucose levels and so minimize the health risks associated with diabetes.

This review covers the latest advances in the development of glucose biosensors, the majority of which are electrochemical in nature but some of which are based on optical sensing. The development of electrochemical sensors is traced from initial beginnings through to the latest sensors being developed incorporating novel organic, biological and nanostructured materials. The development of implantable biosensors and other potential devices for continuous monitoring of blood sugar is also covered. The devices currently on the market and the potential for developing new types of noninvasive biosensors, possibly leading to development of the "artificial pancreas," are also discussed.

6.1 INTRODUCTION

This review provides an up-to-date overview of the most intensively researched area in the biosensor field—the development of glucose biosensors [1–3]. The reason for the continued interest in this field is the prevalence of diabetes mellitus, which has become a world-wide public health problem. The disease is one of the leading causes of death and disability in the world and at the time of writing the incidence is continuing to increase, with estimates of 300 million expected sufferers throughout the world by 2045 [1]. The global market for biosensors in 2004 was approximately $5bn, and of this it was estimated that approximately 85% of the world commercial market for biosensors was for blood glucose monitoring [1]. The development of inexpensive disposable electrochemical biosensors for glucose, incorporating glucose oxidase (GOx), has for these reasons been the focus of sustained and substantive research.

The review will begin with a description of the principle behind glucose biosensors and how biological moieties can be incorporated within sensors so as to lend their unique properties to a detection system. Different types of sensors will be described, with both electrochemical and optical devices being considered. There will then follow a brief history of glucose biosensors starting from its beginnings in the 1960s.

A description will be given of the various classes of enzyme-based electrochemical biosensors, namely the first, second, and third generation type devices. The principles and methods of construction of the most common of these devices will be described. This will be followed by details of the many different methods utilized to immobilize enzymes so as to exploit their specificity and selectivity. A wide variety of materials such as conducting polymers, redox active polymers, nanoparticles, carbon nanotubes, sol-gels, and various porous inorganic compounds have been incorporated within these biosensors so as to improve the "communication" between the biological moiety and the electrode; each of these will be reviewed in turn.

The next section will be dedicated to optically based glucose biosensors. Attempts have been made to miniaturize glucose biosensors and some of the latest results will be reviewed. This will be followed by a section on implantable glucose biosensors and the various problems still to be overcome to permit their routine use. The review will continue with an overview of the commercial glucose sensor field and we will consider the various commercial sensors available. The review will conclude with conclusions and comments on the possible future for glucose biosensing, together with a discussion of the problems that still need to be overcome.

6.2 PRINCIPLES OF GLUCOSE BIOSENSING

The purpose of this section is to introduce the concept of using enzymes as the selective recognition elements within electrochemical biosensors. Most sensors consist of three principle components, as described below and shown in Figure 6.1.

i. The first of these includes a chemical receptor capable of recognizing the anion of interest with a high degree of selectivity; this is usually concurrent with a binding event between the receptor and an analyte.
ii. The second component that must be present is a transducer, where the binding event is recognized via a measurable physical change such as a change in an optical spectrum or a change in electronic state.
iii. Thirdly, there must be inclusion of a method for measuring the change detected at the transducer and converting this into information that can be used.

Many different types of sensors have been constructed using this basic format. Biosensors are a class of sensors which exploit biological materials as active recognition elements within the sensor matrix. A wide variety of biological species have been utilized including enzymes, antibodies, aptamers, DNA, RNA, and even whole cells. Description of all of these types of sensors is outside the scope of this review, although many other reviews and book chapters have been published on this wide ranging field [3–10].

There are several advantages associated with using enzymes as the active recognition entity within a sensor. Enzymes display selectivities that are unsurpassed by many other recognition species; for example glucose oxidase will interact with glucose but not fructose

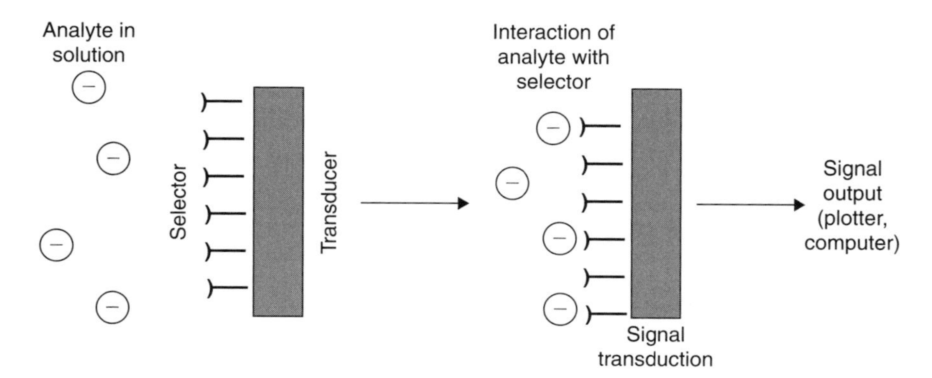

FIGURE 6.1 Schematic of a biosensor.

or sucrose—and in this way enzymes can act as a highly selective receptor. Another advantage that can be exploited is that many enzymes display extremely rapid turnover for their substrates and this is often essential to (a) avoid saturation and (b) to allow sufficient generation of the active species in order to be detectable.

6.3 ELECTROCHEMICAL SENSING OF GLUCOSE

6.3.1 First Generation Biosensors

Glucose oxidase is readily available commercially in high purity, displays high stability, and is relatively inexpensive. All of these factors contribute to making it the material of choice for glucose biosensing. In the case of glucose oxidase, the enzyme catalyzes the conversion of an electrochemically inactive substrate glucose into gluconolactone along with the concurrent generation of hydrogen peroxide. This process is shown in Eq. 6.1.

$$\text{glucose} + O_2 \overset{\text{GOx}}{\leftrightarrow} \text{gluconolactone} + H_2O_2 \qquad (6.1)$$

The first device to directly utilize this reaction was described by Clark in 1962 [11] and can be thought of as the first glucose biosensor. A typical device of this type is shown in Figure 6.2. To construct this device, a solution of glucose oxidase which has been mixed with bovine serum albumin is then crosslinked by glutaraldehyde while being held between two membranes. The resultant laminate is placed on top of a commercial amperometric oxygen electrode. The oxygen electrode is used to measure the concentration of dissolved oxygen in a solution which is placed on the electrode. The presence of glucose leads to consumption of oxygen at the laminate and the resultant measured drop in oxygen is proportional to the concentration of glucose.

This system led to further work by Updike and Hicks [12] who recognized that fluctuations in solution oxygen content could be affected by a variety of factors and therefore utilized two oxygen electrodes, one of which was coated with glucose oxidase. The differential current between these two electrodes was then determined in an attempt to account for and internally compensate for

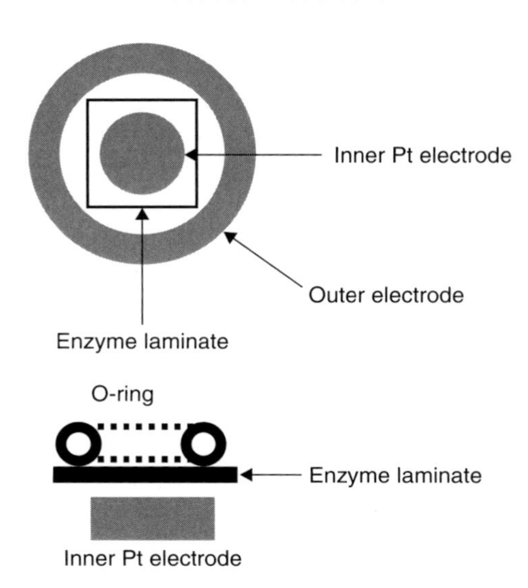

FIGURE 6.2 Schematic of a Rank oxygen electrode and enzyme electrode assembly.

variable oxygen levels [12]. These types of electrodes were utilized in the first commercial glucose analyzer, Model 23 YSI, launched by the Yellow Spring Instrument Company in 1975. This instrument was capable of measuring the glucose level in a 25 ml sample of whole blood. It should be noted in this context that 25 ml represents a relatively large sample for analysis.

An alternative method also exists, where the electrode is polarized to +650 mV vs. Ag/AgCl, so as to amperometrically monitor (Eq. 6.2) the oxidization of the hydrogen peroxide produced enzymatically.

$$H_2O_2 \xrightarrow{+650\ \text{mV vs Ag/AgCl}} 2H^+ + O_2 + 2e^- \tag{6.2}$$

This approach based on the detection of hydrogen peroxide was initially reported in 1973 [13]. The sensors developed using this process exhibited good levels of precision and accuracy when utilized in the analysis of 0.1 ml blood samples.

One major problem associated with these H_2O_2-based sensors, however, is that of erroneous results due to the presence of electroactive species such as ascorbate, acetaminophen, or uric acid that are capable of being oxidized at +650 mV vs. Ag/AgCl. During the 1980s a wide range of permselective and other materials were studied in an attempt to minimize this effect.

One strategy for minimizing interference from these electroactive species is to prevent their access to the electrode via application of a permselective coating to the sensor so that the concentration of interferents at the electrode surface is minimized. During the 1980s a wide range of permselective and other materials were studied in an attempt to minimize this effect. Polymeric materials have been widely utilized, with materials such as the fluorinated ionomer Nafion® [14, 15] and cellulose acetate [16] being two of the most widely studied. Species present in biological fluids will also often bind to and foul many surfaces, and again this can lead to erroneous readings. The polymers mentioned above can also be used in some instances to confer a degree of biocompatibility. Similar biocompatibility can also be conferred by a thin film of diamond-like carbon [17]. An alternative approach has been to electropolymerize suitable monomers to form protective coatings. For example, 1,2-diaminobenzene [18–20], when deposited at the bioelectrode surface, serves to both stabilize the electrode due to its inherent high biocompatibility while also imparting selective exclusion of interferents such as ascorbate. Similar results have also been obtained utilizing overoxidized polypyrrole [21].

With Clark-type electrodes, the results can also be affected by fluctuations in the ambient oxygen concentration. Low levels of oxygen lead to an oxygen deficit and can lead to electrodes giving lower readings than the true value for glucose concentration. Several methods have been utilized to address this: glucose dehydrogenase does not require the presence of oxygen as a cofactor and has been utilized as an alternative to glucose oxidase [22–24]. A series of mutated glucose dehydrogenase enzymes have also been studied with reported improvements in sensitivity and specificity [25]. Polymeric membranes have been utilized which preferably transport oxygen over glucose [26], thereby preventing oxygen deficit from occurring. Similar use has been made of electrodes containing fluorocarbon oils in which oxygen is extremely soluble, with these oils acting as an internal oxygen supply [3]. Another simple method reported has been the development of an electrode which permits oxygen diffusion from both sides of the electrode but glucose from only one [27].

Another approach has been to incorporate catalytic layers within the electrodes which facilitate the detection of the hydrogen peroxide liberated during measurement. Incorporation of these materials can greatly lower the detection potential (down to as low as −0.20 V vs. Ag/AgCl), thereby eliminating many of the interferents common to these systems [3]. Typical materials include carbon paste electrodes incorporating rhodium [28, 29], adsorbing glucose oxidase onto Prussian Blue electrochemically deposited onto glassy carbon [30], screen printed carbon [31], or the electrodeposition of glucose oxidase and Prussian Blue [32]. A glassy carbon electrode could also be coated with a Nafion®/glucose oxidase composite and then rhodium nanoparticles electrodeposited within this matrix to give a biosensor capable of detecting glucose selectively in the presence of common interferents [33]. More recently, entrapped glucose oxidase/poly(1,2-diaminobenzene) films [34] could be coated with copper and palladium. The deposited copper and pallidium layer enhanced the electrode performance due to its better catalytic activity to hydrogen peroxide, and the screening effect of the polymer film led to minimal effects from the interferents. Platinization of carbon paste electrodes has also been reported to improve their performance in glucose oxidase based biosensors [35]. Copper oxide has been incorporated along with glucose oxidase within a carbon paste electrode and shown to have a catalytic effect on peroxide detection [36], again allowing the detection of glucose in the presence of ascorbate, acetaminophen or uric acid. Rhodium oxide has also proved capable of electrocatalysis when incorporated into a Nafion® film, with a very low operating potential of −0.20 V vs. Ag/AgCl, negligible interference and a shelf-life of four months [37].

Similar effects could be observed for macrocyclic ligands, for example when cobalt phthalocyanine [38] was used as an electrocatalytic layer in a glucose electrode, especially when combined with a further coating of Nafion®. A similar material could be adsorbed onto a gold-thiol monolayer and display good electrocatalytic activity [39]. Manganese oxide also catalyzes reactions with hydrogen peroxide and was used as the basis for a glucose oxidase field effect transistor [40]. Combined coatings of manganese oxide with chitosan were also employed within glucose oxidase amperometric biosensors as an oxidant for ascorbate, thereby preventing interference from this material [41]. Nanoparticles of calcium carbonate, because of their high surface area, have also been used to immobilize glucose oxidase, although in this case no electrocatalytic effect is seen [42].

First generation membrane-based sensors are robust and capable of being used for multiple analyses. However, devices such as these are usually utilized within hospitals rather than for home analysis. The trend in recent years has been to develop inexpensive home detection methods where the physiological sample, usually blood, can be analyzed by the patient. The problems of cleaning the sensor are moreover negated by using disposable sensor strips.

6.3.2 Second Generation Biosensors

As described above, there are problems associated with detecting species such as oxygen or peroxide due to the presence of interferents. What would negate this problem would be the transfer of electrons from the active center of glucose oxidase before the reaction with oxygen occurs. However, electrons are unlikely to "jump" from the active center to the electrode because of the substantial protein layer that surrounds the flavin redox center, thereby insulating it from the electrode. For these reasons a new generation of biosensors was conceived which sidestepped this problem by utilizing an artificial electron charge transfer moiety, known as a mediator. The mediator must fulfill several requirements to make it suitable for use within a glucose sensor. Firstly it must react readily with the enzyme to avoid competition by ambient oxygen within the system. It is also essential that in both its reduced and oxidized forms the mediator be nontoxic and stable. Finally, it is preferable that the mediator require as low an over-potential to be oxidized as is feasible. Figure 6.3 shows a typical reaction scheme where a mediator compound is utilized to "shuttle" electrons between the enzyme and the electrode. The advantage of using an artificial mediator is that it lowers the potential required for measurement of the enzyme catalyzed reaction, thereby minimizing the interference by redox active species present within the sample to be studied.

One of the earliest mediators to be utilized was ferrocene [43] which can easily be switched between oxidized and reduced forms (Figure 6.4a). This chemistry led to the development of the first home glucose testing kit, the Exactech® glucose biosensor, a pen sized device produced by Medisense® which utilizes a disposable strip

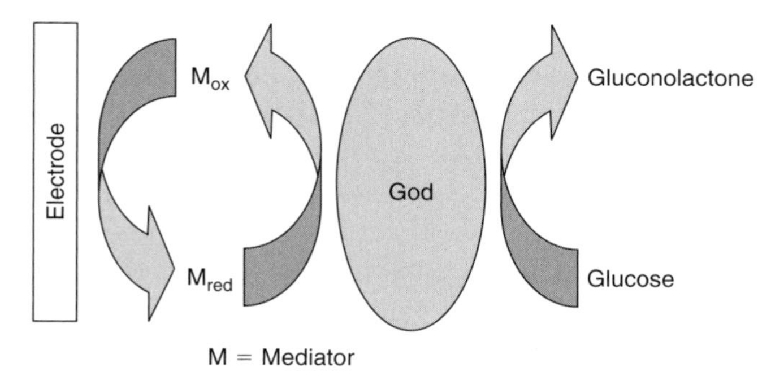

FIGURE 6.3 The oxidation of glucose at an electrode, with electron transfer achieved using a soluble mediator.

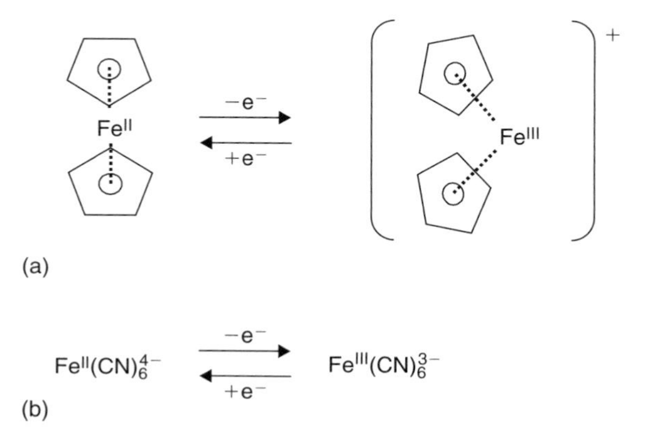

FIGURE 6.4 Oxidation/reduction cycles for (a) ferrocene and (b) ferri/ferrocyanide.

upon which a single drop of blood is placed to allow self-monitoring of blood glucose levels [3]. Usually 15–30 seconds suffices to obtain a blood glucose reading. A wide range of glucose sensors based on this method have since become commercially available. Other materials have also been utilized as mediators, such as, for example, the ferri/ferrocyanide redox couple (Figure 6.4b), which was utilized within the Elite® (Bayer), Surestep® (Lifescan), and Accu-Check® (Roche-Diagnostics) series of glucose biosensors [3]. Further work has produced devices such as the Pelikan® device, which only requires microliter blood volumes [1].

A range of further mediators have also been studied, including, for example, osmium and ruthenium bipyridyl complexes [44], ruthenium cluster compounds [45], viologen compounds [46], vinyl ferrocene [47], and cobalt porphyrin/phthalocyanine complexes [48]. Photocurable polymers containing ferrocene have been studied because of their ease of deposition [49] and the possibility of patterning the film [50]. It has also proved possible to covalently link mediators and enzyme, as seen, for example, in the linking of phenoxazine mediators to GOx with about 20 mediator molecules bound to each enzyme and catalyzing electron transfer [51]. Ferrocene and osmium-based mediators were also used, along with an AC impedance approach to develop interference-free glucose biosensors [52].

6.4 ALTERNATIVE METHODS OF ENZYME IMMOBILIZATION

Although the use of mediators has enabled the production of a wide range of commercial sensors, there is still a great deal of interest in constructing mediator-free glucose biosensors. As mentioned earlier, a major problem is the difficulty of electron transfer between the enzymes and the electrode. A variety of methods have been utilized to immobilize the enzymes on an electrode surface in such a way to facilitate electron transfer. These include novel methods of forming ultrathin films of the enzymes and also via attempts to "wire" the enzyme to the electrode, utilizing such diverse materials as redox active or conducting polymers, metal nanoparticles, and carbon nanotubes. These will be discussed below.

6.4.1 Langmuir-Blodgett (LB) Films

It has long been known that organic materials can spread upon a water surface to form thin films. Classical LB film techniques involve the dissolution of materials in suitable water immiscible solvents and spreading the resultant solution onto a water surface [53, 54]. Classical LB materials are usually substances like stearic acid with a hydrophilic headgroup and a hydrophobic carbon tail. Evaporation of the solvent leaves behind a monolayer on the water surface [53]. Compression of this monolayer with movable barriers can lead to formation of a well-packed monolayer on the water surface.

If a clean substrate (silicon, glass, etc.) is passed through a suitable compressed monolayer into the water subphase, a monolayer of the amphiphile attaches to the substrate. When withdrawn, a

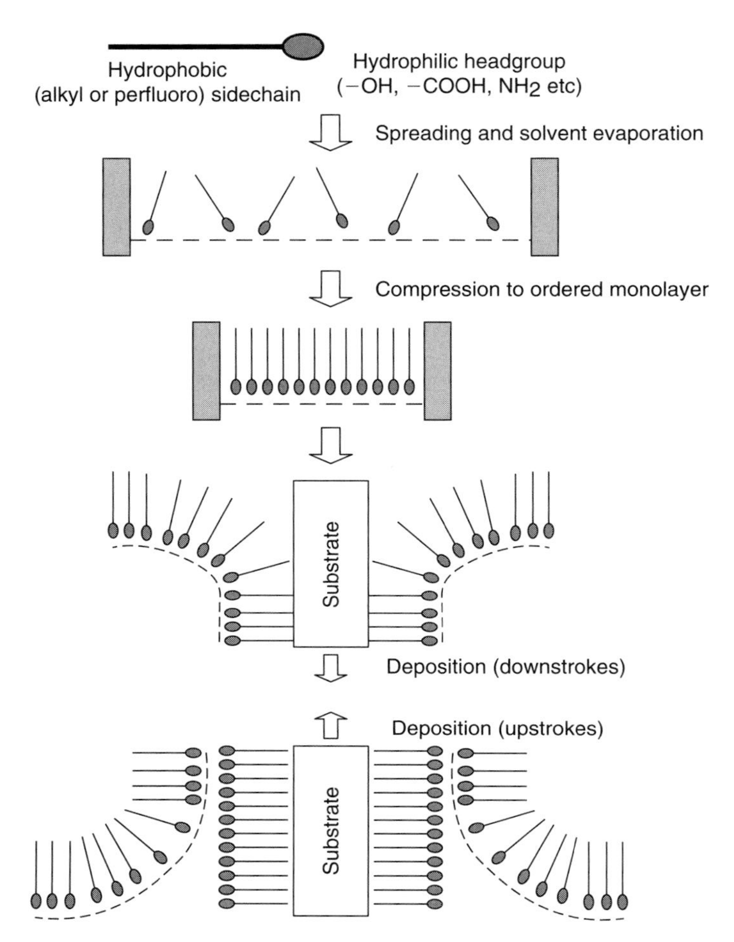

FIGURE 6.5 Formation of an ordered monolayer at the air-water interface and deposition of a Langmuir-Blodgett multilayer.

second layer can attach as shown (Figure 6.5). If the surface pressure is maintained (usually by moving in the barriers automatically via a feedback circuit), this process can be repeated to build up multilayers of any required thickness.

Often species of biological interest are water-soluble, which precludes direct deposition using LB techniques since they would simply dissolve in the subphase. This does not render their incorporation into such films impossible [55]. One method relies on the fact that many biomolecules are electrically charged and capable of being dissolved in the subphase. A layer of an oppositely charged amphiphile can be spread on the water surface and form a composite layer with the biomolecules. This composite can be deposited as an LB film.

Glucose oxidase was utilized in an initial study [56] with a series of different lipids being used to adsorb the enzyme from solution and showed that a glucose sensor could be fabricated by co-depositing GOx with octadecyltrimethylammonium chloride. Similar work [57] involved forming a composite of GOx with a cationic lipid and depositing a bilayer on a platinum electrode. This electrode assembly was shown to respond to glucose with a rapid response time of 5 seconds, probably due to its extreme thinness. Fatty acid/GOx and phospholipids/GOx monolayers were also deposited onto ATR crystals [58] and FTIR studies revealed that the fatty acid incorporated more of the enzyme.

Glucose oxidase was dissolved in the subphase and incorporated into behenic acid bilayers [59] and studied by AFM. Images of LB films of these composites displayed parallel ridges with a periodicity of 6.5 nm, which is compatible with a morphology where the glucose oxidase molecules (which are approximately 6×13 nm in dimension) are aligned in a close packed structure. Later AFM images showed that there is some aggregation of the enzymes which are coated with a behenic acid monolayer [60]. More extensive studies into the behavior of the glucose oxidase/behenic acid monolayer system and its deposition onto silicon were performed [61]. Silinization of the Si/SiO_2 surface and use of high deposition pressures lead to formation of the best quality films as shown by FTIR and AFM.

One major drawback of the spreading and deposition processes used to form LB films is that they can lead to lowering or destruction of the activity of the enzymes used, as can storage. The stability and activity of glucose oxidase bound to lecithin:cholesterol monolayers and deposited (three layers) onto glass, was found to be greatly increased by incorporation of submicron hydrophobic silica particles within the monolayer, probably due to inhibition of leaching of enzyme from the film [62].

A variety of cationic amphiphilic polyelectrolytes based on polyethylenimine or polyvinyl pyridine were spread on subphases containing glucose oxidase [63] and the resultant complexes transferred onto polypropylene membranes. These were incorporated into a Clark oxygen electrode and the resulting biosensors gave linear responses to glucose (1–20 mM); these were found to be reproducible even after up to 100 measurements, or alternatively after 2 months storage in air. Similar work utilized glucose oxidase bound to monolayers of a cationic lipid containing a methacrylate group, deposited onto platinum [64]. UV irradiation, which causes polymerization of the lipid, led to improvements in the stability and reproducibility of the resultant biosensor. A cellulose polymer can also be used as a matrix for the physisorbtion of glucose oxidase while retaining its activity [65]. A composite octadecyl trimethyl ammonium/nanosized Prussian Blue/glucose oxidase composite could also be deposited and gave a clear response current to glucose when polarized at 0 V vs. Ag/AgCl [66].

Conducting polymers can also be incorporated into LB films. Polyaniline/glucose oxidase LB films can be deposited, for example, and used as electrochemical sensors for glucose with linear response from 5–30 mM [67]. Other polymers include poly-3-dodecyl thiophene and this polymer when mixed with stearic acid forms a suitable matrix for the deposition of glucose oxidase [68]. The resultant composite retains its electroactivity and can be utilized for the detection of glucose. Similar work utilized poly 3-hexythiophene [69].

An LB monolayer of a cross-linked polysiloxane [70] could be deposited as a permselective barrier in a glucose sensor. Charged interferents (ascorbate, acetaminophen, etc.) are excluded by the film.

6.4.2 Polyelectrolyte Multilayers

A more recently developed technique for assembly of thin films (shown schematically in Figure 6.6) takes advantage of the strong attraction between oppositely charged polyelectrolytes [71]. A charged solid substrate is placed in a dilute solution of an oppositely charged polyelectrolyte (such as polystyrene sulphonate), with the resultant strong multiple charge interactions causing the deposition of a thin layer of the polymer so as to generate a negatively charged surface. This sample is then rinsed and placed in a second, oppositely charged polyelectrolyte (e.g., polyallylamine hydrochloride) solution, causing adsorption of a second layer regenerating the initial charged surface. This process can be repeated to build up multilayers of any desired thickness.

Besides simple polyelectrolytes, this process can be applied to a wide variety of charged materials, allowing the assembly of

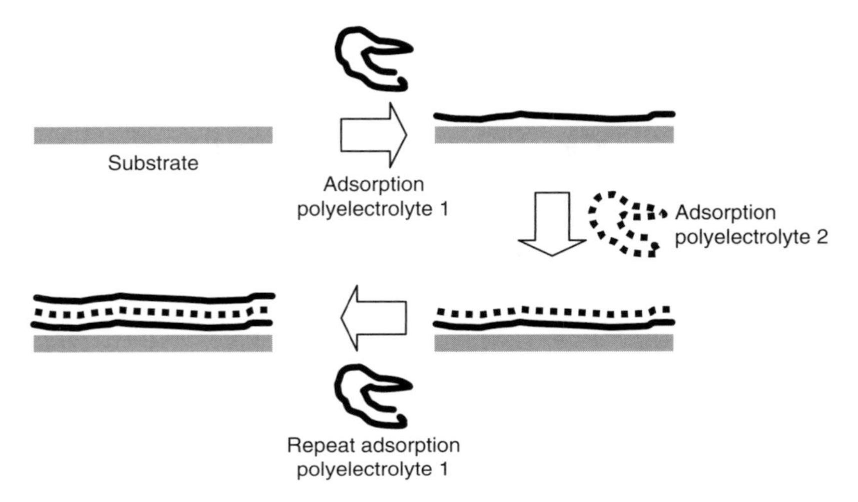

FIGURE 6.6 Deposition of a self-assembled polyelectrolyte multilayer.

layers containing species such as enzymes, proteins, antibodies, polynucleotides, and charged colloidal particles. The formation and structure of these layers has been extensively reviewed elsewhere [72–74]. Whereas LB films are often highly ordered and crystalline, these self-assembled films tend to be more amorphous in nature, with the polyelectrolytes of the individual layers tending to interpenetrate somewhat [72]. The strong multiple charge neutralization interactions were found to lead to the formation of highly stable films and also offer an advantage in that any defects formed in the assembly of a monolayer tend not to propagate through the structure but are "covered over" by overlying layers [71]. The pH and ionic strength of the solutions used to assemble these films and also those in which they are utilized can affect the film structure, allowing for fine-tuning of the multilayer thickness and properties.

Enzymes are often electrically charged, making them suitable for incorporation in these films and, as this method works well from dilute aqueous or buffer solutions, this represents a good approach for easily assembling thin films with enzymatic activity [55]. Initially, simple addition of polyelectrolytes to carbon paste electrodes could be used to stabilize bound enzymes. For example, a polyvinyl pyridine-based cationic material [75] was added to carbon paste and glucose oxidase adsorbed on the paste electrode. This approach was found to give a 25-fold increase in current density over non-polyelectrolyte containing electrodes, as well as with associated improvements in linear range and half-life.

Polyacrylic acid covalently bound to solid surfaces [76] was utilized to bind thin layers of glucose oxidase with retention of its activity. A ferrocene containing polyvinyl pyridine derivative could be deposited on a gold electrode and assembled with glucose oxidase [77]. The polymer in this case acted as both co-assembler and electron mediator, with the resultant film forming the basis of an amperometric enzyme electrode with an extended glucose detection range of 0.01–10 mM glucose as well as experiencing low interference from uric acid. Other redox active materials based on Co incorporated into polyvinyl pyridine could be deposited with glucose, oxidase [78] with surface plasmon resonance studies of the resultant film showing sensitivity to glucose with a linear glucose response from 1–10 mM. Even after 3 weeks in an aqueous environment, no loss of material from the surface was observed, indicating that loss of activity of these systems is due to enzyme inactivation rather than simple "leaching" of active material.

The enzyme does not necessarily need to be in actual solution for self-assembly to occur. Glucose oxidase can be encapsulated in a polyelectrolyte multilayer and when incorporated [79] into a thin film on a carbon electrode has been shown to give electrochemical responses for 1–10 mM glucose, together with a much faster response time than a biosensor prepared by conventional polymer entrapment approaches. Other workers have deposited polyethylenimine on a variety of commercial membranes as supplied or following grafting with polyacrylic acid [80]. These cationically modified materials have been shown to be suitable substrates onto which assembled layers of glucose oxidase and resultant composite membrane were then used in glucose sensing. Polyethylenimine and aminodextran were later used with glucose oxidase [81] and found to give a sensor with excellent stability compared to standard techniques, with no loss of activity after 4 months storage and 50% retention of enzyme activity following 245 hours, continuous operation. GOx could also be deposited as a layer-by-layer film with chitosan and Nafion® to give a highly selective (0% interference from ascorbate and uric acid) sensor for glucose [82]. More recently glucose oxidase has been immobilized in layer-by-layer films containing a variety of polyelectrolytes such as polyethylenimine, Nafion® and DNA to give glucose sensors with fast response times and low sensitivity to ascorbate and uric acid [83]. Polyethylenimine was also utilized to immobilize glucose oxidase onto pyrolytic graphite [84].

Another variation of this technique is especially suitable for biological components; it is known that a very specific strong interaction occurs between the biotin unit (Figure 6.7a) and the protein avidin [85] and this may be used for the site-specific immobilization of enzymes. Since each avidin molecule is capable of binding as many as four biotin units, it can be used in the build-up of multilayers. Since many enzymes can be biotinylated a number of times without great loss of activity, they can be utilized, for example, in the structure in Figure 6.7b which shows a schematic of a multilayer of avidin/a tetra-biotin unit. A biotinylated glucose oxidase and lactate oxidase, for example, could be deposited as alternating layers with avidin to allow good retention of catalytic activity [86].

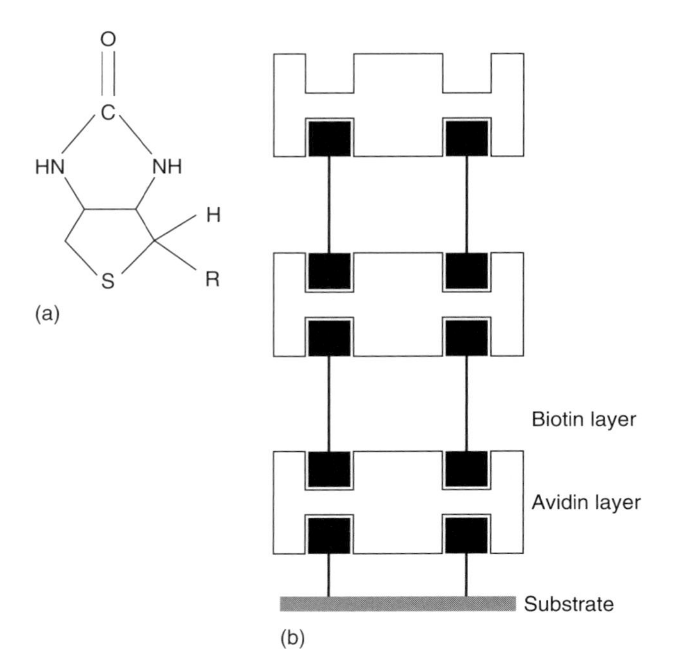

FIGURE 6.7 (a) Structure of the biotin headgroup; (b) Schematic of the structure of an avidin-biotin type multilayer.

6.4.3 Self-Assembled Monolayers (SAMs)

A variety of SAMs have been used to immobilize biomolecules and a number of reviews have been published covering early work in this field [55, 87, 88]. There are several advantages of utilizing SAMs as components of biosensors; they are, for example, extremely thin which leads to minimal diffusion effects, which in turn facilitates rapid detection. Often SAMs are covalently bound to the surface, giving them high stability and a wide variety of chemistries can be utilized, giving us the opportunity to tailor the properties of the surface. They also often provide a membrane-like microenvironment for the biomolecules, similar to environments they occupy in natural systems and since so little material is required to form a monolayer, they only require minimal amounts of what are often expensive biomolecules.

One of the simplest early SAMs involved treating carbon electrodes with an oxygen plasma to generate reactive groups such as alcohols or carboxylic acids, and chemically substituting these with cyanuric chloride groups. These species are then capable of covalently binding enzymes such as glucose oxidase or L-amino acid oxidase [89], which can then be used within amperometric sensors for the target species. However, the field of SAMs (especially for biosensor applications) has recently been dominated by gold-thiol monolayers.

Much early work on these systems has been extensively reviewed by Ulman in the books published in 1991 and 1998 [90, 91]. Sulphur-containing compounds have been shown to display high reactivity towards noble metals and especially gold, which is the most commonly used for several reasons. The gold-sulphur interaction is based on binding between "soft" gold and sulphur atoms, whereas many of the functional groups present in biological species such as acids, amines, etc. are relatively "hard" and do not interact strongly with the gold surface. Therefore, the use of di- or polyfunctional molecules is possible since the gold-sulphur binding interaction frequently dominates and other chemical species tend not to interfere. Gold is also relatively easy to clean, does not oxidize under standard lab conditions, and any weakly physically adsorbed impurities are displaced by the sulphur species. The best binding occurs between gold and thiol groups [90, 91] but other species such as disulphides, thiones, thioesters, etc. have been used. Molecules such as hexadecanethiol lead to well packed quasi-crystalline monolayers, whereas shorter thiols such as hexanethiol give liquid-like monolayers.

The stability and thinness of these layers, plus the versatility of a gold substrate which makes them easily investigated by QCM, FTIR, AFM, SPR, and electrochemical methods, rapidly led to these systems being investigated as possible sensing films. SAMs of simple substituted thiols can be used to generate a surface capable of binding biological moieties. Sulphur containing biological species can alternatively be bound directly to the surface. Both methods have been utilized; for example, glucose oxidase was chemically modified to attach thiol groups to the enzyme. These thiols were then utilized to attach the enzyme to gold microelectrodes [92] to give an amperometric glucose sensor with fast (<20 s) response to glucose in the range 0–50 mM. Alternatively covalent binding of glucose oxidase [93] to previously deposited mercaptopropionic acid monolayers could be utilized, leading to the construction of an amperometric glucose biosensor. In the case of this system, electron transfer could be achieved by use of a mediator (p-benzoquinone). Alternatively, the enzyme electrode could be electrochemically platinized and in this case no mediator is required. A similar system was reported where AC impedance rather than amperometric measurements were used to determine glucose concentration [94].

In most of the sensors reported, the biochemical moiety is adsorbed onto a simple planar electrode or a series of microelectrodes. One advantage of utilizing gold-thiol monolayers is the possibility of these compounds forming patterned systems due to their ability to be easily printed onto gold using a soft polymeric stamp with resolution down to 30 nm [95]. This microcontact printing gives us a method to directly manufacture "circuits" from these monolayers. These layers could then be used to assemble patterns of proteins, enzymes, viruses and cells. Several reviews are available on this subject [96–98].

6.4.4 Conjugated Polymers

A variety of monomers are capable of being electropolymerized on an electrode surface to form stable polymeric films. Typical polymers of this type include polyaniline, poly(1,2-diaminobenzene), polypyrrole, and polythiophene (Figure 6.8). Polymers of these types can often contain a highly conjugated backbone and display

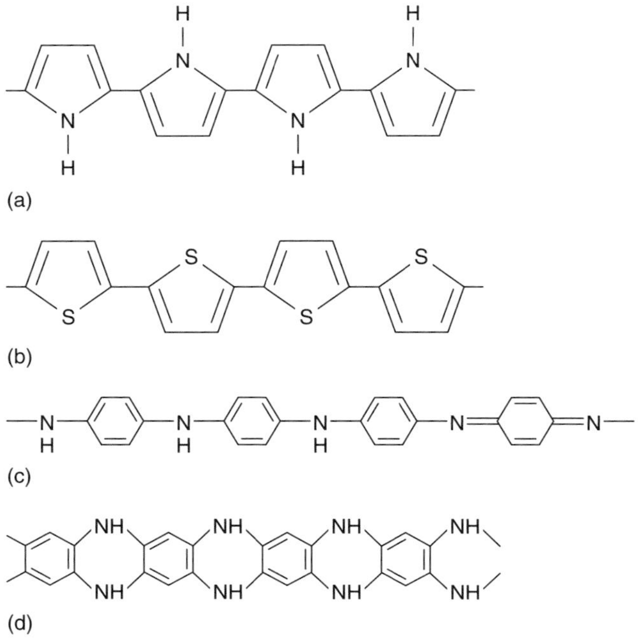

(a)

(b)

(c)

(d)

FIGURE 6.8 Structures of (a) polypyrrole, (b) polythiophene, (c) emeraldine polyaniline and (d) poly(1,2-diaminobenzene).

properties such as electrical conductivity, low energy optical transitions, and a high affinity for electrons. They are, however, not always conductive with, for example, films made from poly(1,2-diaminobenzene) often being insulating depending on the conditions under which the polymer films are grown [18–20].

Electrodeposited polymers are especially suitable for immobilization of enzymes at electrode surfaces and this process has been reviewed in detail elsewhere [99, 100]. Simply having biological species such as enzymes present in solution during the electrochemical polymerization process can lead to their becoming entrapped within the film during the deposition process [101–104]. Alternatively, a polymeric film can be deposited electrochemically and then the biological species can be adsorbed onto or be chemically grafted to the film [99, 100]. These processes both lead to a close association between the polymer and the biomolecule, potentially facilitating rapid electron transfer between the active species and an electrode surface, especially if the polymer is electrically conductive.

Alternatively, should the active species interact with a substrate, changes in its conformation or oxidation state or generation of an active species could lead to a change in the properties of the polymer film. This then may lead to a measurable change in its electrochemical or optical properties. The polymer film can in this way be thought to be acting as the transducer element within the biosensor.

One of the simplest methods involves the entrapment of glucose oxidase within polyaniline films [105]. Aniline was electrochemically polymerized from a solution containing 3 mg/ml glucose oxidase onto platinum electrodes. When exposed to glucose solution, the glucose oxidase led to the formation of hydrogen peroxide [105]. The presence of this active species could then be measured electrochemically. This method is capable of being used for controlled deposition of biological molecules onto electrodes of just about any size and composition. An AC impedance technique was later used to interrogate this system [106] and showed that not only could the hydrogen peroxide produced by oxidation of glucose be detected but, even in anaerobic conditions, the presence of glucose could be detected by the enzyme electrode.

Other workers have also utilized polyaniline as a host for glucose oxidase, for example, within an LB film [67]. More recently, composite materials utilizing polyaniline have been attracting interest. For example, a platinum electrode was coated (by solution casting) with a porous polyacrylonitrile film, into the pores of which aniline was electrodeposited with glucose oxidase [107]. The resultant electrodes showed good performance and no loss of activity following 100 tests or 100 days storage. Polyaniline was deposited electrochemically as a hydrochloride salt and then ion exchanged with either tosylate or ferricyanide ions with concurrent large increases in film porosity and enhanced loading with glucose oxidase [108]. Electrodeposited copolymers of aniline and 2-fluoroaniline [109] were shown to form a suitable substrate for physical adsorption of glucose oxidase and were found to be stable up to 45°C and displayed good linearity from 0.5–22 mM glucose.

A variety of electrodeposited polymers have been studied as hosts for enzymes as reviewed by other authors [99–104] and here we will report more recent work. Polypyrrole is also a popular material since, like polyaniline, it can be deposited in a variety of oxidation states as well as in charged or uncharged, conducting or insulating forms. Glucose oxidase could be co-deposited with polypyrrole onto platinum electrodes of different morphologies to give sensors, with the roughest Pt electrodes displaying higher levels of enzyme adsorption and orders of magnitude greater sensitivity [110]. As a method of eliminating interferents, a four electrode cell was constructed with one polypyrrole/glucose oxidase working electrode and a similar working electrode without the enzyme. Subtraction of the responses of the two electrodes enabled the elimination of the effects of interferents [111].

Again, composites of conducting polymers have been investigated. Polypyrrole was incorporated into an alginate gel [112] to allow a concurrent decrease of permeability and increase in stability over an unsubstituted gel to provide a material suitable as a host for glucose oxidase. Polypyrrole/poly(vinyl sulphonate) films could be electrochemically deposited onto indium tin oxide to give a porous structure capable of adsorbing high levels of glucose oxidase, resulting in increases in stability and lifetime [113]. Similarly, a polypyrrole/poly(vinyl alcohol) composite could be deposited to give an electrically conductive, hydrophilic, and microporous film [114] which was capable of adsorbing ten times as much glucose oxidase (by covalent binding to the film) than a non-porous polypyrrole film. The presence of poly(vinyl alcohol) within the film also minimized protein adsorption upon exposure to blood. Emulsions of polypyrrole and glucose oxidase could be entrapped within polyacrylamide gels and the resultant composites when immobilized on platinum electrodes shown to detect glucose under anaerobic conditions, indicating direct electron transfer between the enzyme and the electrode. These electrodes were moreover free from interference from ascorbate or uric acid [115]. Films of polypyrrole-nickel hexacyanoferrate could be utilized as electrocatalysts for hydrogen peroxide reduction [116], as could polypyrrole composites with a pyridyl substituted hexacyanoferrate [117] where an electron tunneling effect through the aromatic groups is thought to occur. Electrodeposited polypyrrole could also be chemically grafted with viologen groups containing a carboxy group which could then be utilized to covalently bind glucose oxidase [118]. High sensitivities were observed, indicating that the viologen is acting as a mediator. A copolymer of pyrrole and carboxyethyl pyrrole gave a conductive film onto which glucose oxidase could be chemically grafted, with the resultant sensors showing the greatest response when the substituted pyrrole comprised 5% of the monomer units [119]. Glucose oxidase substituted with a polyanion chain could be deposited as a composite with polypyrrole, with the polyanion being strongly bound to cationic polypyrrole, which meant that strong adsorption of the enzyme occurred no matter what its net surface charge [120].

An interesting variation on electrodeposition was reported where glucose, glucose oxidase, and pyrrole were combined in solution [121]. Hydrogen peroxide was produced enzymatically and served as the initiator for the polymerization of pyrrole with the resultant formation of polypyrrole/glucose oxidase nanoparticles (Figure 6.9). Evaporation of the solution on carbon electrodes and glutaraldehyde crosslinking lead to formation of glucose sensors with enhanced stability compared to controls manufactured from crosslinked enzyme only.

Other polymers have also been studied. For example, poly(1,2-diaminobenzene) could be co-deposited with glucose oxidase onto platinized electrodes, with the platinization process being capable of causing a 60-fold increase in sensitivity of the resultant sensor [122]. Nanotubular poly(1,2-diaminobenzene) could also be generated on a platinum electrode and has been used as a support for Prussian Blue, which acted as a catalyst for hydrogen peroxide detection [123]. These assemblies were then used as a base for glucose oxidase sensors, which gave low detection limits (0.05 mM), high selectivity, and were found to be stable for 3 months at 4°C, and 4 weeks if stored at room temperature.

Electropolymerization can be carried out with 2-aminophenol which was used to immobilize glucose oxidase on a copper modified gold electrode, with the copper nanoparticles improving the sensitivity and detection limit when compared to unmodified electrodes [124].

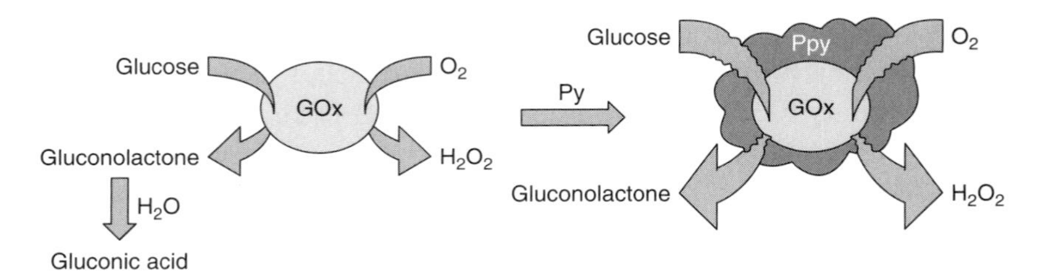

FIGURE 6.9 Synthesis of polypyrrole encapsulated glucose oxidase nanoparticles (reproduced from [121], Figure 1 with permission from Elsevier).

Poly(3,4-ethylenedioxythiophene) (PEDOT) is known as being relatively stable as a conductive polymer and therefore has been investigated as a potential material for incorporation into a biosensor. Electropolymerization of PEDOT could be utilized to entrap glucose oxidase and when combined with a ferricinium mediator was shown to act as a glucose sensor [125]. Comparison of PEDOT with polypyrrole showed that PEDOT retained its electroactivity for much longer and would be preferable in long term sensor applications [126]. A PEDOT-based transistor has been constructed with a platinum gate electrode [127]; when the gate electrode was exposed to a solution containing glucose and GOx, levels of glucose could be determined without need to immobilize the GOx.

Thiophene and thiophene-3-acetic acid could be copolymerized electrochemically on electrodes and then functioned as an immobilized mediator for glucose oxidase, especially when polylysine was used to immobilize glucose oxidase onto the conducting polymer [128]. Polymerization of an indole derivative onto glassy carbon also gave a suitable substrate for glucose oxidase immobilization and construction of a biosensor [129]. Thionine could be electrodeposited onto gold and used as a substrate and mediator for glucose oxidase/glutaraldehyde/Nafion® composites with a reported linear range of 0.005–5 mM glucose [130].

We have within our laboratory utilized a novel method based on combining electrochemistry and sonochemistry, utilizing both nonconductive and conductive polymers to fabricate arrays of conductive microelectrodes with entrapped glucose oxidase [131]. An insulating film of poly(1,2-diaminobenzene) is initially electrochemically deposited on an electrode of either gold or screen printed carbon. Sonochemical ablation may then be used to create an array of pores. Conductive polyaniline containing glucose oxidase is then deposited within the pores, as shown schematically within Figures 6.10a-c. Scanning electron microscopy clearly demonstrates formation of pores within the film and mushroom-like protrusions of polyaniline (Figures 6.10d, e). The resultant electrodes were found to be capable of detection of glucose with high sensitivity, good linearity and stir-independence for glucose responses. Amperometric or AC techniques could both be used to interrogate the sensors.

All these polymers tend to form a "soft" matrix; interestingly, however, when glucose oxidase is electrochemically co-deposited

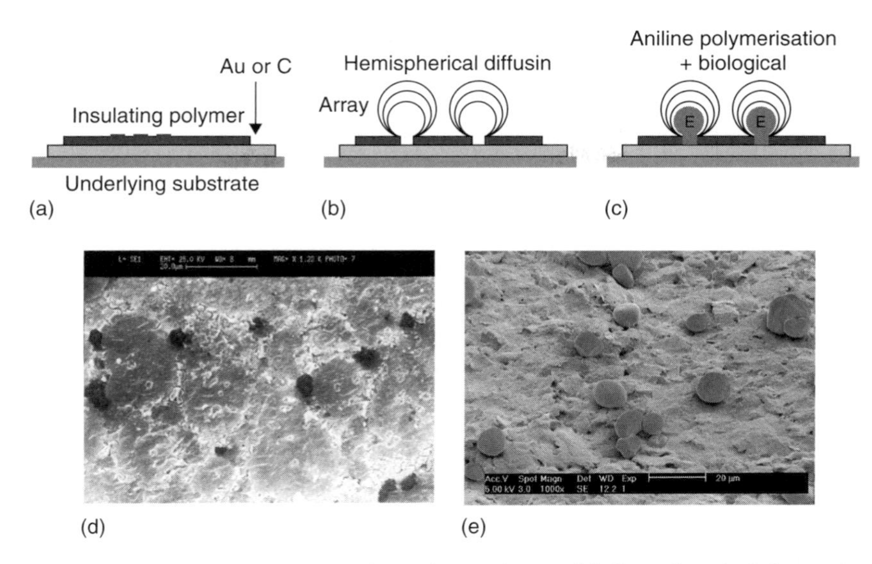

FIGURE 6.10 (a) Deposition of insulating layer; (b) Sonochemical formation of pores; (c) Polymerization of aniline; SEM pictures of (d) pores, (e) polyaniline "mushroom" protrusions.

with a thin film of metallic copper, the enzyme retains its electroactivity and the resultant hydrogen peroxide may be detected electrochemically. Use of crystal violet in the deposition solution appears to increase film porosity and sensitivity [132].

6.4.5 Redox-Active Polymers in Biosensors

As an alternative to the use of mediators, it has been proposed that a suitable polymer could "wire" the enzyme to the electrode. Conducting polymers such as mentioned above have been widely utilized for this purpose; however, another possible method that has been studied utilizes a polymer that does not conduct electrons along the polymer backbone but rather shuttles electrons between electroactive groups bound along the polymer chain. This can be thought of as a solid, polymeric equivalent of the soluble mediators used in second generation biosensors.

Initial work such as that of Heller in 1990 utilized polymers containing electroactive species [133], for example, of a composite material containing polyvinylpyridine and osmium 2,2-bipyridine, as shown in Figure 6.11a. The resultant redox active polymer was

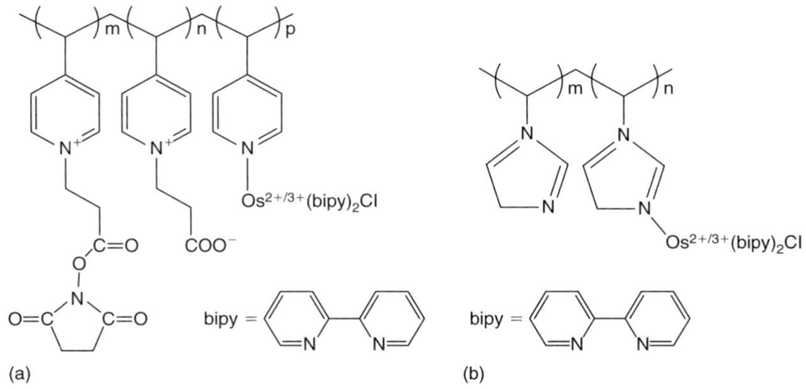

FIGURE 6.11 Structures of redox active polymers based on osmium bipyridyl complexes substituted onto (a) polyvinyl pyridine (b) polyvinyl imidazole.

capable of being deposited at an electrode surface as an electrostatic complex with glucose oxidase and was shown to respond to glucose in the physiological range. An alternative strategy was to immobilize enzymes covalently by use of redox active polymers, which also contained reactive groups such as succinimide [133]. This made possible the construction of films of up to 1 μm thick, which gave electrochemical responses to glucose.

An alternative system was developed based on osmium modified polyvinyl imidazole (Figure 6.11b) which, when mixed with a polyethylene glycol-based crosslinker, could be used to immobilize glucose oxidase onto electrodes to form a glucose sensor [134]. The addition of a second polymer, Nafion® improved the performance of these sensors and allowed constructions of sensors with linear ranges of 6–30 mM (glucose) with only a negligible response being observed to common interferents [135]. A similar polymer was used to immobilize glucose oxidase onto carbon fiber electrodes to give a sensor capable of real-time monitoring of blood glucose levels within rabbits [135].

This technique is highly versatile since the behavior of the polymers can be fine-tuned by variation of their substituents. For example, using a layered enzyme electrode where both glucose oxidase and bilirubin oxidase are "wired" by polyvinyl pyridine/osmium polymers to a glassy carbon electrode, concentrations of glucose as low as 2fM can be detected in the presence of atmospheric oxygen [136].

The photochemically initiated copolymerization of polyethylene glycol dimethacrylate and vinyl ferrocene could be used to synthesize hydrogels containing redox active groups. These materials were utilized to immobilize glucose oxidase on gold electrodes [137] with the resulting glucose sensors showing good linearity between 2–20 mM glucose concentrations. By combining the use of these materials and photolithographic techniques, patterned sensors could be constructed. Other polymers of this nature have been utilized. For example, ferrocene substitution of the polysaccharide chitosan and use of the resultant redox polymer to immobilize glucose oxidase allowed construction of a glucose sensor where the ferrocene units transfer charge between the enzyme and the electrode [138]. Chitosan could also be used to form a composite with hexacyanoferrate and then screen printed carbon electrodes were coated with this composite followed by glucose oxidase. The presence of chitosan was found to lead to a signal enhancement when compared to similar systems without chitosan [139]. A copolymer of vinyl ferrocene and acrylamide, co-deposited with glucose oxidase into a miniature channel on a screen printed electrode, allowed detection of glucose in samples as small as 0.2 μl [140].

Plasma polymerization of vinyl ferrocene onto a needle-type electrode followed by further plasma deposition of acetonitrile allowed the construction of a hydrophilic, redox active layer suitable for the immobilization of glucose oxidase and construction of a glucose sensor [141, 142]. Plasma deposition of a substituted ferrocene over a physisorbed layer of glucose oxidase on a screen printed electrode could also be utilized to develop a glucose sensor with good reproducibility [143].

6.5 NANOTECHNOLOGY IN BIOSENSORS

Much recent work has been performed studying the incorporation of various nano-sized species within biosensors. Two of the most studied species have been metal nanoparticles and carbon nanotubes.

6.5.1 Metal Nanoparticles

It has become possible to synthesize a wide variety of metal nanoparticles with exquisite control of size, composition, and surface coatings. Since these particles often have similar sizes to the enzymes used within this field, attempts have been made to manufacture composites where the small size of the nanoparticles permits intimate contact throughout the composite in which the metal nanoparticles are often conducting in nature, so allowing "wiring" of the enzyme.

Because of its relative inertness and capability to be substituted with thiols, gold has been widely studied. For example, gold electrodes can be modified with a monolayer of gold nanoparticles, which are then used as a substrate to covalently attach glucose oxidase. This led to enhanced electron transfer between enzyme and electrode when compared to direct immobilization of glucose oxidase without the nanoparticle layer [144], with the resulting sensor reported to be stable for up to 30 days. This stability could be doubled by use of a thiol containing silica gel to attach the nanoparticles [145]. A variation on this system deposited Prussian Blue onto a gold electrode, followed by electrodepositing poly(1,2-diaminobenzene), which was then used as a substrate for deposition of gold nanoparticles and then followed by glucose oxidase [146]. This in turn led to formation of a glucose biosensor with linearity for responses between 0.002–6.0 mM glucose together with good stability. Glassy carbon electrodes could also be used as substrates for gold nanoparticles, which could be utilized as the basis for a bienzyme film sensor with glucose oxidase and horseradish peroxidase [147]. A further study was focused towards systems where glucose oxidase was crosslinked onto films of colloidal gold immobilized on gold and carbon electrodes, together with control samples without colloid [148]. The highest sensitivity and stability was observed for the gold colloid modified gold electrode. The procedures described above could also be reversed so that glucose oxidase could be chemically grafted onto a thiol coated gold electrode and then coated with gold nanoparticles. It was reported that this approach also facilitated direct enzyme-electrode transfer [149].

A single-step electrodeposition sufficed for the construction of a chitosan-glucose oxidase-gold nanoparticle composite onto gold electrodes [150], to give a glucose biosensor with a much simpler deposition approach than many of those already described. Formation of this film onto a Prussian Blue modified electrode [151] gave a sensitive glucose biosensor with a wide linear range

(0.001–1.6 mM glucose). A variation of this method utilized a soluble gold salt and allowed formation of the nanoparticles *in situ* during the electrodeposition process [152]. The layer-by-layer polyelectrolyte deposition has also been used to assemble chitosan-gold nanoparticle-glucose oxidase multilayers on platinum electrodes with good electron transfer and glucose sensing properties [153]. Gold nanoparticles and glucose oxidase have also been incorporated in composites with an involatile ionic liquid [154], with dihexadecyl phosphate [155] and with Nafion® [156], each of which showed direct enzyme-electrode transfer. Gold nanoparticles were also combined with a redox active polymer, polyvinyl ferrocene, on platinum with the resultant electrode being used to immobilize glucose oxidase [157]. The resultant sensor was used to detect hydrogen peroxide and shown to have 6.6 times the maximum current, 3.8 times the sensitivity, and 1.6 times the linear range than a similar electrode without the nanoparticles.

Other metals have been utilized in attempts to mediate electron transfer between the electrode and the enzyme. Silver nanoparticles could be entrapped within a crosslinked polymer film along with glucose oxidase, which together were immobilized on a platinum electrode [158]. These electrode assemblies displayed much enhanced electron transfer, as shown by increased current responses in comparison to systems without nanoparticles. Similar results were also found for nanoparticles of an Au-Ag alloy and when utilized in the same system [159], the composites also showed enhanced stability. Silver nanoparticles could be deposited from solutions containing DNA [160], with the DNA preventing particle aggregation and improving the catalytic ability of the nanoparticles towards oxygen and hydrogen peroxide reduction, thus enabling their use in a first-generation glucose sensor.

Platinum nanoparticles have also been studied within these systems. For example, polyaniline could be deposited electrochemically onto stainless steel electrodes followed by electrochemical deposition of platinum nanoparticles. This was then used as a substrate for electrochemical deposition of poly(1,3-diaminobenzene) containing entrapped glucose oxidase [161]. The resultant sensor exhibited linearity from 0.002–12 mM glucose, fast response times (7 s), and good stability with good screening of interferents being in turn provided by the poly(1,3-diaminobenzene). Platinum nanoparticles

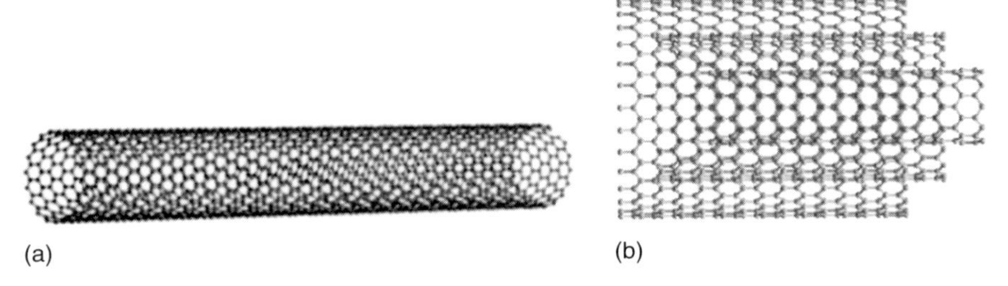

FIGURE 6.12 Structures of (a) single-walled and (b) multi-walled carbon nanotubes.

encapsulated within polymeric dendrimers [162] have also been reported to self-assemble using electrostatic interactions with glucose oxidase onto platinum electrodes to give glucose sensors.

6.5.2 Carbon Nanotubes

Since many of the sensors developed so far use carbon-based electrodes, the use of carbon nanotubes became inevitable once they became available in usable quantities. The structures of single-walled (SWNT) and multi-walled (MWNT) are shown in Figure 6.12. Several properties of carbon nanotubes such as their high aspect ratio, high surface area to volume ratio, ability to be substituted chemically, and abilities to act as electrical conductors make them attractive for use in biosensors. It should also be remembered that carbon nanotubes possess similar dimensions to many biological molecules used within biosensors. Carbon nanotubes are, however, often insoluble and physically intractable and so need to be modified prior to use.

Casting a suspension of MWNTs onto screen-printed carbon electrodes followed by casting of GOx and a ferricyanide mediator gave a glucose sensor with higher sensitivity and linear range than the control samples without nanotubes [163]. MWNTs can be oxidized to give them surface carboxyl groups which can then be modified to allow covalent linking of glucose oxidase [164]. The resultant water-soluble composites can be cast along with Nafion® and ferrocene carboxylic acid onto glassy carbon to give a glucose sensor with a good (0.5–40 mM) linear range for glucose.

A composite of GOx, chitosan, and MWNTs could be simply cast onto glassy carbon [165] to give a glucose sensor which displayed direct electron transfer. A similar composite of GOx, Nafion®, and

MWNTs was cast onto glassy carbon and the resultant H_2O_2 formed in presence of glucose was detected electrochemically [166]. Other polymers used to enable processing of nanotubes have included polydimethyldiallyl ammonium chloride [167] and polyamine dendrimers [168, 169], both of which allow direct electron transfer to occur between GOx and the resultant modified electrodes. Dispersions of MWNTs in aminotriethoxysilane and Nafion® mixtures could be cast onto glassy carbon or carbon fibers [170] and used as substrates for crosslinking glucose oxidase films, so as to allow direct enzyme-electrode electron transfer.

Polypyrrole could be electrodeposited with entrapped GOx and carboxylic acid modified MWNTs to give a sensor capable of detecting the hydrogen peroxide produced in the presence of glucose with linear range up to 50 mM [171]. Poly(1,2-diaminobenzene) [172] and poly(2-aminophenol) [173] could be used as a host in a similar manner and the resultant composite on glassy carbon [172] or gold electrodes [173] gave superior performance as glucose sensors over control samples containing no nanotubes. A Prussian Blue nanodispersion could be cast onto an MWNT/glassy carbon electrode and coated with electrodeposited poly(1,2-diaminobenzene)/GOx [174], with enhanced sensitivity and selectivity.

The layer-by-layer method has also been utilized to assemble multilayers of MWNTs and GOx, with increases in sensitivity with thickness observed for up to five or six bilayers [175, 176]. Similarly a field effect transistor could be modified by MWNTs and layers of cationic polymer and GOx with the resultant H_2O_2 being detected [177]. Layer-by-layer films of MWNTs and electrochemically deposited poly (neutral red) displayed high electrocatalytic activity towards hydrogen peroxide, far above that observed for systems containing only one or neither of the materials [178].

Arrays of vertical carbon nanotubes could be deposited onto a chromium surface and the ends of the nanotubes then selectively modified to allow GOx to be covalently immobilized just on the ends of the nanotubes [179] to give an array of enzyme modified nanoelectrodes. Hydrogen peroxide was produced in the presence of glucose and the nanotubes catalysed its reduction at −0.2 V vs. Ag/AgCl with no interference being observed from ascorbate, acetaminophen or uric acid. Chemical vapor deposition within an alumina substrate could also be utilized to develop arrays of carbon nanotubes, which again could be substituted at the ends with glucose oxidase [180]. The resultant nanoelectrode array showed direct electron transfer rates greater than the rate of oxygen reduction by glucose oxidase.

A recent paper has reported the direct deposition of glucose oxidase onto carbon nanofibers [181] and compared the sensors obtained with sensors based on carbon nanotubes and graphite powder. They report higher sensitivities and longer lifetimes for carbon nanofiber based devices, possibly because of their greater surface area to volume ratio and increased number of active sites. Other novel forms of nanostructured carbon have also been utilized, such as graphite/iron composites where the iron acts as a mediator [182] and allows for the detection of glucose with no effect from common interferents. Glucose oxidase immobilized onto the plane of pyrolytic graphite showed direct electron transfer without the aid of a mediator [183, 184]. Diamond electrodes have been utilized as substrates for glucose sensors [185, 186] as has boron-doped diamond [187] and platinized diamond [188]. Carbon foam [189] has been used to encapsulate glucose oxidase within its pores, leading to construction of a glucose sensor with higher sensitivities than polymer encapsulated GOx, while porous carbon has been similarly used to encapsulate GOx and a ferrocene mediator [190].

As both metal nanoparticles and carbon nanotubes improve the properties of biosensors containing these species, it was inevitable that they would be combined. For example, SWNTs and platinum nanoparticles could be co-dispersed within Nafion® to give GOx sensors with enhanced sensitivity over sensors modified with Pt or SWNTs alone [191]. Improved sensitivity was also observed when carbon paste electrodes were loaded with MWNTs which had platinum nanoparticles attached [192]. Similarly, gold nanoparticle/MWNTs displayed a remarkably higher sensitivity than other GOx-MWCT composites [193]. Carbon nanotubes could also be covalently linked to platinum nanoparticles [194] or alternatively the layer-by-layer technique [195] could be used to assemble layers of GOx and dendrimer modified Pt nanoparticles onto carbon nanotubes to give glucose biosensors. Platinum nanowires could also be encapsulated with carbon nanotubes in a chitosan matrix [196] to provide catalytic activity to hydrogen peroxide and when used in conjunction with GOx were used to form glucose biosensors. Palladium nanoparticles have also been co-deposited with

GOx onto nanotube modified electrodes to allow the catalytic reduction of hydrogen peroxide [197].

6.6 STABILIZATION OF ENZYMES

One major drawback associated with many biosensors is that the activity of the recognition element (usually glucose oxidase in the case of glucose sensors) can diminish with time. Numerous methods have been attempted to stabilize the enzymes, with some such as Nafion® and other polymers having already been mentioned within this chapter. One method that seems to have become more prominent in recent years [3] is via the use of inorganic systems such as stabilizers.

Nanoporous alumina membranes can be prepared by controlled anodization and used to encapsulate glucose oxidase which, when combined with a surface coating of chitosan, were found to prevent enzyme leaching from sensors [198]. Films of glucose oxidase co-deposited with nanosized silica retained bioactivity and due to the high surface area of the silica also displayed high sensitivity [199]. Similar work via the immobilization of GOx/silica composites onto a field-effect transistor displayed enhanced sensitivity and stability over systems without silica [200]. Further stabilization of silica adsorbed GOx (up to 4 months) could be achieved by combining the silica support with modified polyacrylonitrile [201]. Other inorganic species used include magnesium silicate [202], ferrocene intercalated into vanadium pentoxide xerogel [203], and zirconium oxide, which when combined with chitosan led to increases in both sensitivity and storage [204]. Clays such as smectite could be modified with cationic groups and used to immobilize glucose oxidase [205]. In a similar manner anionic layered double hydroxide clays could be used as immobilization substrates for GOx by electrostatic interactions [206], or via electrodeposition followed by glutaraldehyde crosslinking [207] to give glucose sensors.

Hydrogels have been shown to act as stabilizing layers when applied to sensors. A swollen hydrogel of high water content mimics an aqueous environment and helps to prevent denaturing. For example, glucose oxidase could be incorporated within a crosslinked PHEMA membrane and enabled formation of a glucose sensor which only showed a 20% loss in activity following 3 months continuous operation [208]. Alginate microspheres could be used to encapsulate GOx by a variety of methods and when incorporated into polyelectrolyte multilayers allowed enhanced retention of GOx when stored in buffer [209]. Copolymers of acrylonitrile with acrylic acid formed microgels and could be used to entrap GOx which, when used in an amperometric biosensor [210] with the composites, displayed remarkable stability with no loss of initial response from the sensor after 4 months storage. The actual microgel composites were moreover capable of being freeze dried and displayed shelf lives of at least 18 months.

Inorganic sol-gels have been of interest because of their stability and superior physical qualities. Some of the most recent work includes studies where different siloxane compounds were used to host GOx and applied to copper hexacyanoferrate [211], cobalt hexacyanoferrate [212] or polyneutral red [213] modified carbon electrodes to give sensors with lifetimes of several weeks. Another group investigated a wide range of silica sol-gels and conditions of GOx entrapment [214]. Similarly GOx could be encapsulated in silica/titania/Nafion® sol-gel [215], with the Nafion® preventing cracking and improving sensitivity and stability (80% retention after 4 months in phosphate buffer). Prussian Blue modified glassy carbon electrodes could be coated with a GOx/chitosan/silica sol-gel with stability over 60 days [216]. Mediators could also be included in sono-gels; glucose oxidase, for example, in this context could be immobilized at electrodes as a composite with sono-gels containing ferrocene compounds [217, 218].

6.7 OPTICAL METHODS FOR SENSING GLUCOSE

As an alternative to electrochemical approaches, optical methods can be used for a wide variety of analyses. Optical sensors display several advantages over electronic devices: they are, for example, much less sensitive to electronic interference, since the information is carried as photons rather than electrons. Usually the optical components are made from glass chips or fiber-optic cable fibers, which minimizes the weight and the size of the sensors since recent advances originating from the telecommunications industry enable the manufacturing of small sensors. Finally, glass sensors usually display a high chemical stability—i.e., they are not corroded easily and are usually unaffected by organic solvents as well as displaying good thermal and mechanical stability.

A schematic of a fiber-optic sensor is shown in Figure 6.13. The first of these main components of the sensor is an optical chip or fiber (known as the optrode). Usually this will contain an indicator species whose optical properties depend on the presence/absence of an analyte, e.g., adsorption or fluorescence, which is modulated by interaction with the analyte, since the analyte may not give or exhibit changes with readily determined optical properties. This can be thought of as a transduction event.

To follow the transduction event, a light source is used, often with its wavelength specifically matched to the adsorption maximum of the indicator, so as to maximize the sensitivity. Light from the source is passed through the chip and the output, which could be affected by adsorption or fluorescence events that can be measured with a suitable detector. Fluorescence is often measured instead of absorbance (assuming either the analyte is fluorescent or a suitable indicator can be found), due to the inherent higher sensitivity of the technique.

There has been interest in developing optical biosensors for glucose as reviewed recently for fluorescence-based sensors [219]. Commercial fiber-optic oxygen sensitive optrodes are available. Devices such as these may be modified by immobilizing glucose oxidase on its surface and can be utilized to develop a glucose sensor [220]. This sensor, along with an unmodified reference optrode could be used to detect glucose in microdialysis samples. Later versions of this system could be implanted subcutaneously [221] and gave linear response in the physiological range of glucose.

Other recent examples include the immobilization of glucose oxidase onto a novel biomaterial, a swim bladder membrane, followed by glutaraldehyde crosslinking [222]. This was then pressed against a silicone membrane containing a fluorescent ruthenium dye whose fluorescence is quenched by oxygen followed by the sample

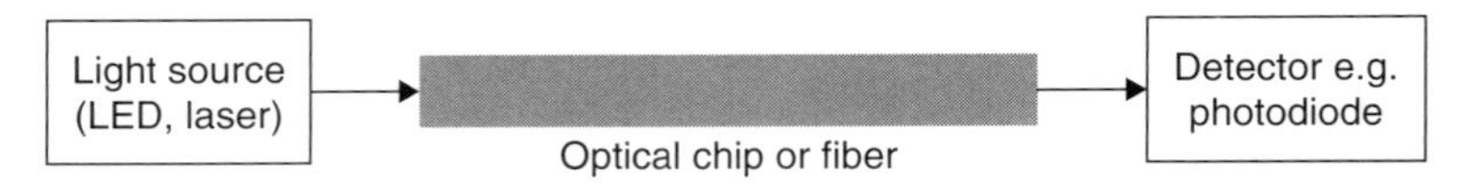

FIGURE 6.13 Schematic of an optical fiber based sensor.

being placed into a flow cell and then exposed to glucose solutions. In the same manner as within a Clark electrode, the presence of glucose causes a drop in oxygen concentration and therefore an increase in fluorescence. This system, when applied to the analysis of urine and serum samples, displayed remarkable stability with 80% retention of enzyme activity after 10 months at 4°C. Use of a bamboo inner shell membrane as the immobilizing medium led to formation of extremely stable glucose biosensors [223], with a 95% retention of enzyme activity remaining after 8 months at 4°C and no effect being observed from common interferents.

Another method involved direct labeling of a glucose binding protein from *E. coli* with two fluorescent probes, one of which, acrylodan, is quenched by the presence of glucose, with the other emission being used as a reference [224]. Concavalin A has also been utilized due to its affinity binding with glucose both in competition assays [225] and when attached to polymer particles in conjunction with a near-infrared fluorescence determination [226]. Glucose binding protein could also be bound to a surface plasmon resonance chip and used to detect glucose with a range of mutant proteins being utilized [227].

6.8 MINIATURIZATION

Miniaturization of fabricated structures has seen great strides in recent years. Microfluidic and micro engineered mechanical systems have shrunk drastically in size while increasing in complexity. The whole subject of microfabrication and its application to sensors would be the subject of a review in itself and is outside the scope of this chapter. A few examples of recent advances will, however, be given.

A microfluidic sensor using an array of hydrogel encapsulated enzymes has been constructed and used to measure glucose, among other analytes [228]. A calorimetric biosensor based on a microfluidic device [229], capable of measuring the heat of reaction caused by enzyme catalyzed oxidation of glucose, has been developed. Microfluidic (micro-reactor) devices based on polydimethyl siloxane containing GOx have been used to measure glucose in serum. Control of the size and placement of the reactor could be used to fine-tune the sensitivity and linear range of the sensor [230]. Gold nanowire arrays modified with GOx have been incorporated into a flow injection device [231] and have been used for measurements of glucose.

Sensors have been constructed with a conventional enzyme electrode contained within a silicon cavity screened by a micromachined porous silicon membrane, the pore size of which could be controlled and so used to set the linear range of the sensor [232] for use with clinical samples. Silicon has also been used as a base for machined microelectrodes, which have then been platinized and coated with polypyrrole for use as a glucose sensor [233]. Similar machining techniques could be used for the construction of microarrays which could then be coated with osmium-polyvinylpyridine complexes upon which GOx could be immobilized within photocurable polymers to make individually addressable multiple electrodes [234]. Microfabricated cantilevers [235] could be functionalized with GOx; in these devices reaction of the enzyme with glucose leads to a change in surface stress, causing bending of the cantilever. Common interferents have been shown to have no effect on the detection process.

The electropolymerization of polypyrrole/GOx composite within a nanoporous alumina membrane has been utilized to generate an array of glucose nanoelectrodes with biocatalytic activity [236]. Electrodeposition of polypyrrole/GOx within containing trenches could be used to generate interdigitated microelectrode arrays which could be used for acute glucose sensing for several hours [237].

6.9 CONTINUOUS MONITORING

One major disadvantage of contemporary commercial biosensors is the requirement of an invasive procedure, requiring frequent withdrawal of blood for testing, which can be both tedious and painful. This aspect of glucose testing could be negated were it possible to implant a sensor within the body which would continuously monitor glucose. Implantable sensors suitable for *in vivo* glucose monitoring would be required to be extremely small and show long-term stability (with minimal drift thereby removing the need for frequent calibration), display no oxygen dependency, and also show high biocompatibility. Although *in vivo* glucose sensors have been developed, the problem of biocompatibility has been an elusive target, leading as yet to only limited lifetimes. Sensor performance can be adversely affected by fouling due to protein deposition on the surface or formation of fibrous tissue around the sensor [238, 239]. Rejection by the immune system or thrombus formation if used intravascularly can degrade device performance and risk harm to the user. The field of *in vivo* glucose monitoring has been recently extensively reviewed [240] so only a brief history and most recent advances will be given here.

Subcutaneously applied glucose sensors with capability of being changed by the user have been developed [241, 242]. However, the effects of biofouling mean their use was limited to periods of 1–2 weeks. Attempts have been made to improve this using polymeric coatings. For example, a needle-based electrochemical probe coated with Nafion® has been developed [243] which is just 0.5 mm in diameter and can be inserted subcutaneously through an 18-gauge needle. Again *in vivo* measurements could be carried out over periods of 2 weeks.

An alternative approach has been to implant a microdialysis fiber into subcutaneous tissue through which an iso-osmotic electrolyte solution may then be pumped. Glucose diffuses from interstitial fluid into the fiber and the electrolyte. Providing flow rates are kept constant the glucose concentration in the fiber outflow can then be directly related to the interstitial fluid glucose concentration. Rapid changes in blood glucose concentration can be determined although a time delay, typically of about 8 minutes [244], is experienced between changes in blood and interstitial fluid glucose concentrations. Use of these probes combined with glucose biosensors to monitor the outflow has achieved 4–7 days continuous use for glucose monitoring in humans [245, 246].

More recently, miniature devices such as a $2 \times 4 \times 0.5$ mm biochip containing two platinum microarray working electrodes have been developed as potential implantable biosensors [247] where the electrode assemblies were used to immobilize glucose and lactate oxidase and were coated with a complex polymer multilayer. Polypyrrole was used to provide screening of interferents while crosslinked glycol polymers and a phosphorylcholine polymer were included so as to provide *in vivo* biocompatibility. Flexible glucose biosensors which could be rolled up into a tube and placed inside a micro-catheter have also been developed as potential *in vivo* monitoring devices [248]. A ribbon-like magnetoelastic sensor could be coated with a pH sensitive polymer and a biolayer of GOx and catalase [249]. Enzyme catalyzed oxidation of glucose to gluconic acid causes a local pH change, leading to shrinkage of the pH sensitive polymer and a decrease in the sensor mass loading. Application of a magnetic field causes vibration of the sensor which can be measured by a pickup coil and is dependent on the mass of the sensor. This leads to the possibility of using these systems as *in vivo* glucose sensors without need for any physical connection between a sensor and a monitoring device [249].

An implantable amperometric glucose sensor has been developed and has been shown to give stable results when implanted in rats for up to 7 days with response times of approximately one minute [250]. Carbon fiber microelectrodes have also been utilized, which were first coated with ruthenium to catalyze H_2O_2 reaction, followed by coating with poly(1,2-diaminobenzene) electrodeposited on top as a screening layer [251]. This was then further coated with GOx and finally a polyurethane/Nafion® composite film to confer biocompatibility. The resultant systems have been shown to detect glucose *in vivo* (rats). An optical glucose sensor [252] has also been developed which will fit inside an 18-gauge needle and has been shown to be capable of measuring glucose *in vivo* (within fish).

As an alternative to implanted sensors, the flexible glucose oxidase sensor based on polydimethylsiloxane has been reported. This type of sensor could be worn on the human body to potentially measure glucose in sweat or tears [253].

6.10 COMMERCIAL BIOSENSORS

At present, the commercial glucose sensors market is dominated by the disposable market, with most devices currently utilizing the mediated electrochemical detection of glucose [1, 3, 254]. A major factor in this dominance is the availability of an inexpensive reliable technology for fabrication of the disposable sensor strips. Screen printing is a mature technology especially suitable for depositing enzyme electrodes [3, 254]. A wide variety of inks containing catalysts, mediators, and reference standards such as Ag/AgCl are commercially available from a variety of suppliers. The ease of scaling up the screen-printing process to mass production levels enables economies of scale to apply and thus reduces the costs of individual electrodes. Medisense (now Abbott), for example, at the time of writing produces over a billion sensor strips per year [3, 254] with other major suppliers including Lifescan and Roche Diagnostics.

The mediated detection of glucose is now a mature technology and the main focus of commercial research appears to be more into the supporting technology. Currently the monitoring of glucose requires lancing of a finger to produce a drop of blood, a process which when repeated on a regular basis can be both painful and distressing. Technologies such as the Pelikan® device, which only require microliter blood volumes [1], have been developed to reduce the amount of blood needed to minimize these problems. However, other ways have also been studied to remove the necessity for invasive procedures.

A variety of techniques have been applied in attempts to measure glucose levels in interstitial fluid. Laser ablation, ultrasonic techniques, and reverse iontophoresis have all been studied and have been shown to be capable of removing interstitial fluid, thereby allowing testing [1, 3, 254]. However, whether this technique actually gives reliable data is still controversial. It is still uncertain whether glucose levels in interstitial fluid follow blood glucose levels closely enough. Should there be a rapid rise or drop in blood glucose and it not be followed quickly enough by corresponding changes in the levels in interstitial fluid, a patient could suffer glycemia with insufficient warning. Other fluids such as urine, tears, or sweat have been suggested; however, as yet no large scale commercial device production has been launched and at the time of writing, the home testing market is still dominated by techniques that require removal of a drop of blood [254].

For patients who require closer monitoring of blood glucose than periodic sampling allows, there is the possibility of using an implanted needle-type sensor for continuous or near continuous glucose monitoring, even when the patient is asleep. Commercial implantable glucose sensors are available, usually consisting of an implanted sensor combined with a pocket-size monitoring and logging device. A subcutaneously implantable device, the CGMS® (Continuous Glucose Monitoring System) has been commercialized by Minimed (www.minimed.com). The glucose biosensor probe is inserted just beneath the skin, usually in the abdomen, and can be used to monitor glucose for up to 72 h and provide a reading every 5 min. Traditional blood sampling of glucose is used to calibrate the device. Other data such as meal times and exercise periods can also be recorded using the device and all data then downloaded to a personal computer.

For many diabetes type 1 sufferers, life involves a constant round of glucose testing and insulin injections. Externally worn insulin pumps have been developed, such as those by Medtronic (www.minimed.com). These have the advantage that they remove the need for injections by introducing insulin subcutaneously through a canella. When coupled together with a glucose biosensor, this provides an opportunity for the device to inject insulin "on demand." This was the first device of its type to receive FDA approval [1]. At present, commercial devices usually can only be used for a few days at a time without maintenance, and injection and sensing sites need to be changed regularly. An implanted device which automatically responds to changes in glucose levels would in effect act as an artificial pancreas so as to enhance the quality of life for many millions of people.

The preferred type of biosensor would be a noninvasive, wearable device which continuously monitored glucose. Devices of this type such as Glucowatch, produced by Cygnus, have been released but have yet to make a successful major impact on the market.

A variety of other methods have been proposed to measure glucose which do not involve use of biological moieties in any way. These lie outside the scope of this chapter but have been reviewed elsewhere [3, 254]. These methods include use of spectroscopic techniques such as near-infrared spectroscopy and use of chemical binders for glucose such as boronic acids or molecularly imprinted polymers. However, as yet none of these methods has been shown to work with sufficient reliability for use in a commercial sensor.

6.11 CONCLUSIONS AND FUTURE

The development of reliable, widely available testing for blood glucose has enhanced the lifestyles and health of many millions of diabetics. A variety of scientific divisions—biology, chemistry, materials science, electrochemistry, and engineering—have all played their part in this success. More research is still progressing to increase reliability, storability, and simplicity of these devices and to remove or minimize the necessary evil of invasive monitoring.

Fields that will be of interest include the removal of mediators by allowing more effective direct communication between the enzyme and the electrode. Other fields include the potential to genetically engineer GOx to maximize selectivity, sensitivity, and stability. Interest is also being shown towards the production of new biocompatible materials to allow implantable monitoring over longer time periods than presently available, along with possible miniaturization to reduce the size of the implanted devices. This could eventually lead to the desirable outcome of an artificial pancreas, a device that responds to glucose levels by delivering a required amount of insulin.

Finally, there are a large number of other factors that we would like to be able to monitor *in vivo* such as lactate, cholesterol, drugs, disease markers, and antibodies. The methods and techniques used for developing high-quality glucose biosensors will also be applied to these fields.

References

[1] J.D. Newman, L.J. Tigwell, A.P.F. Turner, P.J. Warner, Biosensors: A Clearer View, Biosensors 2004 – The 8th World Congress on Biosensors, Elsevier, New York.
[2] J.D. Newman, S.J. Setford, Mol. Biotech. 32 (2006) 249–268.
[3] J. Wang, Electroanalysis 13 (2001) 983–988.
[4] B.R. Eggins, Biosensors, Wiley, Chichester, 1996.
[5] E.A.H. Hall, Biosensors, Open University Press, Maidenhead, 1990.
[6] B.R. Eggins, Chemical Sensors and Biosensors (Analytical Techniques in the Sciences), Wiley, Chichester, 2002.
[7] J. Cooper, A.E.G. Cass, Biosensors (Practical Approach), Oxford University Press, Oxford, 2004.
[8] S. Cosnier, Biosens. Bioelectron. 14 (2003) 443–456.
[9] S. Cosnier, Electroanalysis 17 (2005) 1701–1715.
[10] F. Davis, S.P.J. Higson, in; P.A. Millner (Ed.), Biosensors (Methods Express), Scion Publishing Ltd, Bloxham, 2007, Ch. 8.
[11] L. Clark, C. Lyons, Ann. NY Acad. Sci. 102 (1962) 29–45.
[12] S. Updike, G. Hicks, Nature 214 (1967) 986–988.
[13] G. Guilbault, G. Lubrano, Anal. Chim. Acta 64 (1973) 439–455.
[14] R.B.F. Turner, C.S. Sherwood, ACS Symp. Ser. 556 (1994) 211–221.
[15] Y.N. Zhang, Y.B. Hu, G.S. Wilson, D. Moattisirat, V. Poitout, G. Reach, Anal. Chem. 66 (1994) 1183–1188.

[16] S.P.J. Higson, P. Vadgama, Anal. Chim. Acta 271 (1993) 125–133.
[17] A. Maines, D. Ashworth, P. Vadgama, Anal. Chim. Acta 333 (1997) 223–231.
[18] S. Myler, S. Eaton, S.P.J. Higson, Anal. Chim. Acta 357 (1997) 55–61.
[19] S.A. Emr, A.M. Yacynych, Electroanalysis 7 (1995) 913–923.
[20] S.V. Sasso, R.I. Pierce, R. Walla, A.M. Yacynych, Anal. Chem. 62 (1990) 1111–1117.
[21] F. Palmisano, C. Malitesta, D. Centonze, P.G. Zambonin, Anal. Chem. 67 (1995) 2207–2211.
[22] A. D'Costa, J. Higgins, A.P.F. Turner, Biosensors, 2 (1986) 71–87.
[23] S. Tsujimura, S. Kojima, K. Kano, T. Ikeda, M. Sato, H. Sanada, H. Omura, Biosci. Biotech. Biochem. 70 (2006) 654–659.
[24] M.G. Zhang, A. Smith, W. Gorski, Anal. Chem. (2004) 5045–5050.
[25] S. Igarashi, T. Hirokawa, K. Sode, Biomol. Eng. 21 (2004) 81–86.
[26] G. Reach, G.S. Wilson, Anal. Chem. 64 (1992) A381–A386.
[27] D.A. Gough, J.Y. Lucisano, P.H.S. Tse, Anal. Chem. 57 (1985) 2351–2357.
[28] J. Wang, J. Liu, L. Chen, F. Lu, Anal. Chem. 66 (1994) 3600–3603.
[29] J. Newman, S. White, I. Tothill, A.P.F. Turner, Anal. Chem. 67 (1995) 4594–4599.
[30] A. Karaykin, O. Gitelmacher, E. Karaykina, Anal. Chem. 67 (1995) 2413–2419.
[31] F. Ricci, D. Moscone, C.S. Tuta, G. Palleschi, A. Amine, F. Valgimigli, D. Messeri, Bioelectronics 20 (2005) 1993–2000.
[32] Q. Chi, S. Dong, Anal. Chim. Acta 310 (1995) 429–436.
[33] J. Wang, H. Wu, J. Electroanal. Chem. 395 (1995) 287–291.
[34] Y.Q. Dai, K.K. Shia, Electroanalysis 16 (2004) 1806–1813.
[35] L. Ming, X. Xi, J. Liu, Biotech. Lett. 28 (2006) 1341–1345.
[36] G.L. Luque, M.C. Rodriguez, G.A. Rivas, Talanta 66 (2005) 467–471.
[37] P. Kotsian, P. Brazdilova, S. Reskova, K. Kalcher, K. Vytras, Electroanalysis 18 (2006) 1499–1504.
[38] K. Wang, J.J. Xu, H.Y. Chen, Biosens. Bioelectron. 20 (2005)1388–1396.
[39] P.N. Mashazi, K.I. Ozoemena, T. Nyokong, Electrochim. Acta 52 (2006) 177–186.
[40] X.L. Luo, J.J. Xu, W. Zhao, H.Y. Chen, Biosens. Bioelectron. 19 (2004) 1295–1300.
[41] J.J. Xu, X.L. Luo, Y. Du, H.Y. Chen, Electrochem. Comm. 6 (2004) 1169–1173.
[42] D. Zhan, M.J. Zhu, H.G. Xue, S. Cosnier, Biosens. Bioelectron. 22 (2007) 1612–1617.
[43] A.E.G. Cass, G. Davis, G.D. Francis, H.A.O. Hill, W.J. Aston, I.J. Higgins, E.V. Plotkin, L.D.L. Scott, A.P.F. Turner, Anal. Chem. 56 (1984) 667–671.
[44] S. Warren, T. McCormac, E. Dempsey, Bioelectrochemistry 67 (2005) 23–35.
[45] K. Yamamoto, H.S. Zeng, Y. Shen, M.M. Ahmed, T. Kato, Talanta 66 (2005) 1175–1181.
[46] M.E. Ghica, C.M.A. Brett, Anal. Chim. Acta 532 (2005) 145–151.
[47] R. Kurita, H. Tabei, Y. Iwasaki, K. Hayashi, K. Sunagawa, O. Niwa, Biosens. Bioelectron. 20 (2004) 518–523.
[48] K.I. Ozoemena, T. Nyokong, Electrochim. Acta 51 (2006) 5131–5136.
[49] L.S. Bean, L.Y. Heng, B.M. Yamin, M. Ahmad, Bioelectrochemistry 65 (2005) 157–162.
[50] L.S. Bean, L.Y. Heng, B.M. Yamin, M. Ahmad, Thin Solid Films 477 (2005) 104–110.

[51] K. Krikstopaitis, J. Kulys, L. Tetianec, Electrochem. Comm. 6 (2004) 331–336.
[52] T. Kohma, H. Hasegawa, D. Oyamatsu, S. Kuwabata, Bull. Chem. Soc. Jpn. 80 (2007) 158–165.
[53] G.L. Gaines, Insoluble Monolayers at Liquid-Gas Interfaces, Intersciences, New York, 1966.
[54] R.H. Tredgold, Order in Thin Molecular Films, Cambridge University Press, 1994.
[55] F. Davis, S.P.J. Higson, Biosens. Bioelec. 21 (2005) 1–20.
[56] M. Sriyudthsak, H. Yamagishi, T. Morizumi, Thin Solid Films 160 (1988) 463–469.
[57] Y. Okahata, T. Tsuruta, K. Ijiro. K. Ariga, Langmuir 4 (1988) 1373–1375.
[58] D.G. Zhu, M.C. Petty, H. Ancelin, J. Yarwood, Thin Solid Films 176 (1989) 151–156.
[59] C. Fiol, S. Alexandre, N. Delpire, J.M. Valleton, E. Paris, Thin Solid Films 215 (1992) 88–93.
[60] F. Sommer, S. Alexandre, N. Dubreuil, D. Lair T. M. Duc, J.M. Valleton, Langmuir 13 (1997) 791–795.
[61] J.M. Chovelon, M. Provence, N. Jaffrezic-Renault, S. Alexandre, J.M. Valleton, Mat. Sci. Eng. C. 22 (2002) 79–85.
[62] F.Q. Tang, J.R. Li, L. Zhang, L. Jiang, Biosens. Bioelec. 7 (1992) 503–507.
[63] A. Eremenko, I. Kurochin, S. Chernov, A. Barmin, A. Yaroslavov, T. Moskvitina, Thin Solid Films 260 (1995) 212–216.
[64] S.Y. Zaitsev, Sens. Actuat. B. 24 (1995) 177–179.
[65] A.J. Guiomar, S.D. Evans, J.T. Guthrie, Supramol. Sci. 4 (1997) 279–291.
[66] H. Ohnuki, T. Saiki, A. Kusakari, H. Endo, M. Ichihara, M. Isumi, Langmuir 23 (2007) 4675–4681.
[67] K. Ramanathan, M.K. Ram, B.D. Malholtra, A.S.N. Murthy, Mat. Sci. Eng. C. 3 (1995) 159–163.
[68] R. Singhal, W. Takashima, K. Kaneto, S.B. Samanta, S. Annapoorni, B.D. Malhotra, Sens. Actuat. B. 86 (2002) 42–48.
[69] R. Singhal, A. Chauhey, K. Kaneto, W. Takashima, S. Annapoorni, B.D. Malhotra, Biotech. Bioeng. 85 (2004) 277–282.
[70] D. Kato, M. Masaike, T. Majima, Y. Hirata, F. Mizutani, M. Sakata, C. Hirayama, M. Kunitake, Chem. Commun. 22 (2002) 2616–2617.
[71] G. Decher, J.D. Hong, J. Schmitt, Thin Solid Films 210 (1992) 831–835.
[72] G. Decher, Science 277 (1997) 1232–1237.
[73] P.T. Hammond, Curr. Opin. Coll. Inter. Sci. 4 (1999) 430–442.
[74] M. Schonhoff, Curr. Opin. Coll. Inter. Sci. 8 (2003) 86–95.
[75] J. Parellada, A. Narvaez, E. Dominguez, I. Katakis, Biosens. Bioelectron. 12 (1997) 267–275.
[76] J.G. Franchina, W.M. Lackowsk, D.L. Dermody, R.M. Crooks, D.E. Bergbreiter, K. Sirkar, R.J. Russell, M.V. Pishko, Anal. Chem. 71 (1999) 3133–3139.
[77] S.F. Hou, H.Q. Fang, H.Y. Chen, Anal. Lett. 30 (1997) 163–164.
[78] A.L. Simonian, A. Rezvin, J.R. Wild, J. Elkind, M.V. Pishko, Anal. Chim. Acta 466 (2002) 201–212.
[79] D. Trau, R. Renneberg, Biosens. Bioelectron. 18 (2003) 1491–1499.
[80] Q.T. Nguyen, Z. Ping, T. Nguyen, P. Rigal, J. Memb. Sci. 213 (2003) 85–95.

[81] V.T. Dimakis, V.G. Gavalas, N.A. Chaniotakis, Anal. Chim. Acta 467 (2002) 217–223.
[82] S.A. Miscoria, J. Desbriezes, G.D. Barrera, P. Labbe, G.A. Rivas, Anal. Chim. Acta 578 (2006) 137–144.
[83] M.C. Rodriguez, G.A. Rivas, Electroanalysis 16 (2004) 1717–1722.
[84] W.J. Zhang, Y.X. Huang, H. Dai, X.Y. Wang, C.H. Fan, G.X. Li, Anal. Biochem. 329 (2004) 85–90.
[85] M. Wilchek, E.A. Bayer, Anal. Biochem. 171 (1988) 1–32.
[86] J. Anzai, Y. Kobayashi, Y. Suzuki., H. Takeshita, Q. Chen, T. Osa, T. Hoshi, X. Du, Sens. Actuat. B. 52 (1998) 3–9.
[87] T. Wink, S.J. VanZuilen, A. Bult, W.P. VanBennekom, Analyst 122 (1997) R43–R50.
[88] N.K. Chaki, K. Vijayamohanan, Biosens. Bioelectron. 17 (2002) 1–12.
[89] R.M. Ianello, A.M. Yacynych, Anal. Chem. 53 (1981) 2090–2095.
[90] A. Ulman, An Introduction to Ultrathin Organic Films: From Langmuir-Blodgett to Self-assembly, Academic Press, 1991.
[91] A. Ulman, Thin Films: Self Assembled Monolayers of Thiols (Thin Films), Academic Press, 1998.
[92] M.A. McRipley, R.A. Linsenmeier, J. Electroanal. Chem. 414 (1996) 235–246.
[93] J.J. Gooding, V.G. Praig, E.A.H. Hall, Anal. Chem. 70 (1998) 2396–2402.
[94] R.K. Servedani, A.H. Mehrjardi, N. Zamiri, Bioelectrochemistry 69 (2006) 201–208.
[95] Y. Xia, G.M. Whitesides, Angew. Chem. Int. Ed. 37 (1998) 550–575.
[96] R.S. Kane, S. Takayama, E. Ostuni, D.E. Ingber, G.M. Whitesides, Biomaterials. 20 (1999) 2363–2376.
[97] S. Gaspar, W. Schuhmann, T. Laurell, E. Csoregi, Rev. Anal. Chem. 21 (2002) 245–266.
[98] J.J. Gooding, F. Mearns, W.R. Yang, J.Q. Liu, Electroanalysis 15 (2003) 81–96.
[99] M. Gerard, A. Chaubey, B.D. Malhotra, Biosens. Bioelec. 17 (2002) 345–359.
[100] J.N. Barisci, C. Conn, G.G. Wallace, Trends Polym. Sci. 4 (1996) 301–311.
[101] S. Cosnier, Biosens. Bioelectron. 14 (2003) 443–456.
[102] S. Geetha, C.R.K. Rao, M. Vijayan, D.C. Trivedi, Anal. Chim. Acta 568 (2006) 119–125.
[103] A. Ramanavicius, A. Ramanaviciene, A. Malinauskas, Electrochim. Acta 51 (2006) 6025–6037.
[104] T. Ahuja, I.A. Mir, D. Kumar, Rajesh, Biomaterials. 28 (2007) 791–805.
[105] J.C. Cooper, E.A.H. Hall, Biosens. Bioelectron. 7 (1992) 473–485.
[106] N.G. Skinner, E.A.H. Hall, J. Electroanal. Chem. 420 (1997) 179–188.
[107] H. Xue, Z. Shen, C. Li, Biosens. Bioelectron. 20 (2005) 2330–2334.
[108] M. Gerard, B.D. Malhotra, Curr. Appl. Phys. 5 (2005) 174–177.
[109] A.L. Sharma, R. Singhal, A. Kumar, Rajesh, K.K. Pande, B.D. Malhotra, J. Appl. Poly. Sci. 91 (2004) 3999–4006.
[110] J.J. Wang, N.V. Myang, M.H. Yun, H.G. Monbouquette, J. Electroanal. Chem. 575 (2005) 139–146.
[111] C. Chen, Y. Kiang, J.Q. Kan, Biosens. Bioelectrons. 22 (2006) 639–643.

[112] R.E. Ionescu, K. Abu-Rabeah, S. Cosnier, R.S. Marks, Electrochem. Comm. 7 (2005) 1277–1282.
[113] V.K. Gade, D.J. Shirale, P.D. Gaikwad, P.A. Savale, K.P. Kakde, H.J. Kharat, M.D. Shiraat, React. Funct. Polym. 66 (2006) 1420–1426.
[114] Y.L. Li, K.G. Neoh, E.T. Kang, Langmuir 21 (2005) 10702–10709.
[115] J.R. Retama, E.L. Caberos, D. Mecerreyes, B. Lopez-Ruiz, Biosens. Bioelectron. 20 (2004) 1111–1117.
[116] P.A. Fiorito, S.I.C. de Torresi, J. Electroanal. Chem. 581 (2005) 31–37.
[117] M.A. Alves, P.A. Fiorito, S.I.C. de Torresi, R.M. Torresi, Biosens. Bioelectron. 22 (2006) 298–305.
[118] X. Liu, K.G. Neoh, L. Cen, E.T. Kang, Biosens. Bioelectron. 19 (2004) 823–834.
[119] M. Shimomura, R. Miyataa, T. Kuwaharaa, K. Oshimaa, S. Miyauchi, Eur. Polym. J. 43 (2007) 388–394.
[120] W.J. Sung, T.H. Bae, Sens. Act. B. 114 (2006) 164–169.
[121] A. Ramanavicius, A. Kausaite, A. Ramanaviciene, Sens. Act. B. 111 (2005) 532–539.
[122] J.I. Reyes de Corcuera, R.P. Cavalieri, J.R. Powers, J. Electroanal. Chem. 575 (2005) 229–241.
[123] A. Curulli, F. Valentina, S. Orlanduci, M.L. Terranova, G. Palleschi, Biosens. Bioelectron. 20 (2004) 1223–1232.
[124] D.W. Pan, J.H. Chen, S.Z. Yao, L.H. Nie, J.J. Xia, W.Y. Tao, Sens. Act. B. 104 (2005) 68–74.
[125] P.C. Nien, T.S. Tung, K.C. Ho, Electroanalysis 18 (2006) 1408–1415.
[126] A. Kros, N.A.J.M. Sommerdijk, B.J.M. Nolte, Sens. Act. B. 106 (2005) 289–295.
[127] D.J. Macaya, M.T. Nicolou, S. Takamatsu, J.T. Mabeck, R.M. Owens, G.G. Malliaras, Sens. Act. B. 123 (2007) 374–378.
[128] T. Kuwahara, K. Oshima, M. Shimomura, S. Miyauchi, Polymer 46 (2005) 8091–8097.
[129] S. Yabuki., F. Mizutani, Sens. Act. B. 108 (2005) 651–653.
[130] P.F. Pang, W.Y. Yang, S.J. Huang, Q.Y. Cai, S.Z. Yao, Sens. Act. B. 122 (2007) 148–157.
[131] A.C. Barton, S.D. Collyer, F. Davis, D.D. Gornall, K.A. Law, E.C.D. Lawrence, D.W. Mills, S. Myler, J.A. Pritchard, M. Thompson, S.P.J. Higson, Biosens. Bioelectron. 20 (2004) 328–337.
[132] C.P. Remirez, D.J. Caruana, Electrochem. Comm. 8 (2006) 450–454.
[133] A. Heller, Acc. Chem. Res. 23 (1990) 128–134.
[134] T.J. Ohara, R. Rajagopalan, A. Heller, Anal. Chem. 66 (1994) 2451–2457.
[135] J.J. Fei, K.B. Wu, F. Wang, S.S. Hu, Talanta 65 (2005) 918–924.
[136] N. Mano, A. Heller, Anal. Chem. 77 (2005) 729–732.
[137] K. Sirkar, M.V. Pishko, Anal. Chem. 70 (1998) 2888–2894.
[138] W.W. Yang, M. Zhou, C.Q. Suh, Macromol. Comm. 28 (2007) 265–270.
[139] S.H. Lee, H.Y. Fang, W.C. Chen, Sens. Act. B. 117 (2006) 236–243.
[140] Z.Q. Gao, F. Xie, N. Shariff, M. Arshad, J.Y. Ying, Sens. Act. B. 111 (2005) 339–346.
[141] A. Hiratsuka, H. Muguruma, K.H. Lee, I. Karube, Biosens. Bioelectron. 19 (2004) 1667–1672.

[142] A. Hiratsuka, A. Kojima, H. Muguruma, K.H. Lee, I. Karube, Biosens. Bioelectron. 21 (2005) 957–964.
[143] H. Mugurama, Y. Kase, H. Uehara, Anal. Chem. 77 (2005) 6557–6562.
[144] S.X. Zhang, N. Wang, H.J. Yu, Y.M. Niu, C.Q. Sun, Bioelectrochemistry 67 (2005) 15–22.
[145] S.X. Zhang, N. Wang, Y.M. Niu, C.Q. Sun, Sens. Actuat. B. 109 (2005) 367–374.
[146] X. Zhong, R. Yuan, Y.Q. Chai, J.Y. Dai, Y. Liu, D.P. Tang, Anal. Lett. 38 (2005) 1085–1097.
[147] C.X. Lei, H. Wang, G.L. Shen, R.Q. Yu, Electroanalysis 16 (2004) 736–740.
[148] M.L. Mena, P. Yanez-Sedeno, J.M. Pingarron, Anal. Biochem. 336 (2005) 20–27.
[149] S.X. Zhang, W.W. Yang, Y.M. Niu, Y.C. Li, M. Zhang, C.Q. Sun, Anal. Bioanal. Chem. 384 (2006) 736–741.
[150] X.L. Luo, J.J. Xu, Y. Du, H.Y. Chen, Anal. Biochem. 334 (2004) 284–289.
[151] M.H. Xue, Q. Xu, M. Zhou, J.J. Zhu, Electrochem. Comm. 8 (2006) 1468–1474.
[152] Y. Du, X.L. Luo, J.J. Xu, H.Y. Chen, Bioelectrochemistry 70 (2007) 342–347.
[153] B.Y. Wu, S.H. Hou, F. Yin, J. Li, Z.X. Zhou, J.D. Huang, Q. Chen, Biosens. Bioelectron. 22 (2007) 838–844.
[154] J.W. Li, J.J. Yu, F.Q. Zhao, B.Z. Zeng, Anal. Chim. Acta 587 (2007) 33–40.
[155] Y.H. Wu, S.S. Nu, Bioelectrochemistry 70 (2007) 335–341.
[156] S. Zhao, K. Zhang, Y. Bai, W.W. Yang, C.Q. Sun, Bioelectrochemistry 69 (2006) 158–163.
[157] M.T. Sulak, O. Gokdogan, A. Gulce, H. Gulce, Biosens. Bioelectron. 21 (2006) 1719–1726.
[158] X.L. Ren, X.M. Meng, D. Chen, F.Q. Tang, J. Jiao, Biosens. Bioelectron. 21 (2005) 433–437.
[159] X.L. Ren, X.M. Meng, F.Q. Tang, Sens. Actuat. B. 110 (2005) 358–363.
[160] S. Wu, H.T. Zhao, H.X. Ju, C.G. Shi, J.W. Zhao, Electrochem. Comm. 8 (2006) 1197–1203.
[161] H.H. Zhou, H. Chen, S.L. Luo, J.H. Chen, W.Z. Wei, Y.F. Kuang, Biosens. Bioelectron. 20 (2005) 1305–1311.
[162] L.H. Zhu, H.Y. Zhu, X.L. Yang, L.H. Xu, C.Z. Li, Electroanalysis 19 (2007) 698–703.
[163] W.J. Guan, Y. Li, Y.Q. Chen, X.B. Zhang, G.Q. Hu, Biosens. Bioelectron. 21 (2005) 508–512.
[164] J. Li, Y.B. Wang, J.D. Qiu, D. Sun, X.H. Xia, Anal. Bioanal. Chem. 383 (2005) 918–922.
[165] Y. Liu, M. Wang, F. Zhao, Z. Xu, S. Dong, Biosens. Bioelectron. 21 (2005) 984–988.
[166] Y.C. Tsai, S.C. Li, J.M. Chen, Langmuir 21 (2005) 3653–3658.
[167] D. Wen, Y. Liu, G.C. Wang, S.J. Dong, Electrochim. Acta 52 (2007) 5312–5317.
[168] Y.C. Tsai, S.C. Li, S.W. Liao, Biosens. Bioelectron. 22 (2006) 495–500.
[169] Y.L. Zeng, Y.F. Huang, J.H. Jiang, X.H. Zhang, C.R. Tang, G.L. Shen, H.Q. Yu, Electrochem. Comm. 9 (2007) 185–190.

[170] J.H.T. Luong, S. Hrapovic, D. Wang, F. Bensebaa, B. Simard, Electroanalysis 16 (2004) 132–139.
[171] J. Wang, M. Musameh, Anal. Chim. Acta 539 (2005) 209–213.
[172] Y.Q. Dai, K.K. Shiu, Electroanalysis 16 (2004) 1697.
[173] D.W. Pan, J.H. Chen, S.Z. Yao, W.Y. Tao, L.H. Nie, Anal. Sci. 21 (2005) 367–371.
[174] L.D. Zhu, J.L. Zhai, Y.N. Guo, C.Y. Tian, R.L. Yang, Electroanalysis 18 (2006) 1842–1846.
[175] H.T. Zhao, H.X. Ju, Anal. Biochem. 350 (2006) 138–144.
[176] J.D. Huang, Y. Yang, H.B. Shi, Z. Song, Z.X. Zhao, J. Anzai, T. Osa, Q. Chen, Mat. Sci. Eng. C. 26 (2006) 113–117.
[177] G.D. Liu, Y.H. Lin, Electrochem. Comm. 8 (2006) 251–256.
[178] F.L. Qu, N.H. Yang, J.W. Chen, G.L. Shen, H.Q. Yu, Anal. Lett. 39 (2006) 1785–1799.
[179] Y. Lin, F. Lu, Y. Tu, Z. Ren, Nano Lett. 4 (2004) 191–195.
[180] G.D. Withey, A.D. Lazareck, M.B. Tzolov, A. Yin, P. Aich, J.I. Yeh, J.M. Xu, Biosens. Bioelec. 21 (2006) 1560–1565.
[181] V. Vamvakaki, K. Tsagaraki, N. Chaniotakis, Anal. Chem. 78 (2006) 5538–5542.
[182] J. Wu, Y.H. Zou, N. Gao, J.H. Jiang, G.L. Shen, R.Q. Yu, Talanta 68 (2005) 12–18.
[183] G. Wang, N.M. Thai, S.T. Yau, Biosens. Bioelec. 23 (2007) 2158–2164.
[184] G. Wang, N.M. Thai, S.T. Yau, Electrochem. Comm. 8 (2006) 987–992.
[185] W. Zhao, J.J. Xu, Q.Q. Qiu, N.Y. Chen, Biosens. Bioelectron. 22 (2006) 649–655.
[186] J. Wang, J.A. Carlisle, Diamond Rel. Mat. 15 (2006) 279–284.
[187] W. Jing, Q. Yang, Anal. Bioanal. Chem. 385 (2006) 1330–1335.
[188] H. Olivia, B.V. Sarada, K. Honda, A. Fujishima, Electrochim. Acta 49 (2004) 2069–2076.
[189] D. Lee, J. Lee, J. Kim, J. Kim, H.B. Na, B. Kim, C.N. Shin, J.H. Kwak, A. Dohnalkova, J.W. Grate, T. Hyson, H.S. Kim, Adv. Mat. 17 (2005) 2828.
[190] M.J. Forrow, S.W. Bayliff, Biosens. Bioelectron. 21 (2005) 581–587.
[191] S. Hrapovic, Y.L. Liu, K.B. Male, J.H.T. Luong, Anal. Chem. 76 (2004) 1083–1088.
[192] J.N. Xie, S. Wang, L. Aryasomayajula, V.K. Varadan, Nanotechnology 18 (2007) Art. No. 065503.
[193] J. Manso, M.L. Mena, P. Yáñez-Sedeño, J. Pingarróna, J. Electroanal. Chem. 603 (2007) 1–7.
[194] X. Chu, D.X. Duan, G.L. Shen, R.Q. Yu, Talanta 71 (2007) 2040–2047.
[195] S.J. Geelhood, T.A. Horbett, W.K. Ward, M.D. Wood, M.J. Quinn, Electroanalysis 19 (2007) 717–722.
[196] F.L. Qu, M.H. Yang, G.L. Shen, R.Q. Yu, Biosens. Bioelectron. 22 (2007) 1749–1755.
[197] S.H. Lim, J. Wei, J.Y. Lin, Q.T. Li, J. Kua-You, Biosens. Bioelectron. 20 (2005) 2341–2345.
[198] M. Darder, P. Aranda, M. Hernandez-Valez, E. Manova, E. Ruiz-Hitzky, Thin Solid Films 495 (2006) 321–326.
[199] H.P. Yang, Y.F. Zhu, Anal. Chim. Acta 554 (2005) 92–97.
[200] X.L. Luo, J.J. Xu, W. Zhao, H.Y. Chen, Sens. Act. B. 97 (2004) 249–255.

[201] T. Godjevargova, R. Nenkova, N. Dimova, Macromol. Biosci. 5 (2005) 760–766.
[202] G. Ozyilmaz, S.S. Tukel, O. Alptekin, J. Mol. Cat. B. 35 (2005) 154–160.
[203] C.G. Tsiafoulis, A.B. Florou, P.N. Trikatis, T. Bakas, M.I. Prodromidis, Electrochem. Comm. 7 (2005) 781–788.
[204] Y.H. Yang, H.F. Yang, M.H. Yang, Y.L. Yu, G.L. Shen, R.Q. Yu, Anal. Chim. Acta 525 (2004) 213–220.
[205] J.K. Mbouguen, E. Ngameni, A. Walcarius, Anal. Chim. Acta 578 (2006) 145–155.
[206] D. Shan, W.J. Yoa, H.G. Xue, Electroanalysis 18 (2006) 1485–1491.
[207] A. Mignani, E. Scavetta, D. Tonelli, Anal. Chim. Acta 577 (2006) 98–106.
[208] L. Doretti, D. Ferrara, P. Gattolin, S. Lora, Biosens. Bioelectron. 11 (1996) 365–373.
[209] H.G. Zhu, R. Srivastava, J.Q. Brown, M.J. McShane, Bioconj. Chem. 16 (2005) 1451–1458.
[210] J. Rubio-Betana, E. Lopez-Cabarcos, R. Lopez-Ruis, Talanta 68 (2005) 99–107.
[211] R. Pauliukaite, C.M.A. Brett, Electrochim. Acta 40 (2005) 4973–4980.
[212] M. Florescu, M. Barsan, R. Pauliukaite, C.M.A. Brett, Electroanalysis 19 (2007) 220–226.
[213] R. Pauliukaite, A.M.C. Paquin, A.M.O. Brett, C.M.A. Brett, Electrochim. Acta 52 (2006) 1–8.
[214] K.C. Han, Z.J. Wu, J. Lee, I.S. Ahn, J.W. Park, B.R. Min, K.T. Lee, Biochem. Eng. J. 22 (2005) 161–166.
[215] H.N. Choi, M.A. Kim, W.Y. Lee, Anal. Chim. Acta 537 (2005) 179–187.
[216] X.C. Tan, Y.X. Tian, P.X. Cai, X.Y. Zou, Anal. Bioanal. Chem. 381 (2005) 500–507.
[217] B. Ballarin, M.C. Cassani, R. Mazzoni, E. Scavatta, D. Tonelli, Biosens. Bioelectron. 22 (2007) 1317–1322.
[218] B. Ballarin, M.C. Cassani, R. Mazzoni, E. Scavatta, G. Trioschi, D. Tonelli, Electroanalysis 19 (2007) 200–206.
[219] J.C. Pickup, F. Hussain, N.D. Evans, O.J. Rolinski, D.J.S. Birch, Biosens. Bioelectron. 20 (2005) 2555–2565.
[220] A. Pasic, H. Koehler, L. Schaupp, T.R. Fisher, J. Kliment, Anal. Bioanal. Chem. 386 (2006) 1293–1302.
[221] A. Pasic, H. Koehler, J. Kliment, L. Schaupp, Sens. Act. B. 122 (2007) 60–68.
[222] Z.D. Zhou, L. Qiao, P. Zhang, D. Xiao, M.M.F. Choi, Anal. Bioanal. Chem. 383 (2005) 673–679.
[223] X.F. Yang, Z.D. Zhou, D. Xiao, M.M.F. Choi, Biosens. Bioelectron. 23 (2006) 1613–1620.
[224] X.D. Ge, L. Tolosa, G. Rao, Anal. Chem. 76 (2004) 1403–1410.
[225] S. Mansouri, J.S. Schultz, Biotechnology. 2 (1984) 885–890.
[226] R. Ballerstadt, A. Polak, A. Beuhler, J. Frye, Biosens. Bioelectron. 19 (2004) 905–914.
[227] H.V. Hsieh, Z.A. Pfieffer, T.J. Amis, J.B. Pitner, Biosens. Bioelectron. 19 (2004) 653–660.
[228] J. Heo, R.M. Crooks, Anal. Chem. 77 (2005) 6843–6851.
[229] Y.Y. Zhang, S. Tadagadapa, Biosens. Bioelectron. 19 (2004) 1733–1743.
[230] Q. Zhang, J.J. Xu, H.Y. Chen, J. Chromatogr. A 1135 (2006) 122–126.

[231] M. Delvaux, S. Demoustier-Champagner, A. Walcarius, Electroanalysis 16 (2004) 190–198.
[232] G. Piechotta, J.R. Albers, Hintsche. Biosens. Bioelectron. 21 (2005) 802–808.
[233] J.W. Liu, C. Bian, J.H. Han, S.F. Chen, S.H. Xia, Sens. Act. B. 106 (2005) 591–601.
[234] A. Mugweru, B.L. Clark, M.V. Pishko, Electroanalysis 19 (2007) 453–458.
[235] J.H. Pei, F. Tian, T. Thundat, Anal. Chem. 76 (2004) 292–297.
[236] L. Liu, M.G. Jia, Q. Zhou, M.M. Yan, Z.Y. Jiang, Mat. Sci. Eng. C. 27 (2007) 57–60.
[237] M.Z. Zhu, Z.D. Jiang, W.X. Jing, Sens. Act. B. 110 (2005) 382–389.
[238] P. D'Orazio, Clin. Chim. Acta 334 (2003) 41–69.
[239] S.J. Geelhood, T.A. Horbett, W.K. Ward, M.D. Wood, M.J. Quinn, J. Biomed. Mat. Res. 818 (2007) 251–260.
[240] J.C. Pickup, F. Hussain, N.D. Evans, N. Sachedina, Biosens. Bioelectron. 20 (2005) 1897–1902.
[241] D. Bindra, Y. Zhang, G. Wilson, Anal. Chem. 63 (1991) 1692–1696.
[242] C. Henry, Anal. Chem. 70 (1998) 594A–598A.
[243] F. Moussy, D.J. Harrison, D.W. O'Brien, R.V. Rajotte, Anal. Chem. 65 (1993) 2072–2077.
[244] P.A. Jansson, J. Fowelin, U. Smith, R. Lonnroth, Am. J. Physiol. 255 (1988) E218–E220.
[245] C. Myerhoff, F. Bischof, F. Sternberg, H. Zier, E.F. Pfeiffer, Diabetologia 35 (1992) 1087–1092.
[246] Y. Hashiguchi, M. Sakakida, K. Nishida, T. Uemura, K. Kajiwara, M. Shichiri, Diabetes Care 17 (1994) 387–396.
[247] A. Guiseppi-Elie, S. Brahim, G. Slaughter, K.R. Ward, IEEE Sens. J. 5 (2005) 345–355.
[248] C.Y. Li, J.Y. Han, C.H. Ahn, Biosens. Bioelectron. 22 (2007) 1988–1993.
[249] P. Pang, W. Yang, S. Huang, Q. Cai, S. Yao, Anal. Lett. 40 (2007) 897–904.
[250] S. Woderer, N. Henninger, C.D. Garthe, H.M. Kloetzer, M. Hajnsek, U. Kamecke, N. Gretz, B. Kraenzlin, J. Pill. Anal. Chim. Acta 581 (2007) 7–12.
[251] O. Schuvailo, O. Soldatkin, A. Lefebvre, R. Cespuglio, A.P. Soldatkin, Anal. Chim. Acta 573 (2006) 110–116.
[252] H. Endo, Y. Yonemori, K. Musiya, M. Maita, T. Shibuya, H. Ren, T. Hayashi, K. Mitsubayashi, Anal. Chim. Acta 573 (2006) 117–124.
[253] H. Kudo, T. Sawada, E. Kazawa, H. Yoshida, Y. Iwasaki, K. Mitsubayashi, Biosens. Bioelectron. 22 (2006) 558–562.
[254] J.D. Newman, A.P.F. Turner, Biosens. Bioelectron. 20 (2005) 2435–2453.

Index

F

R

S

Related Fuel Cell Titles from Elsevier

Advances in Fuel Cells, Volume 1 (Advances in Fuel Cells Series)
K.D. Kreuer, Trung Van Van Nguyen, T.S. Zhao

High-temperature Solid Oxide Fuel Cells: Fundamentals, Design and Applications
S.C. Singhal, K. Kendall

PEM Fuel Cell Modeling and Simulation Using Matlab®
Colleen Spiegel

PEM Fuel Cells: Theory and Practice (Sustainable World Series)
Frano Barbir

Fuel Cells ebook Collection: Ultimate CD
S.C. Singhal, Daniel Sperling, Bent Sorensen, Frano Barbir, Nigel Brandon, Colleen Spiegel, T.S. Zhao

Hydrogen and Fuel Cells: Emerging Technologies and Applications (Sustainable World Series)
Bent Sorensen

Stationary Fuel Cells: An Overview
Kerry-Ann Adamson

Fuel Cell Compendium
David Thompsett, Nigel Brandon

DATE DUE

